Tumor
And Tumor Markers
Breast Tumors
Volume (1)

Prof. Dr.Sami AL-Mudhaffar

Introduction

The first tumor marker reported was the Bence – Jones protein. Since its discovery in 1847 by precipitation of a protein in acidified boiled urine, the measurement of Bence – Jones protein has been a diagnostic test for multiple myeloma (a tumor of plasma cells). More than one hundred years after its discovery, Edelman and Poulik identified the Bence–Jones protein as the monoclonal light chain of immunoglobulin secreted by tumor plasma cells. Monoclonal paraproteins appear as sharp bands in the globulin area in electrophoretic patterns of serum. Diagnosis of multiple myeloma is often made based on this finding or on the presence of an elevated level of monoclonal immunoglobulin in the serum.

Definition of Tumor Marker:

A tumor marker is a substance present in or produced by a tumor itself or by the host in response to a tumor that can be used to differentiate a tumor from normal tissue or to determine the presence of a tumor based on measurement in the blood or secretions. Such a substance can be found in cells, tissue or body fluids. It can be measured qualitatively or quantitatively by chemical, Immunological, or molecular biological methods to identify the presence of a cancer.

Morphologically, cancer tissue has been recognized by pathologists as resembling fetal tissue more than normal adult differentiated tissue. Tumors are graded according to their degree of differentiation as being (1) well differentiated (2) poorly differentiated, or (3) an plastic (without forms). Tumor markers are the biochemical or immunological counterparts of the differentiation state of the tumor. In general, tumor markers represent re–expression of substances produced normally by embryogenically closely related tissues.

Few markers are specific for a single individual tumor (tumor – specific markers); most are found with different tumors of the same tissue type (tumor – associated markers). They are present in higher quantities in cancer tissue or in blood from cancer patients than in benign tumors or in the blood of normal subjects

Classification of Tumor Markers:

Tumor markers may be classified to chemical and genetic tumor markers.

Chemical Tumor Markers:

Markers produced by cancers include enzymes and isoenzymes, hormones, oncofetal antigens, carbohydrate epitopes recognized by monoclonal antibodies, receptors, and oncogene products.

Enzymes were of the first groups of tumor markers identified. Their elevated activities were used to indicate the presence of cancer.

Hormones as tumor markers were used for the detection and monitoring of cancer.

Oncofetal antigens were discovered using conventional antisera produced against fluids from cancer – bearing animals or extracts of cancer tissues.

The development of monoclonal antibody techniques allowed more sensitive and specific measurements of tumor antigens. More importantly, new antigens were discovered by developing monoclonal antibodies against tumor cell preparations. They appear to have better clinical sensitivity and specificity than do the oncofetal antigens.

Table (A) summary of the chemical tumor markers.:

Marker type	Associated malignancy	Example
Enzyme	Liver	Alcohol dehydrogenase
	Bone, Liver, Leukemia, Sarcoma	Alkaline phosphatase
	Ovarian, Lung, Trophoplastic, Gastrointestinal, Seminoma, Hodgkin's	Alkaline phosphatase placental
	Pancreas,	Amylase
	Colon, Breast	Aryl Sulfatase B
	Colon, Bladder, Gastrointestinal, Vanous	Galactosyl Transferase
	Lung (small–cell) neuroblastoma, carcinoid, melanoma, Pancreatic	Neuron – Specific enolase
	Prostate, Vanouse (large bowel, lung, ovarian)	Prostate – Specific Antigen (PSA) Ribonuclease
	Colorectal, Breast, etc.	Telomerase
	Colon, Breast, Lung	Sialyl Transferenase
Hormone	cushing s syndrome, Lung (small cell)	ACTH
	Lung (small cell) adrenal cortex, duodenal	Antidiuretic hormone
	Modularl thyroid	Calcitonin
	Pituitary adenoma, Renal, Lung, Embryonal, choriocarcinoma, Testicular (non seminomatous)	Growth hormone hCG
	Torophoblastic, Gonads, Lung, Breast	Human placental lactogen
	Liver, Renal, Breast, Lung, Various	Parathyroid hormone
	Pituitary adenoma, Renal, Lung	Prolactin
	Pancreas, Bronchognic	Vasoactive intestinal
	Pheochromocytoma neuroblastoma	Peptide

Oncofetal Antigen	Hepato cellular, germline (non seminoma)	(α– feto protein)
	Colon	Beta– Oncofetal Antigen (β– oncofetal Antigen)
	Liver	Carcino fetal ferritin
	Colorectal, Gastrointestinal, Pancreatic, Lung, Breast	CEB (Carinoembryonic Antigen)
	Pancreatic	Pancreatic oncofetal
	Cervical, Lung, Skin, Head and neck (Squamous)	Squamous cell Antigen
	Colon, Gastrointestinal, Bladder	Tennessee Antigen
Mucin	Breast, Ovarian	CA 15–3
	Breast	CA 27–29
	Breast, Ovarian	MCA
	Pancreatic, Ovarian, Gastrointestinal, Lung	Du–PAN–2
Protein	Multiple Myeloma, β– –cell lymphoma chronic lymphocytic leukemia, Waldenstrom's macroglubulinnemia	β2 Macroglobulin
	Insulinoma	C– Peptide
	Liver, Lung, , Leukemia	Ferritin
	Multiple myeloma, Lymphomas	Immunoglobulin
	Pancreatic, Stomach	Pancreas associated Antigen
	Trophoplastic, Germ cell	Pregnancy specific protein [2]
	Hepatocellular	Prothrombin precursor
	Ovarian	Tumor associated trypsin inhibitor
	Pancreatic, Hepatic, Gastrointestinal	CA 19–9

	Gastrointestinal, Pancreatic, Ovarian	CA 19-5
	Colon, Gastrointestinal, Pancreatic	CA 50
	Ovarian, Colon, Breast, Gastrointestinal	CA 27,4, CA242
Others	Breast	Estrogen and Progesterone receptors
	Brain, Various	Polyamine
	Bone metastasis, Breast, (multiple myeloma)	Hydroxy Proline
	Neuroblastoma, Pheochromocytoma	Catecholamine metabolites
	Gastrointestinal, Lung, Rheumatoid	Lipid-associated sialic acid

Genetic Tumor Markers

Two classes of genes are implicated in the development of cancer, oncogenes ((cell activation genes and suppressor genes)) involved in the recognition and repair of damaged DNA.

Oncogenes are derived from proto-oncogenes, which may be activated by dominant mutations. The type of mutation could be point mutation, insertion, deletion, translocation, or inversion. Most oncogenes code for proteins that function at same stage of activation of cells for proliferation, and their activation leads to cell division. Most oncogenes are associated with hematological malignancies, such as leukemia and to a lesser extent, solid tumors.

The other class of tumor genes, the suppressor genes, has been isolated from mostly solid tumors. The oncogenicity of suppressor genes is derived from the loss of the gene rather than their activation, as with oncogenes. The major tumor suppressor gene, P_{53}, functions to repair damaged DNA by apoptosis (programmed cell death). Repair is mediated by activation of the production of P_{21}, which blocks the cell cycle in late G_1 to allow repair to take place. The loss of function of this gene due to loss or mutation may result in the inability of the DNA repair process and lead to the development of tumorogensis.

The exciting promise of using detection of oncogenes and suppressor genes for the diagnosis, determining the prognosis, and predicting the response to the chemotherapy remains to be realized. Oncogene detection remains an experimental approach to human cancer, with great expectation not yet fulfilled. The ability to predict susceptibility to develop cancer by the detection of mutations in tumor suppressor genes raises ethical questions that remain to be resolved.

Ideal Tumor Markers:

The ideal tumor markers have the following properties :–
* High clinical sensitivity.
* High clinical specificity.
* Tumor markers levels proportional to tumor volume.
* Reflect tumor heterogeneity.
* Low levels in healthy population.
* Low levels in benign diseases.
* Discriminatory to identify tumor and metastasis from benign to healthy states.
* Provide adequate times lead for early diagnosis and early treatment.
* Assay sensitivity to detect stage I cancer.

Clinical Applications of Tumor Markers:

The potential clinical applications of tumor marker assay are listed below.

Screening:

Some tumor markers have been used in mass screening programs of asymptomatic individuals, with limited success, in high – risk sectors of the population. However, it is to be emphasized that no biochemical tumor marker is yet specific and sensitive enough to be recommended as a definitive screening test for cancer.

Diagnosis:

Almost every tumor marker has been investigated for its suitability as a primary diagnostic test for cancer in symptomatic individuals. However, sufficient false – positive and false – negatives have been encountered with every marker so far discovered to preclude their use in distinguishing malignant and nonmalignant conditions. The ultimate goal of identifying tumor – specific antigens has so far eluded oncologists because most tumor markers have been found in some normal tissue and the serum of some non – cancerous individuals and in many benign diseases. For this reason, these antigens are often referred to as tumor associated antigens. Nevertheless, a number of tumor markers have proved to be useful in confirming diagnosis, often in conjunction with a battery of other clinical methods. Another approach attempts to use multiple tumor markers to diagnose tumors and to identify the primary origin of metastic disease

Differential Diagnosis and Classification:

Immunoassay for some tumor markers are used in clinics to distinguish between clinical conditions with similar symptoms , where one or both could be cancerous, For example, the measurement of neuron– specific enolase levels allows differentiation between neuroblastoma and Wilm's tumor when a child present with a palpable abdominal mass .

Staging and Grading:

The degree of elevation in the concentration of several tumor markers can help to stage tumors. In general the mean circulating levels of these tumor markers increase with the stage of the cancer. In contrast, placental alkaline phosphatase is a tumor marker related to the grade of cancer, and serum levels of this anylate are higher in grade 1 and 2 tumors than in grade 3 ovarian carcinomas

Prognosis

Prognosis is the probability of cure of a cancer patient. Positive lymph node detection is a classical method of determining prognosis invasively. The magnitude of tumor marker levels in several cancers corresponds to the mass of tumor. Moderate elevations are suggestive of better prognosis than persistent high levels. Tumor aggressiveness resulting in widespread metastasis; precipitates very high serum tumor marker levels indicating poor prognosis. Generally, well– differentiated tumors tend to be less aggressive than undifferentiated or anaplastic tumors. Whereas most tumor marker over expression indicates poor prognosis, the increased levels of progesterone and estrogen receptors in breast cancers determine the type of treatment (hormone) as well as good prognosis.

Monitoring and Recurrence:

The profile of tumor marker concentration against time can mirror the condition of patients diagnosed to have cancer.

Tumor markers profiles usually reflect one of the following classical patterns:–

* A rapid decline in tumor marker level to normal concentrations following surgery or other forms of first– line therapy suggests that treatment has been successful.

* The lack of a decline to basal levels following first– line therapy may indicate that treatment has been only partially successful.

* Continued low levels of the tumor marker indicate that remission has been maintained as a result of treatment.

* A subsequent rise in the concentration of the tumor marker (from the basal level) suggests a recurrence of the disease. Tumor markers can warn of renewed tumor growth or recurrence 3 – 12 months before other methods provide confirmation.

* Decline of the marker levels after an increase has been associated with a recurrence, is suggestive of the responsiveness of a tumor to second– line or subsequent treatment.

* If tumor marker concentrations remain elevated after treatment, the tumor may be resistant to the therapeutic method employed and

prognosis of the patient is poor unless alternative therapeutic modalities are available.

Table (B) Refers to the up to date clinical usefulness of tumor markers :

Type of cancer	Tumor Marker	Clinical use
GI (colorectal, pancreas, stomach)	CEA *CA19–9, CA195, CA72.4* CA50	Prognosis and Monitoring therapy.
Liver	AFP	Screening, diagnosis, Prognosis, Monitoring therapy.
	CEA	Monitoring therapy.
Breast	CEA, CA15–3, CA549, CAM2, M29, CA27, 29 MCA	Monitoring therapy.
Prostate	Estrogen /progesterone receptors PSA–total, Free, Complexes	Screening, Prognosis,
Ovarian	CA125	Prognosis, therapy.
Lung (small cell carcinoma)	NSE, CK–ββ	Monitoring therapy
Neuroblastoma	VMA,	Screening, diagnosis,

	Catecholamine	Prognosis, Monitoring therapy.
	NSE	Monitoring therapy.
Testicular (gem cell tumors)	AFP, β–hCG	Diagnosis, Prognosis and Monitoring therapy.
Myeloma	Immunoglobulin	Diagnosis, Prognosis.
Thyroid	Thyroglobulin	Screening, Monitoring therapy.
	Calcitonin	Diagnosis, Monitoring therapy.
Neuron endocrine	variety of hormones	Diagnosis.
Bone	ALP, 5–NT	Diagnosis, Monitoring therapy.
Bladder	BTA, NMP22	Recurrence Detection.
Liver metastases	ALP	Diagnosis, Prognosis, Monitoring therapy.
Trophoblastic diseases	hCG	Diagnosis, Prognosis, Monitoring therapy.

Chapter one
The Breast Tumors

Clinical & Chemical Aspects of Breast Tumors Human Breast Development

Two important concepts, in breast development, which has been and continued being a major biological puzzle are that, this organ is one of few that is not completely developed at birth and it reaches its full differentiation only through the hormonal stimuli induced by pregnancy and lactation.

The development of the breast, which is vigorously controlled by the ovary, placenta and pituitary hormones, can be defined by several parameters, such as its external appearance, total area, volume, degree of branching, number of structures present in the mammary gland, and degree of differentiation of individual structures, i.e., lobules and veoli. The study of breast development reveals that the breast is composed of lobular structures reflecting different stages of development. It composes of lobules type 1, which are the most undifferentiated ones and lobules type 2 evolve from the previous ones and have a more complex morphology. During pregnancy, lobules type 1 and type 2 progress to lobules type 3. Lobules type 4, which are present only during the lactational period of the mammary gland, regress to type 3 after weaning

There is a significant difference in the content and relative percentage of lobular structures present in the breast according to the parity status of a woman. In nulliparous women, lobules type 1 and 2 are almost constantly present throughout their lifespan; the lobules type 2 decrease in number after menopause. In the parous woman's breast the lobules type 3 are the most frequent structures present. Only after the fourth decade of life there is an increase in the number of type 1 due to the regression of the more differentiated lobules type 3 at the end of the fifth decade of life, the breast of both nulliparous and parous women contains predominately lobules type 1.

The Breast Tumors

The breast is constantly responding to changes in hormonal, nutritional, genetic, psychological, and environmental stimuli such as radiation that cause continual cellular changes . As a result of these changes, breast tumors (abnormal breast tissue) may develop either benign (noncancerous) or malignant (cancerous) . The major significance of the benign processes less in the need to separate them from malignancies. The World Health Organization (WHO) classifies tumors of the breast (1981) according to histological aspects (Table 1).

Table (1): Histological classification of breast tumors

I. Epithelial Tumors		
A. Benign		
	1.	Intraductal papilloma
	2.	Adenoma of the nipple
	3.	Adenoma
	a.	Tublar
	b.	Lactating
B. Malignant		
	1.	Non invasive
	a.	Intraductal carcinoma
	b.	Lobular carcinoma
	2.	Invasive
	a.	Invasive ductal carcinoma
	b.	Invasive lobular carcinoma
	3.	Paget's disease of the nipples
II. Mixed Connective Tissue and Epithelial Tumors		
	a.	Fibroadenoma

Table (1): Continued.

	b.	Phyllodes tumor (cystosarcoma phyllodes)
	c.	Carcinosarcoma
III. Miscellaneous Tumors		

a.	**Soft tissue tumors**
b.	**Skin tumors**
c.	**Tumors of haemopcietia and lymph tissues**
IV. Unclassified Tumors	
V. Mammary Dysplasia / Fibrocystic Change	
VI. Tumor like Lesion	
a.	**Duct ectasia**
b.	**Inflammatory pseudotumors**

Benign Breast Tumors

Fibroadenoma

This is the most common benign tumor of the female breast. It is a new growth composed of both fibrous and glandular tissue . These tumors are commonly found in younger women between the ages 20-35 years . It increases in size, during pregnancy . It is less likely to develop after menopause. An epidemiological study suggests that fibroadenoma represents a long-term risk for breast carcinomas and that risk is increased in women with ductal hyperplasias, or a family history of breast carcinoma .

Fibrocystic disease

This is an ill-defined condition of the breast where palpable lumps can be felt and is usually associated with pain and tenderness that fluctuates with the menstrual cycle . Fibrocystic changes are the most common, occurring in approximately 60% of premenopausal women. Women with this disease usually have a freely movable, palpable mass, at time it may cause pain, particularly when women are in the premenopausal phase of the menstrual cycle, however, breast pain can be caused by lesions other than fibrocystic changes. The palpable lesion may appear to increase and decrease is size cyclically; usually achieving its maximum size in the premenstrual phase of the menstrual cycle. Cystic disease is frequently accompanied by varying degrees of epithelial hyperplasia in adjacent ducts and lobules . In patients who have one particular form of fibrocystic disease (the proliferate form), the incidence of cancer increased very slightly .

Malignant Breast Tumors
Incidence

Breast cancer is the most common malignant tumors in women, and it is the leading malignancy affecting women in North America and Europe. In 2000, approximately 184200 new cases of invasive breast cancer were diagnosed in the United States. The number of noninvasive breast cancer is hard to verify, but it probably account for and an additional 20000 to 30000 new cases; thus, the number of invasive and noninvasive breast cancer treated in 2000 approximately 200000.

In Iraq, according to the results of Iraqi Cancer Registry (ICR), breast cancer accounting for (31.11%) and remained the commonest tumors in the year 2000, it was also shown that breast cancer was the first among the commonest ten cancers in Iraq. Figure 1 represents the population of breast cancer in Iraq for the last nine years. As shown in the same figure, the population of breast cancer increased more than double in the last nine years.

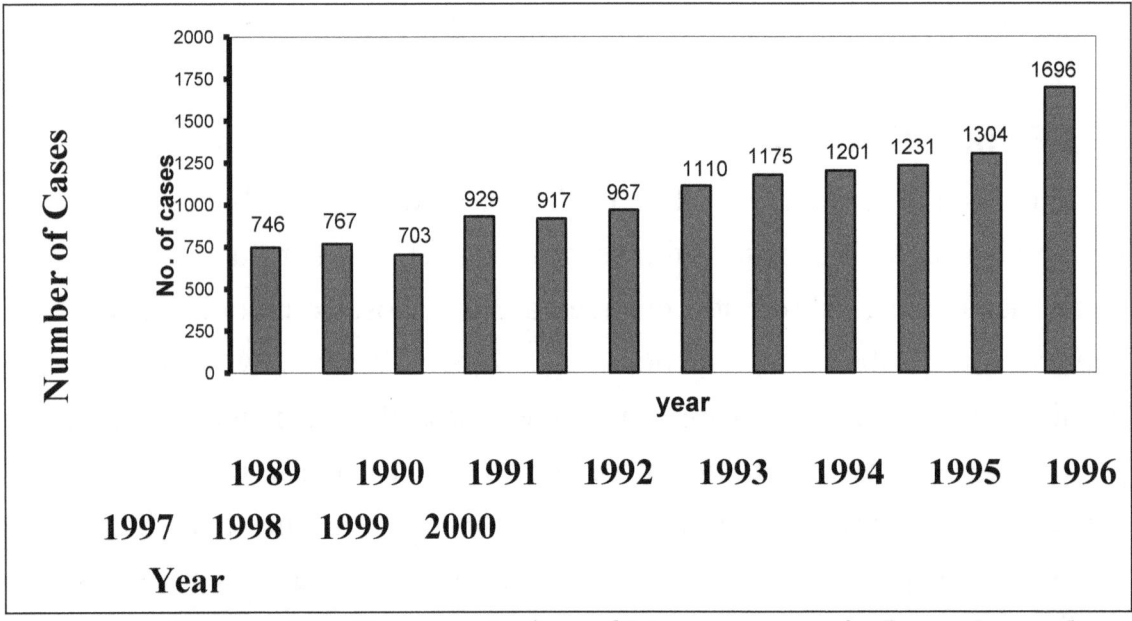

Figure (1): The population of breast cancer in Iraq through (1989-2000).

Pathogenesis and Site of Origin of Breast Cancer

It is not known when in the lifetime of a woman breast tumor initiates. The term tumor is applied indistinctly to either benign or malignant lesions, with notifying that the breast is the most frequently diagnosed malignancy in the female population. Studies of chemically induced carcinogenesis in an experimental animal model and primary cultures of human breast epithelial cells have shown that the initiation of the neoplastic process is inversely related to the degree of differentiation and *in vivo* cell proliferation of the mammary gland. An important concept that emerged from these studies is that the lobules type 1 have been identified as the site of origin of preneoplastic lesions such as atypical ductal hyperplasia, which evolve to ductal carcinoma *in situ*, progressing to invasive carcinoma. Lobulars type 2 give rise to the origin of atypical lobular hyperplasia and lobular carcinoma in situ, whereas lobules type 3 and 4 originate more in benign breast lesions.

Effecting Factors on Breast Cancer

Despite the numerous uncertainties surrounding the origin of cancer and no clear understanding of the cause of the worldwide breast cancer incidence increase, several studies indicated several factors that cause or prevent breast cancer.

Several experimental studies on some of these factors postulate that they seem to affect the/or effect by architectural pattern of the breast.

Although there is no explanation as yet for higher breast cancer risk exhibited by nulliparous and late parous women, experimentally fact induced rat mammary carcinomas model develop only when the carcinogen interacts with the undifferentiated and highly proliferating mammary apithelium of young nulliparous rats. These observation also support the hypothesis that the presence of lob1 explains the higher breast cancer risk of nulliparous women, as they represent the population with highest concentration of undifferentiated structures in the breast. Suggesting that these lobules are biologically different form those of early parous women.

The direct association of breast cancer risk with nulliparity, as well as the protection afforded by early first full-term pregnancy, has been in great part explained in many studies figure(2). It has been observed in the rodent experimental model and in human, that cells derived from the differentiated lob 3 are resistant to growth in vitro and do not express transformation phenotype upon carcinogen treatment, as do cells from lob1. These observations support the hypothesis, which postulated that the induction of differentiation of the breast by the reproductive process is responsible for the inhibition of carcinogenic initiation, figure (3).

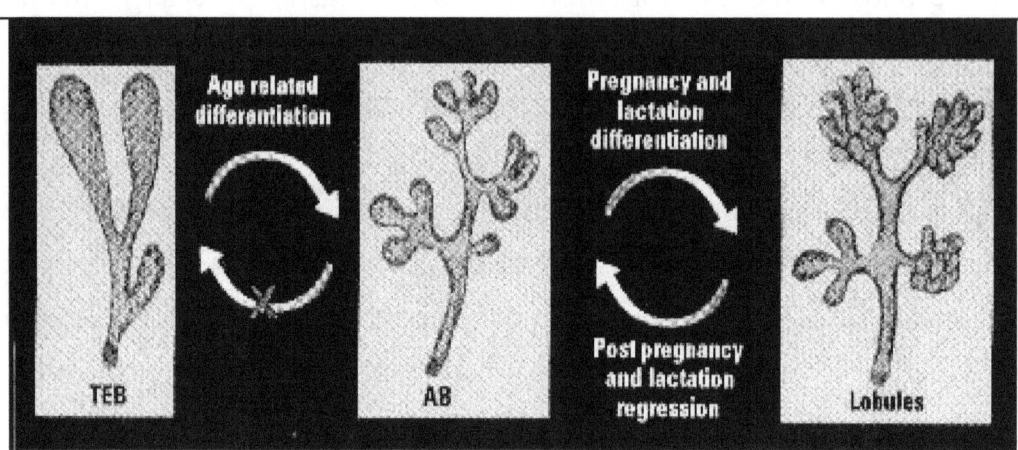

Figure (2):Terminal end buds (TEB) differentiation to alveolar buds(AB) in the nulliparous female under the regular hormonal stimuli of the menstrual cycle, pregnancy and lactation.

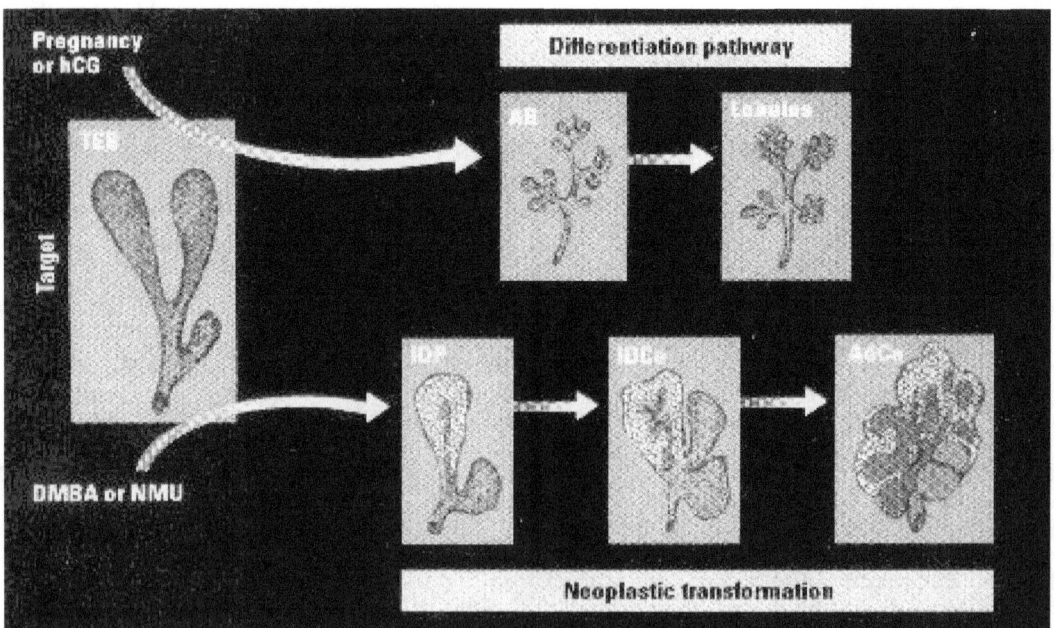

Figure (3): Terminal end buds (TEB) evolve to alveolar duds (AB) or lobules if pregnancy or hCG stimulate them towards the differentiation pathway. If a carcinogen reaches the target (TEB) during the susceptibility period, it diverts this evolution to yhe neoplastic transformation pathway, developing instead into intraductal proliferation (IDP), intraductal

carcinoma (IDCa), and invasive adenocarcinoma (AdCa).

Etiology and Risk Factors in Breast Cancer

Numerous risk factors have been associated with the development of breast cancer, such as genetic, environmental, hormonal, and nutritional.

Despite all available data on breast cancer risk factors, 75% of women with this cancer have not exposed to any risk factors.

Genetic Factors

Breast cancer is the result of mutations in one or more critical genes. Two genes in women on chromosome 17 have been implicated. The most important gene is called BRCA-1; the other is the P53 gene. A third gene is BRCA-2 on chromosome 13.

Increasing Ages

Breast cancer is uncommon before age 25 years, but then there is a steady rise to the time of menopause, followed by a slower rise throughout life. The average age at the diagnosis is 64 years.

Family History

The overall relative risk of breast cancer in women with a positive family history in a first –degree relative (mother, daughter, or sister) is 1.7. Premenopausal onset of the disease in a first–degree relative is associated with three-fold increase in breast cancer risk, whereas postmenopausal diagnosis increases relative risk by only 1.5. When the first degree has relative bilateral disease, there is fivefold increase in risk. The relative risk for a woman whose first-degree relative developed both bilateral and premenopausal breast cancer is nearly nine. No increased risk has been demonstrated when only a second-degree relative (aunt, cousin, or grandmother) has had breast cancer.

Proliferative Breast Disease

The diagnosis of certain condition after breast biopsy is also associated with an increased risk for the subsequent development of invasive breast cancer . Women with proliferative disease of the breast with a typical hyperplasia (atypia) are at increased risk for developing breast cancer (five-fold increase), however. The risk for atypia is greater in patient with a strong family history of breast cancer (11-fold increase) . Personal Cancer History

A personal history of breast cancer is significant risk factor for the subsequent development of a second, new breast cancer. This risk has been estimated to be as high as 1% per year from the time of diagnosis of the initial cancer. Women with a history of endometerial, ovarian, or colon cancer also have a higher likelihood of developing breast cancer than those with no history of these malignancies .

Menstrual and Reproductive Factors

Early onset of menarche (<12 years old) has been associated with a modest increase in breast cancer risk (two fold or less). Women who undergo menopause before age 30 have a twofold reduction in breast cancer risk when compared to women who undergo menopause after age 55. A first full-term pregnancy before age 30 appears to have a protective effect against breast cancer, whereas a late first full-term pregnancy or nulliparity may be associated with higher risk, There is also suggestion that lactation protects against breast cancer development

Environmental Factors

Women exposed to therapeutic radiation or after atom bomb exposure have a higher rate of breast cancer. Risk increases with younger age and higher radiation doses. Radiation is believed to cause 1-2% of all cancer deaths. In respect to pollution studies, which have shown that there is a well-established correlation between many pollutants and cancer, it has been estimated, that 1% of cancer deaths is due to air, water and land pollution .

Hormonal Influences

Endogenous estrogen excess, or more accurately, hormonal imbalance, clearly plays a significant role. Many risk factors mentioned-long duration of reproductive life, nulliparity, and late age at first child-imply increased exposure to estrogen peaks during the menstrual cycle.

Dietary Factors

Diets that are high in fat have been associated with an increased risk factor for breast cancer. It has been suggested that differences in dietary fat content may account for the variations in breast cancer incidence observed among different countries. Sala et.al, illustrated that certain macronuterients and food such as protein, carbohydrate and meat intake influence of risk of breast cancer through their effects on breast tissue morphology . Data from prospective studies have confirmed that the relationship exists between alcohol intake and risk of developing breast cancer . Alterations in endrogenous estrogen levels secondary to obesity may enhance breast cancer risk .

Lactation

In the search of practical methods to prevent breast cancer, lactation has strong evidence as a potentially modifying factor especially at early age and for long period. There is a significant reduction in the risk of breast cancer associated with lactation for more than two years. This effect appeared to be limited to premenopausal women . Lactation may reduce the risk of breast cancer by interrupting ovulation or by modifying pituitary and ovarian hormone secretions. Direct physical changes in the breast that accompany milk production may also contribute to prevent the effect .

Histopathology of Breast Cancer

Most cancer of the breast is a carcinoma of the epithelial cells that line breast ducts and lobules. Rarer forms of cancer occurring in the breast arise from the stromal cells that surround the epithelial glands . Breast cancer is a complex, devastating diseases and the most frequently diagnosed cancer in women. It is the single leading cause death for women of age 20-59 years .

There are a various type of breast carcinomas according to (WHO) classification:

In Situ Carcinoma (Non-Spreading Type)

A. Ductal Carcinoma In Situ (DCIS)

The malignant cells in this disease are confined to the ductal basement membrane . DCIS usually occurs without forming a mass because there is no scirrhous component . DCIS is also known as intraductal carcinoma.

B. Lobular Carcinoma In Situ (LCIS)

LCIS is composed of smaller lobular or acinar cells and fills the terminal breast lobule with a homogenous proliferation; most clinicians currently regard LCIS as a risk factor for the development of invasive breast cancer. LCIS is usually not treated, but affected women are placed under frequent surveillance

. Infiltrating Carcinoma (Invasive Carcinoma)

A. Infiltrating Ductal Carcinoma (IDC)

Most invasive carcinoma of the breast is ductal in origin . Infiltrating (invasive) breast carcinoma differs from intraductal carcinoma (ductal carcinoma in situ) by the presence of stromal invasion, through which tumor cells spread not only locally but also regionally and distantly via vascular lymphatic space.

B. Infiltrating Lobular Carcinoma (ILC)

This type is rare, from about (5%-10%) of breast cancer. It is begins in the milk-secreting glands of the breast. It is often multicentirc, several areas of thickening may occur in one or both breasts. It is characterized by the presence of small and relatively uniform tumors cell growing singly around lobules involved by in situ lobular neoplasia

Staging System of Breast Cancer

The standard staging system for breast cancer is the TNM system table . TheTNM classification devised by the International Union Against Cancer (UICC) and accepted by the American Joint Commission on Cancer Staging a world standard . Another system is the Colombia Clinical Classification (CCC), formulated by Haagensen and Stout. Although this system was a valuable precursor and is easier to remember than the TNM, it is a less precise classification where stage A represents a tumor confined to the breast; stage B include tumors with clinical axillary lymph node enlargement; stage C represents the presence of grave prognostic sings in the breast; and stage D indicates metastatic disease. The TNM based on the clinical features of tumor (T), the regional lymph nodes (N), and the presence or absences of distant metastases (M) . The purposes of staging are the following :

- Plane a therapeutic strategy that most appropriate for the patient.
- Allow for more intelligent prognostication of the disease statues of the patient.
- Permit comparison of therapeutic results obtained from different sources by different means.

Table (2): TNM System and Stage Grouping.

	T	Primary tumor
	T0	No evidence of primary tumor
	Tis	Carcinoma in situ
	T1	Tumor 2 cm or less in greatest diameter
	T2	Tumor more than 2 cm but less than 5 cm in greatest diameter
	T3	Tumor more than 5 cm in greatest diameter
	T4	Tumor of any size with direct extension to chest wall or skin
	N	Regional lymph nodes

Table (2): Continued.

	N0	No regional lymph node metastases
	N1	Metastasis to movable ipsilateral axillary lymph node(s)
	N2	Metastasis to ipsilateral lymph node(s) fixed to one another or to other structures
	N3	Metastasis to ipsilateral internal mammary lymph node(s)
	M	Distant metastasis
	Mo	No distant metastasis
	M1	Distant metastasis (including metastasis to ipsilateral supraclavicular lymph nodes)
Stage 0	Tis	No	Mo
Stage I	T1	No	Mo
Stage	T0	N1	Mo

IIA	T1	N1	Mo
	T2	N0	Mo
Stage IIB	T2	N1	Mo
	T3	N0	Mo
Stage IIIA	T0	N2	Mo
	T1	N2	Mo
	T2	N2	Mo
	T3	N1/N2	Mo
Stage IIIB	T4	Any N	Mo
	Any T	N3	Mo
Stage IV	Any T	Any N	M1

Treatment of the Breast Tumors

The goal of any oncologic treatment is to maximize the cure and at the same time optimize the quality of life.

Surgical Therapy

Surgical treatment represents most frequently used and the most successful sign method of cancer therapy currently available. More patients are cured of cancer by surgery than by any other therapeutic modality.

Conservation Breast Cancer Surgery

It is aimed at removing the tumor plus a rim of at least (1 cm) of normal breast tissue. This is commonly referred to as a wide local excision or lymphectomy.

Mastectomy

Removal of all breast tissue, choice may not be offered if the lesion is too large, multi-focal, and lobular or, in the surgeon's opinion, so close the nipple that it is likely to cause distortion.

Radiotherapy

Palliative radiotherapy may be advised for locally advanced cancers with distant metastasis in order to control ulceration, pain, and other manifestation in the breast and regional nodes. Radiotherapy is especially useful in the treatment of the isolated bony metastasis, chest wall recurrence and brain metastasis .

Chemotherapy

Breast cancer is responsive to all major classes of cytoxic drugs: alkylating agents, antimetabolites, mitotic inhibitor, and the antitumor antibiotic. Among the most active are alkylating agent including cyclophosphamide and thaitepa, and anthracyclines such as doxorubicin. The antimetabolites methotrexate (MTX) and 5–flourouracil (5-FU) are also active. Numerous combination of chemotherabutic agents have been evaluated in the treatment of metastatic breast cancers such as: CMF, CAF .

Endocrine Therapy

Hormonal therapy is the initial treatment of metastatic disease in patients with ER or PR positive tumors. Tamoxifen, a nonsteroidel antiestragon was approved first for the treatment of metastic breast cancer over 20 years ago, is usually the first agents of choice because of its favorable toxicity profile.

Detection and Diagnosis

Although an accurate history and clinical examination are still the most important method of detecting breast disease, there are a number of investigations that can assist in the diagnosis as follows:

Self Examination

All women over age 20 should be advised to examine their breast monthly–premenopausal women should perform the examination 7-8 days after the menstrual period. The breast should be inspected initially while standing before a mirror with the hands at the side, overhead, and pressed firmly on the hips to contract the pectoralis muscles. Masses asymmetry of breasts and slight dimpling of the skin may become apparent as a result of these maneuvers. Next, in a supine position, each breast should be carefully palpated with the fingers of the opposite hand. Some women discover small breast lumps more readily when their skin is moist while bathing or showering. Most women do not practice self-examination, and its value is controversial. Clearly, however, it is not harmful, it is inexpensive, and it may be beneficial .

Mammography

Soft tissue x-rays are taken by placing the breast in direct contact with ultrasensitive film and exposing it to low-voltage, high-amperage x-rays. The dose of radiation in approximately 0.1 Gy and therefore mammography is a very safe investigation .

Ultrasound

Ultrasound is particularly useful in young women with dense breasts in whom mammograms are difficult to interpret, and in distinguishing cysts for solid lesions. It can also be used to localize impalpable breast lumps .

Magnetic Resonance Imaging (MRI)

MRI is of increasing interest to breast surgeons in a number of settings, it can be useful to distinguish scar from recurrence in women who have had previous breast conservation therapy for cancer (although it is not accurate within 9 months of radiotherapy because of abnormal enhancement); it is the gold standard for imaging the breast of women with implants; it may prove useful as a screening tool in high-risk women; and it is being evaluated in the management of the axilla in both primary breast cancer and recurrent disease .

Needle Biopsy/Cytology

Histology can be obtained by using a fine needle such as a trucut or corecut biopsy device under local anesthesia. Cytology is obtained by using a 21 or 23 gauge needle and 10 mL syringe with multiple passes throughout the lump without releasing the negative pressure in the syringe. The aspirate is then smeared on to a slid, which is air-dried. Fine needle aspiration cytology (FNAC) is the least invasive technique to obtain a cell diagnosis and is very accurate if both operator and cytologist are experienced. However, false negatives do occur mainly through sampling error, and invasive cancer cannot be distinguished from in situ disease .

Triple Assessment

In any patients who presents with a breast lump or other symptoms suspicious of carcinoma, the diagnosis should be made by a combination of clinical assessment, radiological imaging and tissue sample taken for either cytological or histological analysis .

Chapter Two
Tumor Markers

Definition of Tumor Marker

Several definitions of tumor markers exist in the literature. A restricted and classical definition of tumor markers pertains to the measurement of certain analytes either using chemical, biochemical or immunochemical methods in conveniently obtainable body fluids such as urine and blood. However, this narrow definition excludes the important class of tissue and cellular markers. Not all tumor markers are secreted into body fluids such as (CA170, CA174) membrane bound antigens and several intracellular antigens including aberrant nucleic acid sequences. However, at the fifth international conference on tumor markers (Stockholm, 1988) a consensus definition of one category of tumor markers was adopted: "Biochemical tumor markers are substances developed in tumor cells and secreted into body fluids in which they can be quantitated by non-invasive analysis. Because of a correlation between marker concentration and active tumor mass, tumor markers are useful in the management of cancer patients. Markers, which are available for most cancer cases, are additional, valuable tools in patient prognosis, surveillance and therapy monitoring whereas they are presently not applicable for screening. Serodiagnostic measurements of markers should emphasize relative trends instead of absolute values and cut-off levels". The general potential uses of tumor markers include: Screening, diagnosis, differential diagnosis and classification, staging and grading, prognosis and monitoring treatment .Finally A tumor marker is a substance present in or produced by a tumor or by the tumor's host in response to the tumor's presence that can be used to differentiate a tumor from normal tissue or to determine the presence of a tumor based on measurement in the blood or secretions. Such a substance can be found in cells, tissue or body fluids. It can be measured qualitatively or quantitatively by chemical, immunological, or molecular biological methods to identify the presence of a cancer. Tumor markers are the biochemical for immunological counterparts of the differentiation state of the tumor. In general, tumor markers represents re-expression of substances produced normally in embryoginically closely related tissues. Few markers are specific for a single individual tumor (tumor-specific markers); most are found with different tumor of the same tissue type (tumor-associated markers). They are present in higher quantities in cancer tissue or in blood from cancer patients than in benign tumors or in the blood of normal subjects.

Measurement of Tumor Markers:

Different assay techniques have been used for the measurement of tumor markers. The in vitro measurement of tumor markers is in two realms: histochemical and body fluid analysis. Current probes for such measurements are largely based on monoclonal antibodies and polyclonal antibodies although some assays based on the total absence of immune reactions still exists. While immunohistochemical analysis of tumor markers generally are qualitative, the availability of image analysis techniques, software and digital signal acquisition methods provide a degree of quantitation of the intensity of staining and the degree of heterogenicity of tumor marker expression in a tumor mass. Almost all in vitro immunoassays for body fluid analytes are quantitative.

Today, most immunoassays of tumor markers are based on monoclonal antibodies using the dual monoclonal sandwich assay technique.

Further refinements in assay technology include the development of homogeneous assays wherein there is no separation of the bound and free reactants: new non-isotopic signal generation such as time resolved, fluorescence, bioluminescence and chemiluminescence and dye based particles.

Recently a new technique called immuno-PCR (polymerase chain reaction) has been introduced that uses a short piece of DNA as the tag in an immunoassay.

The in vivo identification of tumor mass needs a special mention here. Anti-tumor polyclonal antibodies and now monoclonal antibodies have been radiolabeled with a variely of diagnostic isotopes such as ^{131}I, ^{111}In and ^{99m}Tc and infused into cancer patients for in situ detection by and external gamma camera of primary and recurrent metastasis exploiting tumor markers as targets. Although this technique of scintigraphy is largely a qualitative in vivo measurement of the tumor marker, some estimate of the quantitative uptake in tumor markers has been made for dosimetry and eventual radiotherapeutic applications.

Routes of Tumor Markers Production

Benign tumors are generally well differentiated. The cells in a benign tumor are similar to the cells of the normal tissue, and the tumor markers produced are the products found in the normal tissue. They may be found in increased amounts in the circulation depending on the size of the tumor.

Malignant tumors may produce substance may associated with normal cell, or they may be different. As a zygote is transformed into an embryo, which then evolves into a fetus, the rapidly dividing cells became differentiated into specialized tissues by selective gene expression. The genes expressed are responsible for the production of hormones, enzymes, receptors, structural proteins, and cell metabolism. When a normal cell is transformed into tumor cell, gene expression changes. The affected cell may lose its ability to synthesize some specific cell products, or it may manufacture greatly increased amounts. The cell may be less specialized than the tissue it evolved from and assume the characteristics of the less well-differentiated cells of the embryo, synthesizing proteins found in the embryo but not in a normal adult. Cell proliferation rates change as the metabolic rate of the cells increases. After the cell is transformed, it loses growth control and begin to divide rapidly. The cells lose contact inhibition and invade the primary site. They then invade the adjacent organs and blood and lymph system, which may carry the cells to distance organs. The cell may then lodge in a capillary bed and begin to invade the new site. As this invasion process takes place, new proteins are produced that actively aid in the invasion. These proteins can also be used as markers.

Classification of Tumor Markers

Tumor markers may be classified into chemical and genetic tumor markers. Tumor markers can be broadly classified into tumor specific antigens and tumor-associated markers. Most tumor markers were often heralded as highly tumor specific, but subsequent studies demonstrated their presence in normal tissues of the adult or in various stages of ontogeny.

As a result, very few tumor-specific antigens can be recognized. The idiotypes of immunoglobulins of B cell tumors and certain neo-antigens of virus induced tumors are two examples that are strictly tumor specific. The vast majority of tumor markers are in reality tumor-associated antigens and can be classified into two types based on their size. The low-molecular weight tumor markers (- < 1000 Daltons) include some nucleosides, lipid associated sialic acid, polyamines, pseudouridine, pigment derivatives, and other metabolites. The macromolecular tumor antigens are the most important sub-type useful in the clinical management of cancer patients.

The large cancer antigens are either enzymes, growth factors, hormones, receptors, biological response modifiers, oncogenes and their products, or glucoconjugates which include glycoproteins and glycolipids. The classification of tumor markers into its various categories are summarized in (Table 3).

Chemical Tumor Markers

Table (3) summarizes the chemical tumor markers classified according to biochemical characteristics, and their associated malignancy. The table shows the low specificity of tumor marker for cancer.

Table (3): Classification of tumor markers.

Category	Examples
I. Tumor –specific markers	B-cell tumor immunoglobuline idiotype: virus induced antigens e.g.: SV40T antigen, T-cell receptor of T-cell leukemia.
II. Tumor-associated markers a. Low molecular weight markers	Polyamines, nucleoside derivatives; sialic acid (lipid associated) Vanillylmandelic acid and catechoamine. Van metabolites.
b. Macromolecular markers 1. Enzymes, Isoenzymes 2. Hormones cyctokines, growth factors soluble and receptors.	Placental alkaline phosphatase: prostate specific antigen, prostatic acid phosphatase, Thymidine kinase: Neuorone specific enolase. HCG: 11.2: EGI: estrogen and progesterone receptors.

3. Oncogenes and oncoproteins
4. Oncofetal proteins
5. Complex glycoconjugates:
   i- Glycoproteins and glycosamino-glycans.
   ii- Glycolipids
6. Cellular markers
	C-myc: src: ras: crb: neu: sis.
	CEA, AFP, lewis X; lewis Y.
	
	CA125; CA15-3; SIA; CA19-9; TAG72; Matrix protein; N-CAM.
	
	Lewis X; Lewis Y; GM2; GD2; CA19-9 glycolipid
	Philadelphia chromosome; pre-cancerous cells in PAP smear

Table (4): Chemical tumor markers.

Marker	Example	Associated Malignancy
Enzyme	Alcohol dehydrogenas	Liver
	Alkaline phosphatase	Bone, Liver, Leukemia, Sarcoma
	Alkaline phosphatase Placental	Ovarian, Lung, trophoplastic gastrointestinal, seminoma, Hodgkin's
	Amylase	Pancreas, Various
	Aryl Sulfatase B	Colon, breast
	Galactosyi transferase	Colon, bladder, gastrointestinal, Various
	Neuron-Specific enolase	Lung (small-cell), neuroblastoma, carcinoid, melanoma, Pheochromocytoma,
	Prostate-specific antigen (PSA)	Prostate Various (Large bowel. Lung, ovarian)
	Telomerase	Colorectal, Breast, etc.
	Sialyl transferenase	Colon, Breast, Lung
Hormone	ACTH	Gushing's syndrome, lung (small – cell)
	Antidiuretic hormone	Lung (small - cell) adrenal cortex,
	Calcitonin	Medullary thyroid
	Growth hormone hCG	Pituitary adenoma, renal, lung, Embryonal choriocarcinoma, testicular
	Human placental	Trophoblastic, gonads, lung, breast
	Parathyroid hormone	Liver, renal, breast, lung, various
	Prolactin	Pituitary adenoma, renal, lung. Breast
	Vasoactive intestinal	Pancreas, bronchogenic,
	Peptide 5	Pheochroinocytom neuroblastoma

Table (4): Continued.

Oncofetal Antigen	α-Feto protein	Hepato cellular, germ line (non-
	β-oncofeta antigen	Colon
	Carcino fetal ferritin	Liver
	CEA	Colorectal, gastrointestinal,
	Pancreatic oncofetal	Pancreatic
	Sequamous cell antigen	Cervical, lung, skin, head and neck
	Tennessee antigen	Colon, gastrointestinal, bladder
Mucin	CA 125	Ovarian, endometrial
	CA 15-3 (Episialin)	Breast, Ovarian
	CA 27-29	Breast
	MCA	Breast, ovarian
	Du-PAN-2	Pancreatic, ovarian, gastrointestinal,
Blood group related antigen	CA 19-9	Pancreatic, hepatic; gastrointestinal
	CA 19-5	GastrointestinaT, pancreatic, ovarian
	CA50	Pancreatic, colon, gastrointestinal
	CA 27.4	Ovarian, breast, colon,
	CA 242 1	Ovarian, breast, colon,
Protein	β2-Microglobulin	Multiple myeloma, β-cell
	C-peptide	Insulinoma
	Ferritin	Liver, lung, breast, leukemia
	Immunoglobuin	Multiple myeloma, lymphomas
	Melanoma associated	Melanoma
	Pancreas associated	Pancreatic, stomach
	Pregnancy specific	Trophopiastic, germ cell
	Prothrombin precursor	Meato cellular
	Tumor associated trypsin inhibitor	Ovarian
Others	Estrogen and progesteron	Breast
	Catecholamine	Neuroblastoma, pheochromocytoma
	Hydroxy proline	Bone metastasis (breast) multiple
	Lipid-associated sialic	Gastrointestinal, lung. Rheumatoid
	Polyamine	Brain, various

Genetic Tumor Markers

Two classes of genes are implicated in the development of cancer: Oncogenes (Cell activation genes–table (5) and suppressor genes (genes involved in the recognition and repair of damaged DNA-table (5). Oncogenes are derived from proto-oncogenes, which may be activated by dominate mutations. The type of mutation could be point mutation, insertion, deletion, translocation, and inversion. Most oncogenes code for proteins that function at the same stage of activation of cells for proliferation, and there activation leads to cell division. Most oncogenes are associated with hematological malignancies, such as Leukemia and to a lesser extent, solid tumors.

The other class of tumor genes the suppressor genes has been isolated from mostly solid tumors. The oncogenicity of suppressor genes is derived from the loss of the gene rather than their activation, as with oncogenes. The major tumor suppressor gene, P53, functions to repair damaged DNA by apoptosis (programmed cell death).

Repair is mediated by activation of the production of P21, which blocks the cell cycle in late G1 to allow repair to take place. The loss of function of this gene may result in the inability of the DNA repair process and lead to the development of tumorgensis.

The exciting promise of using detection of oncogenes and suppressor genes, for the diagnosis, determining the prognosis, and predicting the response to chemotherapy remains to be realized. However, oncogenes detection remains an experimental approach to human cancer, with great expectations not yet fulfilled. The ability to develop cancer by detection of mutations in tumor suppressor genes raises ethical questions that remain to be resolved.

Table (5): Classification of genetic tumor markers .

Marker	Example	Associated Malignancy
Oncogene	N-ras mutation	Acute myeloid leukemia neuroblastoma
	K-ras mutation	Leukemia, lymphoma

	C-myc	b-and T-cell lymphoma., small cell
	C-erb B-2	Breast, ovarian, gastrointestinal
	C-abllber	Chronic myelocytic leukemia
	N-myc	Neuroendocrine
	bcL-2	Leukemia, iymphoma
Suppressor Gene	VHL mutation	Kidney
	APC mutation	Colorectal
	PI 6 (cd Kn2)	Bladder, glioblastoma, melanoma
	WT1 mutation	Wilms', tumor
	Loss of	Wilms', breast, hepatoblastoma
	BRCA2, PB1	Breast
	RB1 mutation	Retinoblastoma, osteosarcoma small-
	PI 6 E-cadheim	Breast
	BRCA1 mutation	Neurofibromatosis 1 Melanoma, breast
	P53 mutation	Breast, colorectal, lung, liver, renal cell,
	DCC mutation	Colorectal
	NF2 mutation	Neurofibro matosis2, meningioma

Clinical Applications of Tumor Mrkers

The potential uses of tumor markers are summarized in table (6). In general, tumor markers may be used for diagnosis and prognosis of carcinomas and for monitoring the effects of therapy as well as targets for localization and therapy. Ideally, a tumor marker should be produced by tumor cells and be detectable in body which fluids should not be present in healthy people or in benign conditions. Therefore, it could be used for screening for the presence of cancer in symptomatic individuals in general population.

Most tumor markers are present in normal, benign, and cancer tissues. They are not specific enough to be used for screening cancer. In situations where the incidence of cancer is high among certain populations, screening might be possible.

Table (6): Clinical Usefulnees of tumor markers.

	Biochemical properties	Molecular weight	Primary clinical applications
Alpha-fetoprotein (AFP)	Glycoprotein, 4% carbohydrate; considerable homology with albumin	~70 KD	Diagnosis and monitoring of primary hepatocellular carcinoma and germ cell tumors. Prognosis of germ cell tumors.
Cancer antigen 125 (CA 125)	Mucin identified by monoclonal antibodies	~200 KD	Monitoring ovarian carcinoma. Prognosis after chemotherapy
Cancer antigen 15-3 (CA 15.3, BR 27.29)	Mucin identified by monoclonal antibodies	>250 KD	Monitoring breast cancer
Cancer antigen 72.4 (CA 72.4)	Glycoprotein identified by monoclonal antibodies	~48 KD	Monitoring gastric carcinoma
Cancer antigen 19-9 (CA 19-9)	Glycolipid carring the Lewisa blood group determinate	~1,000 KD	Monitoring pancreatic carcinoma
Carcinoembri-yonic antigen	Family of glycoproteins, 45%-60%	~180 KD	Monitoring gastrointestinal and other adenocarcinomas

(CEA)	carbohydrate		
CYFRA 21-1	Fragments of cytokeratin	~30 KD	Monitoring bladder and lung carcinoma
Estrogen receptor	Nuclear transcription	65 KD	Predicting response to endocrine therapy in breast cancer

Table (6): Continued.

Human chorionic gonadotrophin (hCG)	Glycoprotein hormone consisting of tow non-covalently bound subunits (α and β)	~36 KD	Diagnosis and monitoring non-seminomatous germ cell tumors, choriocarcinomas, hydtidiform moles, seminomas. Prognosis of germ cell tumors.
Neuron specific enolase (NSE)	Dimer of the enzyme enolase	~87 KD	Monitoring small cell lung carcinoma, neuroblastoma, apudoma.
Placental alkaline phosphatase (PLAP)	Heat-stable isoenzyme of alkaline phosphatase	~86 KD	Monitoring of germ cell tumors (seminomas)
Progesterone receptor	Nuclear transcription factor	A from: 94 KD B from: 120 KD	Predicting response to endocrine therapy in breast cancer.
Prostate specific antigen (PSA)	Glycoprotein serine protease	~36 KD	Diagnosis, screening and monitoring prostatic carcinoma
Squamous	Glycoprotein	48 KD	Monitoring

cell carcinoma antigen (SCC)	sub-fradion of tumor antigen T4		squamous cell carcinomas
Tissue polypeptide antigen (TPA)	Fragments of cytokeratin 8,18 and 19	~22 KD	Monitoring bladder and lung carcinoma
Tissue polypeptide specific antigen (TPS)	Fragment of cytokeratins 18	~22 KD	Monitoring metastatic breast carcinoma

CA 125

CA 125 is a high-molecular mass (>200 KD) glycoprotein recognized by the monoclonal antibody OC 125. The level of CA 125 is measured quantitatively by using immunoradiometric assay (70). In healthy population, the upper limit of CA 125 level is 35 KU.L-1. CA 125 is elevated in ovarian carcinoma, endometerial, pancreatic, Lung, breast, colorectal and other gastrointestinal tumors . CA 125 is useful to detecting residual disease in cancer patients following initial therapy .

TPA

Tissue polypeptide antigen is not a specific tumor marker . Antibodies that react with cytokeratin 8,18 and 19 identify it. TPA is a heterogeneous group of molecules with molecular weight range 20-45 KD . Both normal and cancerous cells produce TPA; it is useful in the monitoring of metastic diseases.

TPS

Tissue polypeptide-specific antigen (TPS) is a new tumor marker defined by monoclonal antibody against the soluble tissue polypeptide antigen (TPA) . First described as specific tumor marker by Bjorklund in 1957 . In breast cancer patients TPS was especially useful in monitoring response to treatment and effectiveness of therapy in metastatic disease .

MAM-6

MAM-6 an epithelial membrane antigen present on ductal and alveoli epithelial cells that is detected by monoclonal antibody raised against human milk-fat globule membranes . Partial characterization of the antigen by SDS-PAGE showed that the antigen is a polymorphic epithelial sialomucin with a molecular mass over 400 KD .

Estrogen receptors and Progesterone Receptors

The estrogen and progesterone receptors are intracellular receptors that are measured directly in tumor tissue. These receptors are polypeptides that bind their respective hormones translocate to the nucleus , and induce specific gene expression. There are three domains on these polypeptides: a C-terminal hormone binding domain, a central DNA binding domain, and an N-terminal domain that is important for transcription .

When the hormone binds to its respective receptor, the DNA binding domain is modified so it binds to DNA quite avidly and initiates transcription. This process clearly affects the growth of the cell. However, the intricacies of how the hormone and receptor complex affect cell growth are only partially understood.

DNA Flow Cytometrically Derived Parameters

DNA diploid tumors are those in which a single peak containing an amount of DNA similar to normal control cells is generated by flow cytometry. DNA aneuploid tumors have additional peaks on DNA histogram. Presumably representing cells containing more or less nucleic acid than is found in 46 normal chromosomes .

C-erbB-2 (HER-2/neu)

The C-erbB-2 gene encodes a transmembrane tyrosine kinase that is the receptor for a family of peptide hormones.

P53

P53 is a tumor suppressor gene on the short arm of chromosome 17 that encodes a protein that is important in the regulation of cell division. The P53 gene product appears to regulate transcription of several other genes. The full role of P53 in the normal and neoplastic cell is unknown. There is evidence that the gene product is important in preventing the division of cells containing damaged DNA. P53 gene deletion or mutation is a frequent event along with other molecular abnormalities in colorectal carcinogenesis.

carbohydrate Antigen 19-9 (CA 19-9)

CA 19-9 is a carbohydrate antigen identified as a glycolipid-that is, sialylated lacto-N-fucopentose II ganglioside, which is a sialylated derivative of the Lewis a blood group antigen and is denoted as Le a. CA19-9 is synthesized by normal human pancreatic and biliary ductular cells and by gastric, colonic, endometerial, kidny, salivary gland, sweat gland and present in ductal epithelium of breast . In serum it exists as a mucin, a high-molecular weight (200-1000 KD) glycoprotein complete . The monoclonal antibody against CA19-9 was developed from a human colon carcinoma cell line, SW-1116 by Koprowski and associates .

Monoclonal antibody 19-9 derived from spleen cells of a mouse immunized with human colon adenocarcinoma cell line SW-1116 . The epitope of this antibody is carbohydrate with the sugar sequence

NeuNAcα 2-3 Gal β 1-3 GlcNAc β 1-3 Gal…

Fuc α 1

As described by Magnani et.al. .

. Methodology

CA 19-9 is measured with a double monoclonal immunoradiometric assay, using monoclonal antibodies raised against the SW-1116 cell line . The antibody reacts with CA19-9 found at low concentrations in sera from healthy individuals, but frequently increased in sera from patients with adenocarcinomas . The upper limit of normal for healthy subjects has been defined by the cutoff value of 37.0 (U.mL-1) . CA 19-9 has become an established marker for pancreatic cancer , but it must still be regarded as a research test for colorectal cancer.

Another methods to determinate CA 19-9 were enzyme-linked immunosorbent assay. Both the capture and the enzyme-conjugated antibody use the CA 19-9 monoclonal antibody. It should be noted that this antibody is useless for cancer diagnosis when a patient is lacking the enzyme for the synthesis of sialyl Le a. In Japanese, about 5-10% of the population lacks this enzyme. Determination carbohydrate antigen CA 19-9 levels in serum were also measured by radioimmunoassay (RIA) . Immunohistochemical technique used for the distribution of CA19-9 in tissues using an immunoperoxidase assay .By this technique the CA 19-9 can be detected not only in cancerous tissues but also in non cancerous normal tissues.

Screening

Numerous studies have addressed the potential utility of CA 19-9 in adenocarcinoma of the colon and rectum.

The reported incidence of elevated serum CA 19-9 in colorectal cancer ranges from 20% to 40% . The incidence of elevated CA 19-9 in stage-related, with the highest sensitivity occurring in patients with metastases . However, the sensitivity of CA 19-9 was always less than that of the CEA test for all stages of disease . The false-positive rate (>37.0 U.mL-1) is 15% to 30% in patients with non-neoplastic diseases of the pancreas, liver and biliary tract . Consequently, CA 19-9 cannot be used for screening asymptomatic populations.

Monitoring Response to Treatment

Kouri et.al. compared CEA and CA 19-9 for predicting response to chemotherapy in patients. Decreases in CEA more accurately reflect the response to therapy than did the decreases of CA 19-9. The pretreatment CA 19-9 value was, however, an important prognostic factor. Median survival was 30 months for patients with normal CA 19-9 values and 10.3 months for patients with elevated CA 19-9 values. CA 19-9 used to examined the serum levels and immunohistochemistry during the clinical course of female patient treatment with idiopathic interstitial pneumonia (IIp) that had elevated serum levels of CA 19-9 .

Clinical Application

Elevated levels (>37 U.mL-1) were seen in patients with pancreatic (80%), hepatobiliary (67%), gastric (40-50%), hepatocellular (30-50%), colorectal (30%), and breast (15%) cancer. Pancreatits and other benign gastrointestinal diseases show a 10 to 20% elevation; however, the levels are usually lower than 120 (U.mL-1). CA 19-9 levels correlate with pancreatic cancer staging . CA19-9 is useful in monitoring pancreatic and colorectal cancer. Elevated levels can indicate the recurrence before clinical finding by 1 to 7 months . Unfortunately, early detection of relapse may not be useful because of the lack of effective therapy for pancreatic cancer.

ntroduction

The role of tumor markers in breast cancer is to enhance the clinicians, ability to provide more effective management of the disease . Serum CA15-3 concentration was determined by using sandwich enzyme immunoassay of a double monoclonal antibody , automated chemiluminescent immunoanalyzer , immunoradiometric assay and radioimmunoassay, in women with benign breast tumor and breast cancer.

Tumor Markers in Breast Cancer

Several tumor markers have been investigated for one or more clinical use in breast cancer (Table 7). Tumor markers may be used to indicate the risk presence status, or future behavior of cancer. However, it is more valuable to use tumor markers for discrimination between malignant and benign tumors. New markers are frequently introduced into clinical practice without rigorous analysis, with the assumption that any information available to the clinician will help the patient.

Tumor-associated antigens (TAAs) that have been associated with breast cancer include carcinoembryonic antigen (CEA); tissue polypeptide antigen (TPA), tissue polypeptide-specific antigen (TPS), gross cystic disease protein (GCDP); prostate specific antigen (PSA); and the products of the MUC-1 gene. The MUC-1 gene encodes a cell-associated mucin-like protein. Secretary epithelial cells such as breast epithelial cells express this antigen. Several assays detect the MUC-1 gene products, but they are not identical. These proteins have been identified by monoclonal antibodies to breast cancer cell lines, breast cancer tissue, or human milk fat globule membranes. Assays that detect circulating MUC-1 products include CA 15-3, CA 27-29, CA 549, breast cancer mucin (BCM), mammary serum antigen (MSA), and mucin-like carcinoma-associated antigen (MCA) .

The results obtained with these assays may not be identical, presumably due to reactivity of different antibodies to different epitopes, and/or different sensitivities and specificities that result from different assay configurations .

More recently, markers of tumor biology have been investigated in breast cancer (Table7), and molecules related to angiogenesis, adhesion, invasion, and metastases. Several, but not all of these are indeed, detected with immunologic assays, and could arguably be designated as TAAs.

Table (7): Tumor markers that have been investigated in breast cancer

Tumor-associated antigens
Carcinoembryonic antigen (CEA)
Products of or related to products of the MUC-1 gene
CA 15-3
CA 27-29
CA 549
Breast cancer mucin (BCM)
Mammary serum antigen (MSA)

Table (7): Continued.

Mucinous carcinoma antigen (MCA)
Tissue polypeptide antigen (TPA)

Tissue polypeptide-specific antigen (TPS)
Gross cystic disease protein (GCDP)
Prostate-specific antigen (PSA)
Markers of tumor biology
Extra-ceullular domain (ECD) of c-erbB-2/HER2/neu
Molecules of adhesion and invasion
E-selectin
Soluble urokinase plasminogen activator receptor (SuPAR)
Intercellular adhesion molecule-1 (ICAM-1)
Molecules associated with angiogenesis
Vascular endothelial growth factor (VEGF)
Basic fibroblast growth factor (bFGF)
Hepatocyte growth factor (HGF)
HUVEC assay
Antibody response against TAAs
c-erbB-2/HER2/neu

P53

There are several tumor markers correlate with the incidence of breast cancer, but the most important markers are: Tumor-associated antigens (TAAs) that have been associated with breast cancer include carcinoembryonic antigen (CEA); tissue polypeptide antigen (TPA), tissue polypeptide-specific antigen (TPS), gross cystic disease protein (GCDP); prostate specific antigen (PSA); and the products of the MUC-1 gene. The MUC-1 gene encodes a cell-associated mucin-like protein. Secretary epithelial cells such as breast epithelial cells express this antigen. Several assays detect the MUC-1 gene products, but they are not identical. These proteins have been identified by monoclonal antibodies to breast cancer cell lines, breast cancer tissue, or human milk fat globule membranes. Assays that detect circulating MUC-1 products include CA 15-3, CA 27-29, CA 549, breast cancer mucin (BCM), mammary serum antigen (MSA), and mucin-like carcinoma-associated antigen (MCA) .

. CEA

Carcinoembryonic antigen is a marker for breast carcinoma , lung, gastrointestinal and colorectal . CEA is one of the older oncofetal protenis in use. CEA is a large family of related cell-surface glycoproteins with a high molecular mass of 150 to 300 KD, it contains 45 to 55% carbohydrate with increase expression found in a variety of malignancies, including breast cancer .CEA is not recommended for screening, diagnosis, staging or routine surveillance of breast cancer patients following primary therapy .CEA belongs to family of cell-surface glycoproteins with increased expression found in a variety of malignancies, including breast cancer. CEA is not recommended for screening, diagnosis, staging, or routine surveillance of breast cancer patients following primary therapy. Routine use of CEA for monitoring response of metastatic disease to treatment is not recommended. But in the absence of readily measurable disease, an increasing CEA level may be used to suggest treatment failure.

TPA

Tissue polypeptide antigen is not a specific tumor marker. Antibodies that react with cytokeratin 8,18 and 19 identify it. TPA is a heterogeneous group of molecules with molecular weight range 20-45 KD Both normal and cancerous cells produce TPA; it is useful in the monitoring of metastic diseases.

TPS

Tissue polypeptide-specific antigen (TPS) is a new tumor marker defined by monoclonal antibody against the soluble tissue polypeptide antigen (TPA). First described as specific tumor marker by Bjorklund in 1957. In breast cancer patients TPS was especially useful in monitoring response to treatment and effectiveness of therapy in metastatic disease.

CA 549

CA 549 is an acidic glycoprotein and it is a marker for breast carcinoma. CA 549 is not useful in detecting early breast carcinoma but it is useful is detecting recurrence of breast cancer in patients after initial therapy followed by adjuvant therapy.

CA 27.29

CA 27.29 is detected by a monoclonal antibody B 27.29, this is produced against antigen in ascites of patients with metastatic breast carcinoma. CA 27.29 test above 37.7 $KU.L^{-1}$ were considered positive, its most useful in monitoring metastatic breast carcinoma.

CA 125

CA 125 is a high-molecular mass (>200 KD) glycoprotein recognized by the monoclonal antibody OC 125. The level of CA 125 is measured quantitatively by using immunoradiometric assay. In healthy population, the upper limit of CA 125 level is 35 $KU.L^{-1}$. CA 125 is elevated in ovarian carcinoma, endometerial, pancreatic, Lung, breast, colorectal and other gastrointestinal tumors. CA 125 is useful to detecting residual disease in cancer patients following initial therapy.

Mammary Antigen

Several new antigens have been recognized by monoclonal antibodies. Which have been identified in patients with breast cancer. They have been proposed as "tumor markers":

MCA

Mucin-like carcinoma associated antigen (MCA) is a mucin glycoprotein with a molecular mass of 350 KD. MCA was identified on the surface of a breast carcinoma cell line by the monoclonal antibody b-12. MCA level is elevated in 60% of metastatic breast cancer patients.

- **MAM-6**

MAM-6 an epithelial membrane antigen present on ductal and alveoli epithelial cells that is detected by monoclonal antibody raised against human milk-fat globule membranes. Partial characterization of the antigen by SDS-PAGE showed that the antigen is a polymorphic epithelial sialomucin with a molecular mass over 400 KD.

- **MSA**

Mammary serum antigen (MSA) was detected by an antibody raised against a whole cell suspension of intraductal breast cancer.

Galectin-4

A protein Galectin-4 is expressed in non-invasive and invasive breast cancer but not in normal cell. An anti-Galectin-4 antibody was able to detect the presence of Galectin-4 very specifically. Galectin-4 is specific diagnostic marker of breast cancer whose patterns of expression at early stages of disease could identify those patients with a high risk of progression to aggressive cancer.

Cathepsin-D

Cathepsin-D is a glycoprotein with molecular weight M.wt: 52 KD. It was discovered in 1979 in the culture medium of hormone dependent human breast cancer. It is a precursor to lysosomal acidic protease. This proteolytic enzyme can react against basement membranes .

Cathapsin-D may facilitate cellular actions such as migration, metastasis, and an invasion of other tissues. Estrogen has been shown to stimulate secretion of this tumor marker in certain hormone-dependent breast cancer cell lines. This antigen has been found to have potential application in breast cancer prognosis as its concentration appears to be related to the patients overall change for survival . .

. Cathepsin-D, this proteolytic enzyme can react against basement membranes. Cathepsin-D also has mitogenic activity on MCF-7 cells that are estrogen-depleted. Further studies showed that cathepsin-D was relatively low in resting mammary cells but was elevated in malignant and benign proliferative breast diseases. These findings raised the suspicion that the Cathepsin-D could both promote abnormal growth of cells as well as contribute to the metastatic potential of malignant cells through its disruption of the basement membrane and therefore might be a marker for a poor prognosis.

Carbohydrate Antigen 15-3 (CA 15-3)

CA 15-3 is a breast-associated antigen identified on the apical side of alveoli and ducts of mammary glands and as a circulating antigen [82]. Distinct epitopes of this high molecular-weight mucin-like glycoprotein of 300-400 KD, which carbohydrate side chain account for about 50% .

Also known as polymorphic epithelial mucin (PEM), epithelial membrane antigen (EMA) or episialin . CA 15-3 can be identified by two monoclonal antibodies DF3 and 115 D8, in a double-determinate or sandwich-type immunoassay . The 115 D8 antibody was prepared against human milk-fat globulin membrane while the DF3 antibody was raised against a membrane-enriched fraction of a human breast carcinoma .

Chapter Three

CA15-3 in Breast Tumors

Carbohydrate Antigen 15-3 (CA 15-3)

CA 15-3 is a breast-associated antigen identified on the apical side of alveoli and ducts of mammary glands and as a circulating antigen . Distinct epitopes of this high molecular-weight mucin-like glycoprotein of 300-400 KD(83-85), which carbohydrate side chain account for about 50% .

Also known as polymorphic epithelial mucin (PEM), epithelial membrane antigen (EMA) or episialin . CA 15-3 can be identified by two monoclonal antibodies DF3 and 115 D8, in a double-determinate or sandwich-type immunoassay . The 115 D8 antibody was prepared against human milk-fat globulin membrane while the DF3 antibody was raised against a membrane-enriched fraction of a human breast carcinoma. CA 15-3 measures the serum level of a mucin-like membrane glycoprotein, which is shed from tumor cells into the bloodstream. The CA15-3 epitope is recognized by two monoclonal antibodies in a double-determinant or sandwich radioimmunoassay.

Structure of CA 15-3

CA 15-3 (Episialin) is synthesized as transmembrane molecule with a relatively large extracellular domain and cytoplasmic domain of 69 amino acids. The extracellular domain mainly consists of region of nearly identical repeats population, leading to substantial differences in molecular weights of the CA 15-3 molecules from different individuals .

The repeats together with adjacent degenerated repeats contain many serins and threonines that are potential attachment sites for O-liked glycans and constitute the mucin-like domain, which comprises more than half of the polypeptide backbone. The mucin domain of CA15-3 contains many prolines and other helix-breaking amino acids, resulting in a molecule with an extended structure and many β-turns . The extended structure is very rigid as aresult of the numerous O-linked glycans attached to the molecule . The CA15-3 extends 200 to 500 nm above the cell membrane .

CA 15-3 Expression

- ### CA 15-3 Expression in Normal Tissues

CA 15-3 is predominatly found at the apical side of epithelial cells lining the acini alveoli, or lumens in various organs, i.e. in the mammary glands, salivary glands, sebacious glands, sweat glands, esophagus, stomach, pancreas, bile ducts, lungs, kidney, bladder, prostate, uterus, and rete testis .

- ### CA 15-3 Expression In Malignant Tissues

Relative to the expression levels of CA 15-3 found in normal tissues, CA15-3 is often overexpressed several-fold in many types of carcinomas derived from these tissues . In these tumors, polarization of the cells is often lost, resulting in the presence of CA 15-3 at the entire cell surface. High levels of CA 15-3 are also detected on carcinoma cells present in pleural effusions on ascites from patients with breast or ovary carcinoma and on many breast carcinoma cell lines.

Biosynthesis of CA 15-3

CA15-3 is synthesized as a large single polypeptide, in most cell lines approximately 200 KD or more . This precursor is rapidly cleaved by proteolysis in a small moiety, which contains the transmembrane and cytoplasmic domains, and a larger part, which comprises most of the extra cellular domain. Both moieties remain non-covalently associated . This proteolytic processing step occurs in the endoplasmic reticulum and may be essential for further maturation. CA 15-3 is mainly processed by adding numerous O-linked glycans, which increases the apparent molecular weight on SDS-polyacrylamide gels to more than 400 KD. The extensive glycosylation protects the molecule against proteolytic degradation, since the precursors without O-linked sugars are degarded rapidly, while the mature molecule is extremely resistant to the action of proteases. The glycosylation also determines the rigidity of the molecule. The last step in the processing of CA15-3 is the addition of sialic acid to the glycans, which increases the mobility of the molecule on SDS-gels .

The early proteolytic cleavage step is not directly responsible for the release of CA15-3 for the membrane, which suggests that CA15-3 is most likely released from the membrane by a second proteolytic cleavage step after arrival at the cell surface. The second proteolytic cleavage seems to be a slow and probably a random process, allowing the mucin to remain associated with the cell surface with a half-life of 16-24 hrs.

Methodology

The CA 15-3 test from all sources uses both DF3 and 115-D8 antibodies. Serum is initially incubated with a polystyrene bead to which 115-D8 antibody has been attached. This antibody binds to antigenic sites on the glycoprotein, pulling it out of solution. The beads are then washed to remove unbound meterial and incubated with the radioiodine ($125I$)-labeled DF3 antibody. The radiolabeled DF3 antibody binds its antigenic sites and then the amount of radioactivity is quantitated. This is called Immunoradiometric Assay (IRMA).

Biology of CA15-3

- CA15-3 and Cell Adhesion

Similar to mucins in mucus, membran-associated mucins might act as barrier molecules to protect cells against toxic substances, as in pancreatic and bladder ducts. The high densities of CA15-3, due to its extended and relatively rigid structure, might also interfere with the function of the adhesion molecules. In this way, CA15-3 might prevent interactions between opposing apical membrane of polarized normal cells and facilitate the formation and maintenance of the ducts during development.

In carcinomas, the combination of overproduction and loss of polarization of CA15-3 expression might reduce cell adhesion and facilitate the invasion of tumor cells because CA15-3 might now interfere with the function of molecules required from tissue integrity.

CA15-3 and Immune System

The putative function of CA15-3 in tumor progression may not only be restricted to inhibition of adhesion which will probably result in an increased invasive potential of cells, but CA15-3 overexpression may well be critical to the survival of tumor cells during dissemination. A completely different aspect of CA15-3 is its ability to act as a tumor-specific antigen. The underglycosylation of CA15-3 in various tumor cells exposes the protein backbone, leading to the generation of novel epitopes. This could elicit an immune response.

Clinical Application

In healthy subjects, the upper limit of CA 15-3 concentration is 25 (KU.L-1). At this level, (5.5%) of 1050 normal subjects, (23%) of patients with primary breast cancer, and (69%) of those with metastatic breast cancer show elevated CA 15-3 levels .

Elevated CA15-3 levels are also found in other malignancies, including pancreatic (80%), lung (71%), breast (69%), ovarian (64%), colorectal (63%) and liver (28%) cancer. It is also reported to be elevated in benign diseases, although with less frequency (e.g., in benign liver [42%] and benign breast diseases [16%]).

CA15-3 should be used to diagnose primary breast cancer, because the incidence of elevation (23%) is fairly low. CA 15-3 is most useful in monitoring therapy and disease progression in metastatic breast cancer patients. A significant change must be at least (25%) and correlates with disease progression in (90%) of patients, with its regression in (78%). No change correlates with disease stability in (60%). CA 15-3 could replace CEA in metastatic breast cancer owing to its sensitivity and specificity. CA 15-3 has been used in management of patients, with breast cancer. CA 15-3 has been evaluated for its ability to determine diagnosis, prognosis, monitor therapy and predict recurrence of breast cancer following curative surgery and radiation therapy . Low incidences of CA 15-3 elevation in early stage cancer (stage I and stage II) have been observed .

Incidence of abnormal values of CA 15-3 in stage III and stage IV, and a very high CA 15-3 level have been correlated with metastases of breast cancer.

Therefore the development of immunoradiometric was planned to carry out the determination of the optimum conditions of 125I-anti CA 15-3 antibody binding with CA 15-3 in breast tumor tissue homogenate, hence determination of CA 15-3.

Materials and Methods

Patients

Three groups of breast tumors patients were included in this study.

Group I   : Consisted of 40 patients with benign breast tumors

Group II : Consisted of 32 premenopausal patients with breast cancer.

Group III : Consisted of 15 postmenopausal patients with breast cancer.

Group IV : Consisted of 10 controls.

Patients suffered from any disease that may interfere with this study were excluded. All surgical operation of breast tumors were carried out under the supervision of the following surgeons:

The host information of all patients and normal healthy subjects is summarized in table (8).

Table (8): The host information of breast tumors patients and healthy subjects studied.

Group	Patients	No.	Age	Type of tumor	Metastases
I	Benign breast tumor	40	18-42	23 fibroepithelial tumor (fibroadenoma) 17 fibrocystic changes (adenosis)	– –
II	Premenopausal malignant breast tumor	32	34-52	22 Infiltrative Ductal carcinoma 10 Ductal carcinoma	2 lymph nodes
III	Postmenopausal malignant breast tumor	15	55-73	Infiltrative Ductal carcinoma	4 lumph nodes
IV	Control	10	25-40		

Preparation of Blood Samples

Five milliliters of blood samples were obtained from patients by venipuncture just before surgery. Ten physically normal age volunteers were used as controls. Blood samples were left for 20 min. at room temperature. After coagulation, sera were separated centrifugation at 3000r.p.m for 10 min., and then sera were aspirated and stored at –20oC until time analysis. The samples were not thawed and refrozen before testing.

Collections of Specimens

The tumors tissues were surgically removed from breast tumor patients by either mastectomy (cancer patients) or lumpectomy (benign tumor patients). The specimens were cut off and immediately rinsed with ice-cold isotonic saline solution. They were collected individually in plastic receptacle and stored at –20 oC until homogenization.

Preparation of Breast Tumors Tissues Homogenates

The frozen tissue were weighed, sliced finely and scalped in petri dish standing on ice bath, and then homogenized with fivefold volumes of PBS buffer pH7.2, using manual homogenizer . The homogenate was filtered through four layers of nylon gauze in order to eliminate fibers connective tissues, and then centrifuged at 4000 r.p.m for 45 min. at 4 oC in order to precipitate the remaining intact cells and the intact nucleus. The supernatant fraction at this speed was separated, divided in aliquots and freezed-20 oC until use.

Statistical Analyses

Students' t-test was used to determine if the mean values of studied parameters were significant different in the individual groups included in this work. $P<0.05$ were considered significant .

Methods
Protein Determinations

Total homogenate protein content was determined by the method of Lowry, using bovine serum albumin (BSA) as the standard.

Figure (3) represents the standard curve of protein, which was constructed by plotting the absorbance at 600 nm against standard protein concentrations

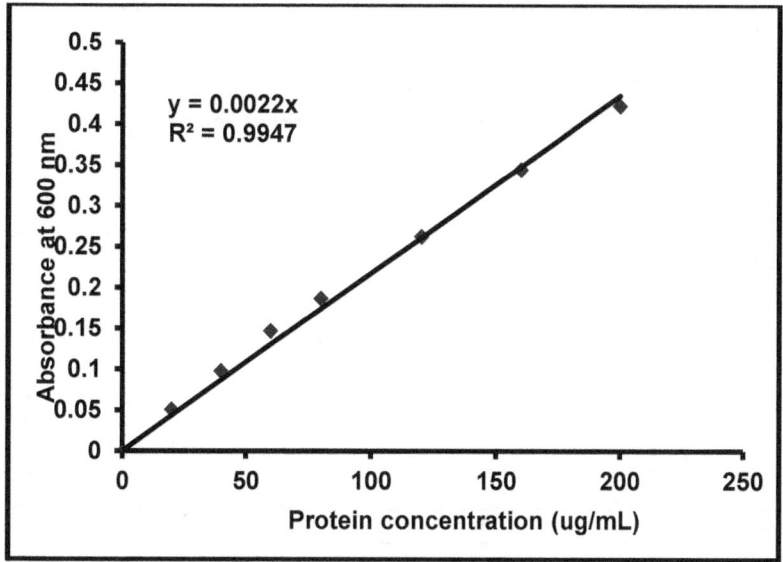

Figure (3): Standard curve of protein concentration. (All other details are explained in the text).

Determination of CA 15-3 Levels in Sera of Breast Tumors Patients

Reagents
- Tracer: two vials each one contained 1.0 μ Ci/mL (37.1 KBa /mL). CA15-3 antibody labeled with 125I in 10 mL / Tris buffer with protein stabilizer and preservative.
- CA15-3 standards: The vial contained 100 mL, which represented 0 U.mL-1. There are four vials, 1.2 mL in each vial with different concentrations of human CA15-3 (25, 50, 100, 200) U.mL-1 in Tris. buffer with protein stabilizer and preservative.
- One bottle contained 100-coated beads, Anti-CA15-3-mouse, monoclonal.
- One vial contained 0.5mL CA15-3 control, CA15-3 in re-calcified human plasma with preservative.

Procedure

The assay protocol is described in table (9).

Table (9): IRMA protocol of serum CA 15-3 (U.mL-1) (All other details are explained in the text).

	CA 15-3 standard (U.mL-1)					Control	Unknown samples	
	0	25	50	100	200		1	2-etc.
Reaction trays no.	1.2	3.4	5.6	7.8	9.10	11.12	13.14	15.16-etc.
Standard (µL)	200	200	200	200	200	–	–	–
Control serum or samples (µL)	–	–	–	–	–	200	200	200
125I-anti CA15-3	200	200	200	200	200	200	200	200

The specimens and reagents must be brought to room temperature (20-30 oC) before opening. The reaction trays and data sheets were marked.

First Incubation

The specimens and the control were diluted to (1:15) prepared by adding 20 µL of specimen or control to 1000 µL CA15-3 standard, 0 U.mL-1 in a tube marked proper identification of specimen. Two hundred microliters of diluted specimen and control were pipette to their assigned wells. Two hundred microliters of each standard was pipette to its assigned well (standards are not to be diluted). One bead was dispensed into each well and the adhesive cover sealer was applied. After incubation for 2hrs at room temperature, the adhesive cover sealer was removed and the liquid was aspirated, then each bead was washed three times with 5 mL distilled water.

Second Incubation

Two hundred microliters of 125I-antiCA15-3 was pipetted on each bead. The adhesive cover sealer was applied again. After incubation time for 3hrs at room temperature, the cover was removed and the liquid was aspirated from wells, then the beads were washed as it is above. The beads were transferred to the counting tubes, and then the tubes were counted for 1 min.

Calculations

The standard curve was constructed by plotting counts per min. (Y axis) versus concentration for CA15-3 standard (X axis), figure 4). Then the points were connected with straight-line segments.

The CA15-3 concentration of specimens and control were determined directly from the standard curve.

Preliminary Test of CA 15-3 Binding to 125I -Anti CA15-3 Antibody in Breast Tumor Homogenate

The supernatant and pellet were centrifuged and detected by using ordinary tubes. In order to detect CA 15-3, 100 µL of the supernatant breast homogenate having (900µg protein) were incubated with 50 µL (0.35 mg.mL-1) of 125I -anti CA15-3.The volume of reaction was completed to 500 µL with PBS buffer pH7.2, then incubated at 37 oC for 2hrs. The assay tubes were centrifuged at 4000 r.p.m. for 45 min. at 4 oC.

The supernatant was discarded, the rim at each tube was swabbed with cotton, and then gamma counter counted the complex formed for one minute. The pellet of CA 15-3 was estimated by dissolving the sediment in PBS-buffer pH 7.2 with the ratio 1:5 (weight: volume) shaking was then carried out. Hundred microliters of the supernatant fraction of the sediment having (540 µg.mL-1 protein) was added to 50 µL (0.35 mg.mL-1) of 125I -anti CA 15-3 antibody. The same steps mentioned in this experiment were followed to determine the radioactivity of the complex formed. For total count two additional tubes with 50µL of 125I –anti CA 15-3 antibody were counted in gamma counter.

Calculations

1. The counted radioactivity in each tube (expressed in c.p.m.) represents the bound fraction (B), (i.e., 125I antiCA15-3 antibody/CA 15-3 complex).

2. The counted radioactivity in the tubes containing 125I-anti CA15-3 antibody only represents the total count (T).

3. The (B/T) ratio for each tube counted as follows:

$$(B/T)\% = \frac{\text{Sample Counts}(B)}{\text{Total Counts}(T)} \times 100$$

Factors Effecting of 125I-Anti CA-3 Antibody Binding to CA 15-3 in Breast Tumors Homogenates

The Effect of Different Amounts of Protein Concentration of the Tumor Homogenate on the Binding with 125I-Anti CA 15-3 Antibody

Procedure

1. Fifty microliters (0.35 mg.mL-1) of 125I -anti CA 15-3 antibody were added to 100µL of the supernatant (benign Fibroadenoma, pre-and post-menopausal malignant breast tumors (IDC) respectively) containing increasing amounts of protein (50, 100, 150, 200, 250 µg.mL-1) then completed to a final volume of reaction to 500 µL with 0.15 M PBS pH 7.2.

2. The assay tubes were then incubated for 2 hrs at 37oC.

3. Two additional tubes, containing 50µL (0.35 mg.mL-1) of 125I –anti CAB-3 antibody only, for total counts were set-aside until counting.

4. At the end of incubation, the assay tubes were centrifuged at 4000 r.p.m for 45 min at 4oC.

5. The supernatant were decanted, the rims at the tube were swabbed with cotton piece.

6. The radioactivity of the complex were counted using gamma counter.

Calculations

1. The B/T percent were determined according to the text.

2. The percent of binding values B/T were plotted versus the increasing amount of protein of the breast tissue homogenate.

The Effect of 125I -Anti CA15-3 Antibody Concentration on the Binding

Procedure

1. Fifty microliters of increasing concentration (0.070, 0.140, 0.175, 0.350, 0.701 mg.mL-1) of 125I -anti CA 15-3 antibody were added to 100µL of

homogenate (benign breast tumor (Fibroadenoma), pre-and post-menopausal malignant breast tumors) (IDC) containing (100, 100, 200 μg.mL-1 protein) respectively.

2. The volume of reaction was made up to 500 μL with PBS pH 7.2.

3. Steps 2,3,4,5 and 6 of the experiment were repeated.

Calculations

1. The same mathematical equation mentioned in the text was used to calculate (B/T)%.

2. Values of (B/T)% were plotted versus concentration of labeled antibody (125I -anti CA15-3 antibody).

The Effect of pH on the Binding

Procedure

1. One hundred microliters of human homogenate (benign breast tumor (Fibroadenoma), pre-and post-menopausal malignant breast tumors (IDC)) containing (100,100,200, μg.mL-1 protein) were added to 50μL (0.175, 0.175,0.140mg.mL-1) of 125I -anti CA15-3 antibody respectively.

2. Each mixture was completed to 500 μL with PBS of different pH ranging (6.8-8.0).

3. Step 2,3,4,5 and 6 of the experiment were repeated.

Calculations

1. Values of (B/T) % were calculated as described in the text

2. (B/T)% were plotted against their corresponding pH.

Time Course of the Binding of 125I -Anti CA15-3 Antibody to CA 15-3 in Breast Tumors Homogenates

Procedure

1. One hundred microliters of homogenate (benign breast tumor (Fibroadenoma), pre-and post-menopausal malignant breast tumors (IDC)) containing (100,100,and 200 μg.mL-1 protein) were incubated with 50μL of 125I -anti CA15-3 antibody concentration (0.175,0.175 and 0.140 gm.mL-1).

2. The volume of reaction was made up to 500 µL with PBS pH (7.0, 7.6 and 7.8).

3. All tubes were incubated at 37°C at different time intervals (30, 60, 90, 120, 150 and 180) min.

4. Step 3,4,5,6 of the experiment was repeated.

5. To determine the time course of CA 15-3 binding to 125I –anti CA 15-3 antibody at different temperatures. Steps 1, 2, 3 and 4 in the same experiment were repeated at different temperatures (5, 15, 25, 45°C).

Calculations

1. The values of (B/T)% were calculated as described in text at each time and temperature used.

2. The values (B/T)% was plotted against the time of incubation at different temperatures.

The Effect of Different Halides on the Binding

The breast tumors homogenates (benign breast tumor (Fibroadenoma)) were prepared as described in the text except using PB-buffer instead of PBS at the same pH and concentration carried out the homogenization.

Procedure

1. One hundred microliters of each group homogenate (benign breast tumors (Fibroadenoma) and pre-post menopausal malignant breast tumors (IDC)) containing (100, 100 and 200 µg.mL-1 protein) were incubated with 50 µL of 125I -anti CA 15-3 antibody concentration (0.175, 0.175 and 0.140 gm.mL-1). The volume was made up to 500 µL with PB pH (7.0, 7.6 and 7.8) containing 0.01 M of the following halides: NaF, NaCl, NaBr and NaI in each assay tube. (A sample without the addition of any salt was used as a control).

2. The assay tubes were then incubated for 90min. at 45, 15 and 45°C for the three groups individually.

3. Steps 3, 4, 5 and 6 of the experiment were repeated.

Calculations

1. The values of (B/T) % were calculated (B/T)% was plotted against halides concentrations.

The Effect of Monovalent and Divalent Cations on the Binding

Procedure

1. The same steps mentioned in the text were followed to determinate the effect of monovalent and divalent of CA 15-3 in the tissues homogenates of (benign breast tumors (Fibroadenoma) and pre-and postmenopausal malignant breast tumors (IDC)) with 125I -anti CA 15-3 antibody, except the PB buffer containing (0.025M) of the following salts: KCl, NH4.Cl, MgCl2.6H2O, CaCl2.2H2O, MnCl2.4H2O, CuSO4.5H2O, ZnCl2.

2. A sample without the addition of any salt was used as control.

Calculations

1. The values of (B/T)% were calculated as described (B/T)%was plotted against monovalent and divalent cations salts concentrations.

Recovery of CA 15-3

Known concentration of CA15-3 (200 U.mL-1) was added to the three groups of tissues homogenates (benign breast tumors (Fibroadenoma), and pre-and post-menopausal malignant breast tumors (IDC)). The experiment was carried out at optimum conditions that were obtained in experiments of . The CA15-3 was determined according to the experiment in the text.

Calculations

1. The bound (c.p.m) of the reaction mixture (standard CA 15-3 was added to tissue homogenate) with 125I -antiCA15-3 antibody, represent the measured value.

2. The bound (c.p.m.) of CA 15-3 in tissue homogenate with 125I – antibody CA 15-3 antibody only, represent the expected value.

3. The recovery % (yield%) was calculated as follows:

$$\text{Recovery\%} = \frac{\text{Measured values}}{\text{Expected values}} \times 100$$

Results and Discussion

Human breast tissues in this study were classified according to type of breast tumors (benign and malignant) and the malignant breast tumors were again classified into sub groups (premenopausal and postmenopausal). Each type was examined histologically according to WHO classification.

Homogenization was carried out in a cold medium (i.e.4°C) to avoid protein denaturation , by proteolytic enzymes . The filtration of the tissue homogenate through several layers of nylon gauze was used to remove any suspended pieces unhomogenized fragments and blood vessels.

Determination of CA 15-3 Levels in Sera of Breast Tumors Patients

CA 15-3 levels in sera of patients with benign breast tumors (group I) and pre and post-menopausal malignant breast tumors (group II and group III) were measured by IRMA method. These three groups were matched with a group of control subjects.

Table (10) summarizes the groups and the mean concentrations of CA15-3 for the control women and patients with benign breast tumors and pre-and post-menopausal malignant breast tumors.

Table (10) showes that CA15-3 levels in three different groups (benign breast tumors and pre-and post-menopausal malignant breast tumors) were significantly elevated ($p<0.05$) for benign breast tumors and highly significantly elevation ($p<0.0001$) for pre-and post menopausal malignant breast tumors respectively, as compared with the control.

The mean serum CA15-3 level of the control was found to be (17.26 ± 4.06 U.mL-1) as shown in table (2-5), and the cutoff values was (25 U.mL-1) that obtained from (mean +2 SD). This cutoff value is in agreement with Geraghty J.G(151), other study obtained that cutoff value of 40 U.mL-1 (152), 22 U.mL-1 (153), 30 U.mL-1 .

It has shown that widely different cutoff value which was described ranging from 20-40 U.mL-1 in different reference .

According to Bon et al the upper limit of CA 15-3 of normal may be method-dependent. No association between the CA 15-3 and either age or menopausal status was found in the control group. Therefore , the cutoff values do not require adjustments related to these variables. These results were in agreements with Gion M.et.al. . The distribution of the individual values of CA15-3 in sera of patients with benign breast tumors and pre-and post menopausal malignant breast tumors and control, were determined by using the standared curve in figure (4).

It was found that the mean of serum CA 15-3 concentration in 20 patients with benign breast tumors was 21.9 ± 6.6 U.mL-1 (mean ± SD). These results are in agreement with Hayes D.F. et.al . The results show there was highly significant correlation between serum connections of CA15-3 in both groups pre-and post-menopausal status with control, while it was significantly lower in benign breast tumors status.

This is in agreement with Ichihara S. et.al . Therefore all of the cases used in the binding studies were concentrated to this type of carcinoma (IDC) and this type is the common type of breast cancer. In Iraq very high levels of CA15-3 advanced disease and the value 5 to 10 times of normal suggest the presence of metastasis. Increasing numbers of metastatic sites correlate with increasing CA15-3 levels .

These findings suggest that higher levels of CA15-3 represent the breast cancer extent and reflect the cell differentiation and aggressiveness of the tumor. Therefore, it could be concluded that the determination of CA15-3 before surgical operative may be useful as a prognostic factor in breast cancer.

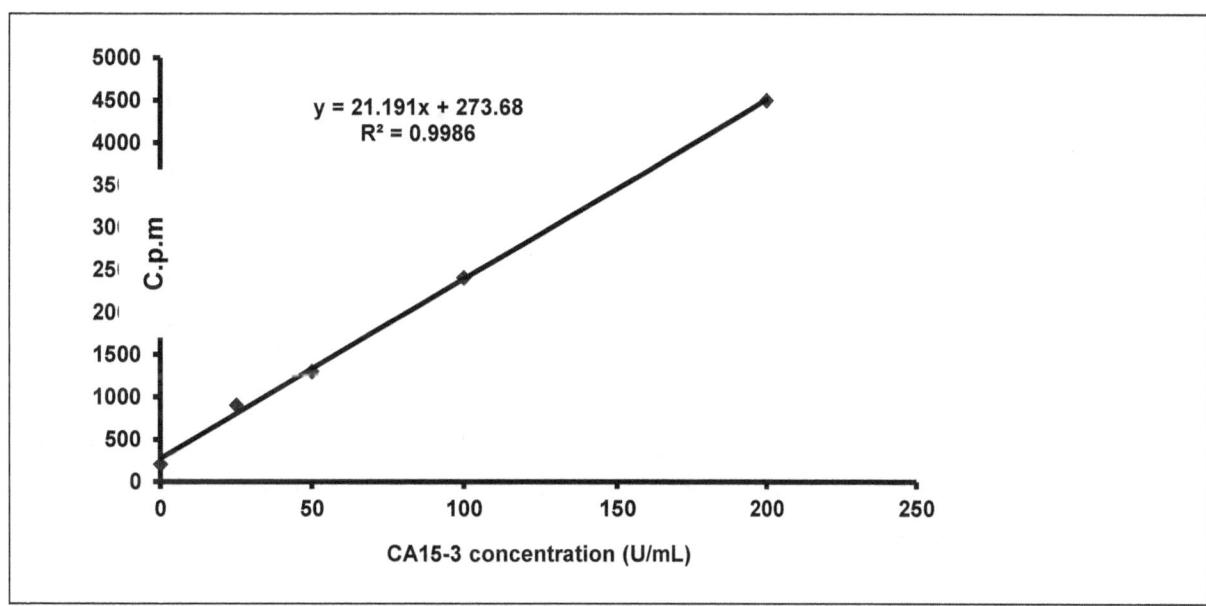

Figure (4): Standard curve of CA 15-3 determination in human sera by IRMA method.(All other details are explained in the text).

Table (10): Sera CA15-3 levels (U.mL-1) in patients with benign and malignant breast tumors. (All details are explained in the text).

Group	Patients	No. of cases	Age range (Year)	Sera CA15-3 U.mL-1 (mean ± SD)	P values
I	Benign	20	18-42	21.9 ± 6.6	P<0.05

	breast tumors				
II	Premenopausal malignant breast tumors	16	34-52	37.3 ± 6.8	P<0.0001
III	Postmenopausal malignant breast tumors	12	55-73	60.3 ± 10.9	P<0.0001
IV	Control	10	25-40	17.3 ± 4.06	--

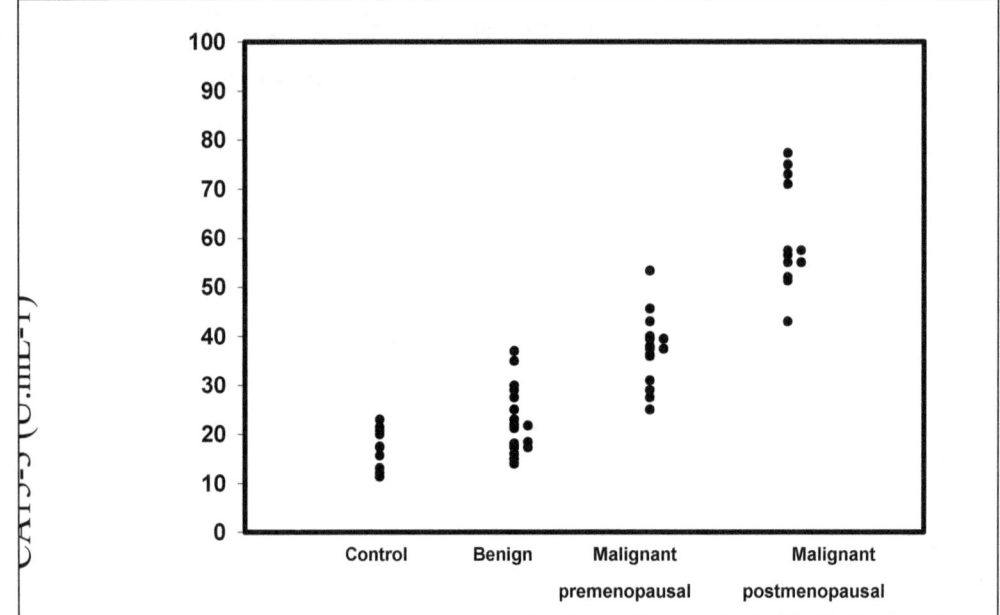

Figure (5): Distribution of the individual value of CA15-3 U.mL-1 in the sera of benign and malignant breast tumors patients. (All other details are explained in the text).

Binding Studies of 125I-Anti CA15-3 Antibody with CA15-3 in Benign and Malignant Breast Tumors Homogenates

Preliminary Test of the Binding of 125I-Anti CA15-3 Antibody with CA 15-3 in Breast tumor homogenate

Supernatant and pellet formed at speed (4000 r.p.m.) in three groups of human breast tumor homogenate (benign breast tumors, pre-and post-menopausal malignant breast tumors) were used in this experiment. In each fraction CA 15-3 was detected through the incubation of 125I-anti CA15-3 antibody with supernatant fraction and pellet individually for 2hrs at 37°C in PBS buffer as a medium to complete the reaction. The separation of the bound antibody from the unbound was carried out at 4000 r.p.m. for 45 min. to precipitate the (125I-anti CA15-3 antibody/CA15-3) complex formed. Preliminary experimental conditions used in Table (11), which is show, the amount of binding B/T% in both fractions. The data revealed that CA15-3 was higher in incidence according to B/T%.

Table (11): Incidence of CA15-3 in supernatant and pellet fractions in three different breast homogenate.

Groups	(B/T)%		CA15-3 U.mL-1 in supernatant fraction kit
	Supernatant Fraction	Pellet Fraction	
Benign	6.20	3.40	90
Premenopausal	8.04	5.53	356
Postmenopausal	6.31	4.64	144

B/T% in supernatant is more than in pellet fractions of this speed (4000 r.p.m.). According to these results supernatant fractions was collected and the pellet was then discarded. The CA15-3 levels in the supernatant of breast tumors homogenate were determined according to IRMA method.

In general, results show that CA15-3 concentration in pre-and post-menopausal malignant breast tumors homogenates is more than benign breast tumors homogenates. These results are in agreement with the result obtained from B/T% from IRMA developed method.

From these results, it can be said that developed method was useful for determination CA15-3 in breast tumors homogenate using 125I-anti CA15-3 antibody.

Factors Effecting of 125I-Anti CA15-3 Antibody Binding to CA15-3 in Breast Tumors Homogenates

The Effect of Different Amounts of Protein Concentration of the Tumor Homogenate on the Binding with 125I-Anti CA 15-3 Antibody

To obtain the optimum protein of homogenate for the binding of CA15-3 with 125I-anti CA15-3 antibody, the supernatant homogenate containing increasing amount of CA15-3 in the presence of fixed amount of 125I-anti CA15-3 antibody as it was mentioned in the text

Figure represents the quantitative precipitation curve in which the amount of (125I-anti CA15-3 antibody/CA15-3) complex in three groups (benign breast tumors and pre-and-post menopausal malignant breast tumors) was plotted as a function of CA15-3 concentration.

As shown in this figure, in the first phase of the reaction no precipitate was formed. The amount of precipitate increased until a point of maximum binding was reached. After this point as the amount of CA15-3 increased the amount of precipitate diminished; thus the increase in protein concentration which would increase the number of binding site and hence increase the percent of binding until the saturation state at (100, 100, and 200 µg.mL-1) homogenate concentration for (benign breast tumors, pre-and post menopausal malignant breast tumors respectively).

The complex precipitate out of solution because of the multivalent nature of both molecules . The radioactive antibody has two binding sites, it can cross-link antigenic sites of two different CA15-3 molecules and can produce maximum complex formation and therefore maximum precipitate will occur. When CA15-3 is in greater excess, large complex are again less probable.

In all subsequent experiments the amonts of (100, 100 and 200 µg.mL-1 protein) of tissue homogenate in benign breast tumors and pre-and post menopausal malignant breast tumors were used according to the result obtained in this experiment.

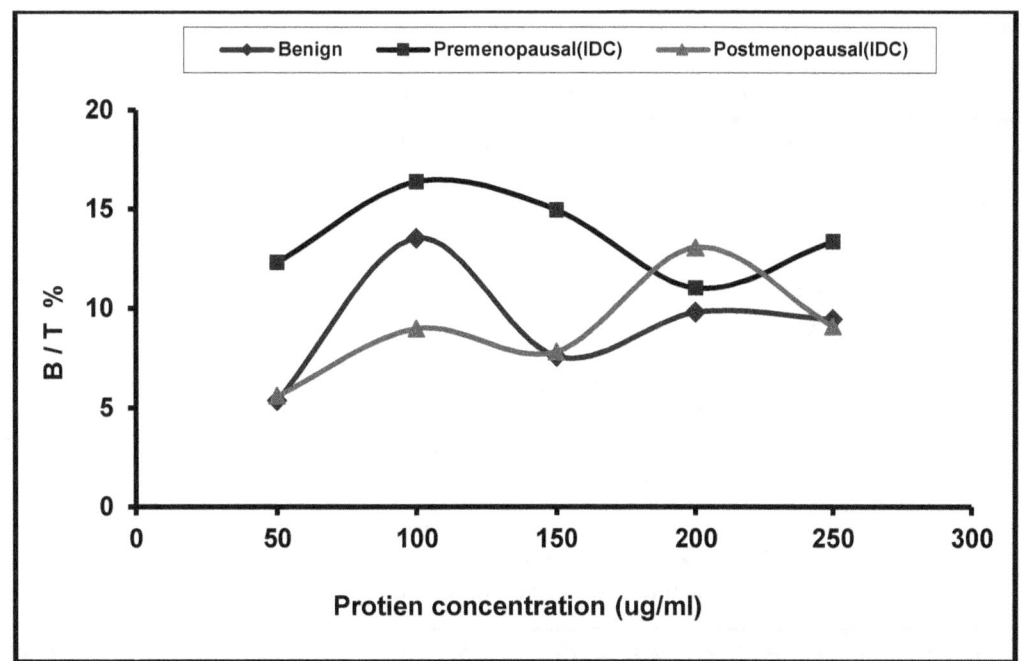

Figure (6): Influence of increasing protein concentration on the binding with 125I-anti CA15-3 antibody. (All other details are explained in the text).

The Effect of 125I-Anti CA15-3 Antibody Concentration on the Binding

The experiment was carried out in the presence of fixed amount of protein concentration of the homogenate and increasing concentration of 125I-anti CA15-3 antibody.

The results are illustrated in figure (7). Which represent 125I-anti CA15-3 antibody binding curve with supernatant fraction of benign breast tumor, pre- and post-menopausal malignant breast tumors. As shown in figure (7) it is obvious that the amount of (125I-anti CA15-3 antibody/CA15-3) complex rises gradually, and then the breast tumor protein was saturated with 125I-anti CA15-3 antibody. When the amount of antibody is in moderate excess, the probability of cross-linking of Ag by Ab in the incubation mixture is more likely, and hence large complex formation is favored. Then the maximum B/T percent was detected. The presence of (0.175, 0.175, 0.14 mg.mL-1) of 125I-anti CA15-3 antibody in benign, pre-and post-menopausal breast tumors homogenates give the optimum concentration of 125I-anti CA15-3 antibody in three groups. Then the binding percent decreased as the amount of 125I-anti CA15-3 antibody increased.

This is because all antigenic sites are covered with antibody and complex formation is inhibited . These results indicate that the binding is principally dependent on the amount of the antibody in the reaction mixture .

According to the results of this experiment the above concentration of 125I-anti CA15-3 antibody was used in the subsequent experiments.

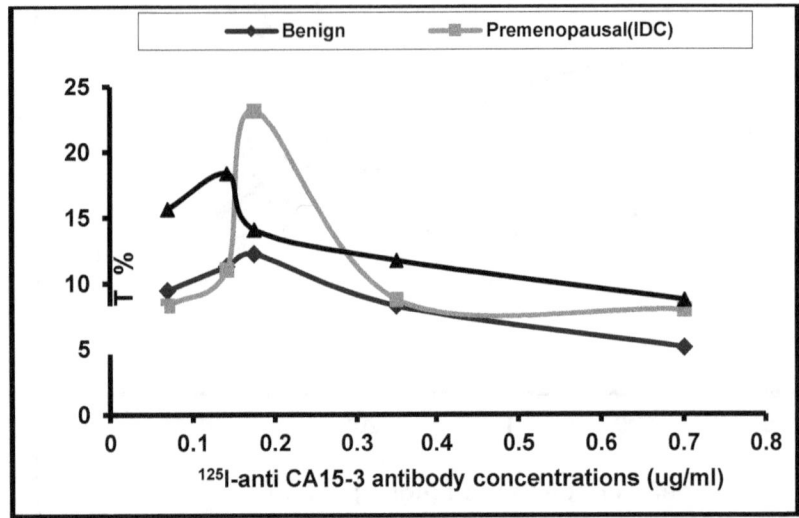

Figure (7): Effect of different concentrations of 125I-anti CA15-3 antibody on the binding of with CA15-3. (All other details are explained in the text).

The Effect of pH on the Binding

Figure (8) shows the values of the binding of 125I-anti CA 15-3 antibody to CA 15-3 in benign breast tumor, pre- and post-menopausal malignant breast tumors, at different pH values. Maximum value of the binding occurs at (pH 7, pH 7.6, pH 7.8) for benign breast tumor, pre-and post-menopausal malignant breast tumors respectively.

The formation of (125I-anti CA 15-3 antibody/CA 15-3) complex is usually performed at pH between 6.8-8.0; the results indicate that the shift in the pH of the environment may affect the properties of CA 15-13 molecules involved in the binding. This effect may include the protonation deprotonation processes occurring within the possible ionizable groups of the amino acids present in the binding domain of these molecules.

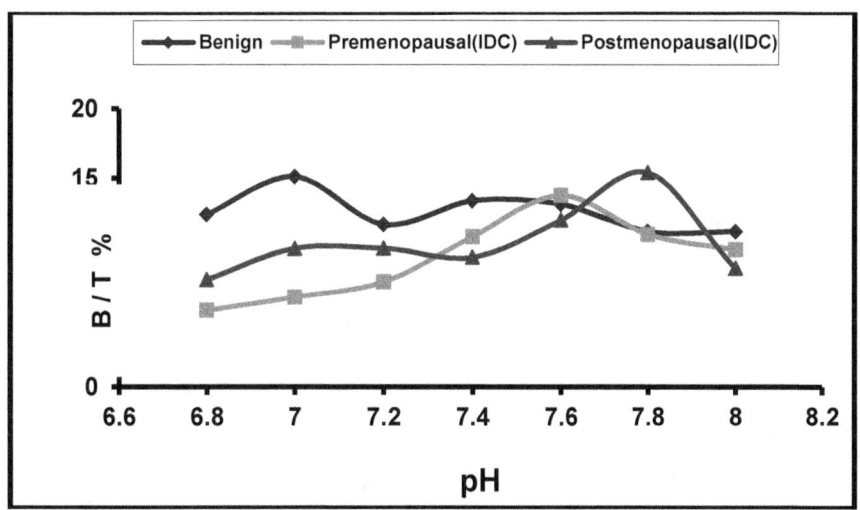

Figure (8): Effect of pH on the binding of 125I-anti CA 15-3 antibody with CA 15-3 in breast tumors homogenates. (All other details are explained in the text).

Time Course of the Binding of 125I -Anti CA15-3 Antibody to CA 15-3 in Breast Tumors Homogenates

The results of time course pattern at different temperatures (5, 15, 25, 37, 45oC) indicate the 125I-anti CA 15-3 antibody binding to crude fractions of CA 15-3 is temperature and time dependent process, as shown in figures (9),. (The maximum binding was obtained at 45oC after incubation for 90 min. in crude fractions of benign breast tumors and postmenopausal malignant breast tumors respectively, whereas the binding in crude fractions of premenopausal malignant breast tumors occurs at 15oC after incubation for 30 min.

The decrease of the binding activity may be due to reversible dissociation of (125I-anti CA 15-3 antibody/CA 15-3) complex after reaching the equilibrium state.

At 45oC the CA 15-3 molecule preserve the nature of protein structure and gave the maximum binding, but at higher temperature than 45oC denaturation may occur.

In the premenopausal malignant breast tumors the maximum binding occurs at 15oC for 30 min., in this state the energy is enough to overcome the energy barrier and give the maximum binding , the decrease in binding after 15oC may be due to proteolytic enzyme.

The difference in incubation time to give the maximum binding may be due to the different source of CA 15-3. According to this results, the binding studies of the subsequent experiments were carried out a 45oC for 90 min incubation for benign and postmenopausal breast tumors homogenate, whereas 15oC for 30 min. incubation for premenopausal malignant breast tumors homogenate.

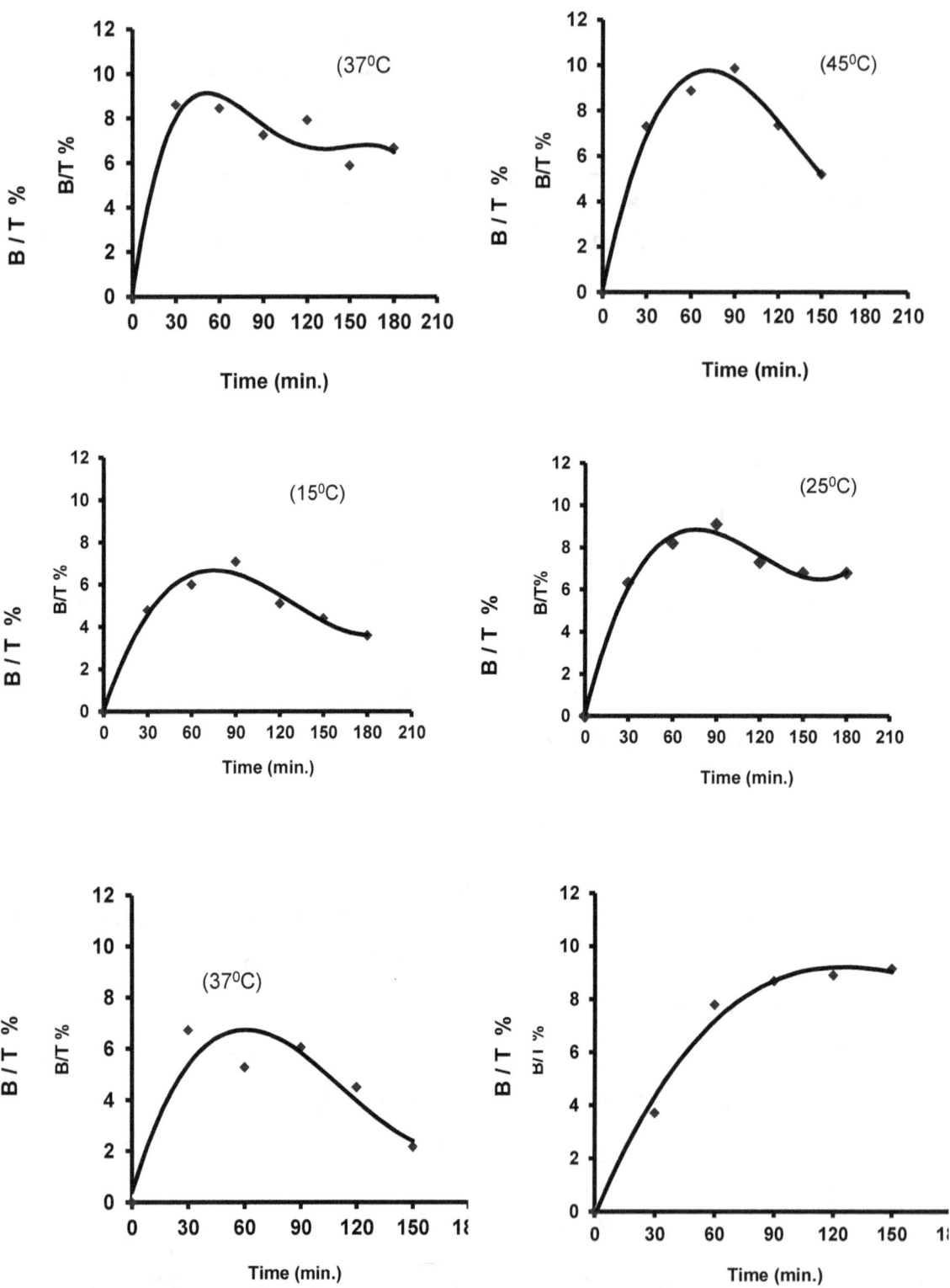

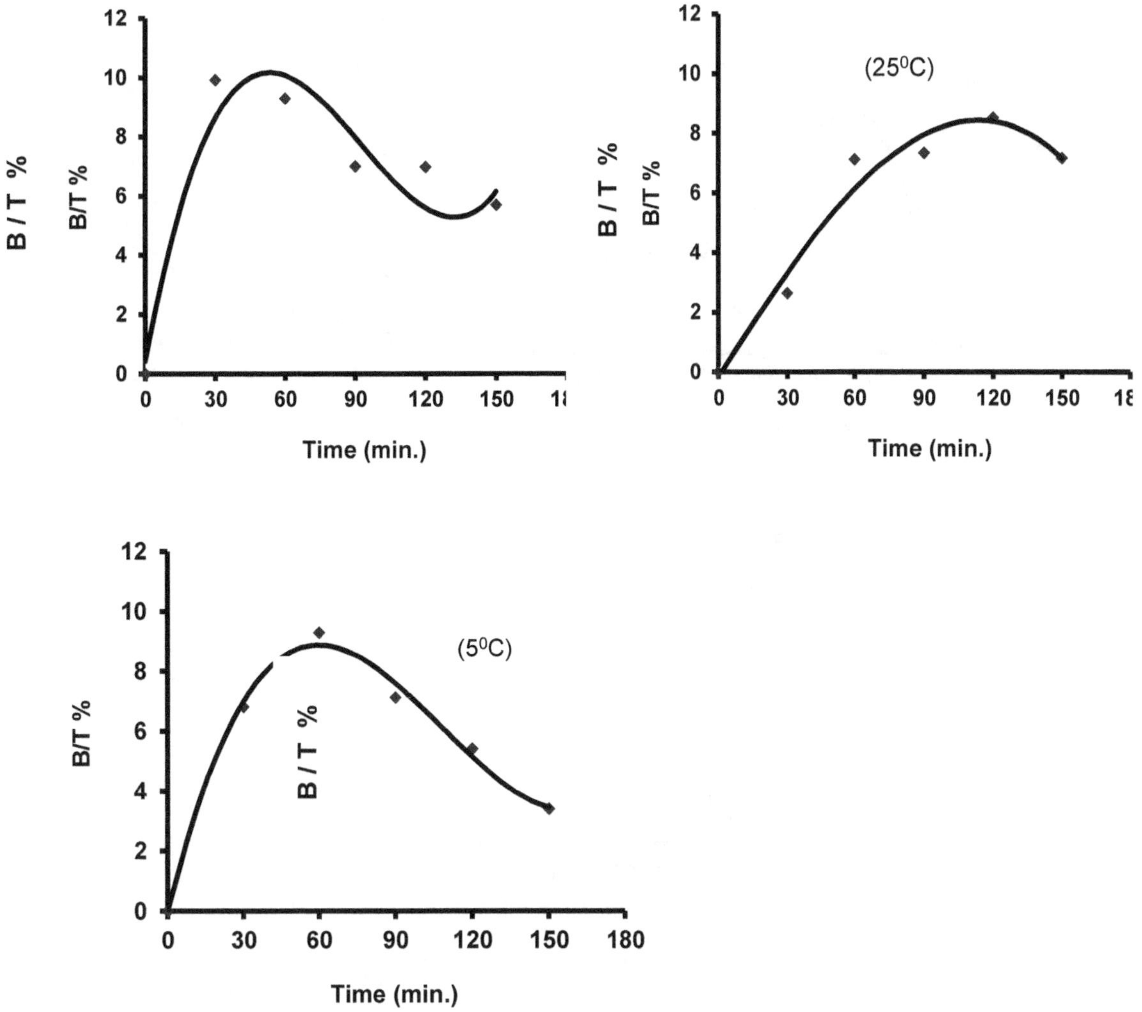

Figure (9): Time course of the binding of 125I-antiCA15-3 antibody with CA15-3 in premenopausal malignant breast tumor. (All other details are explained in the text).

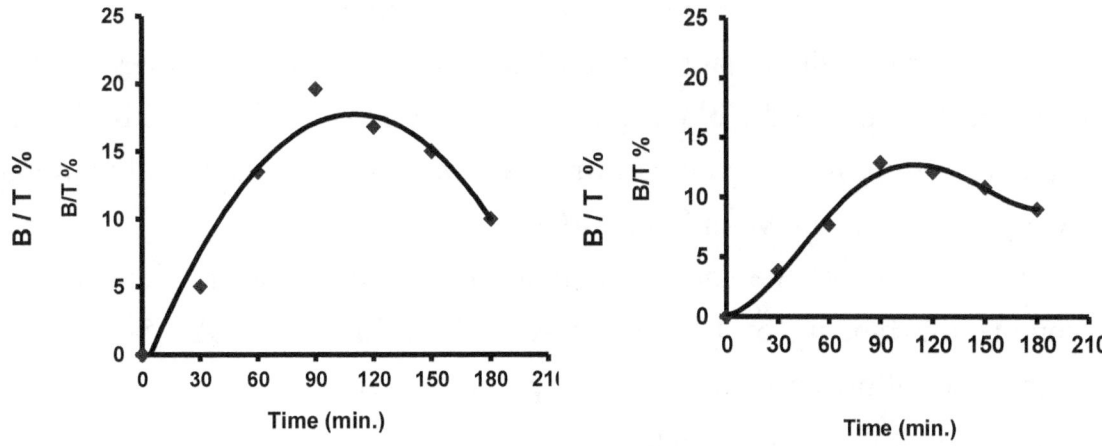

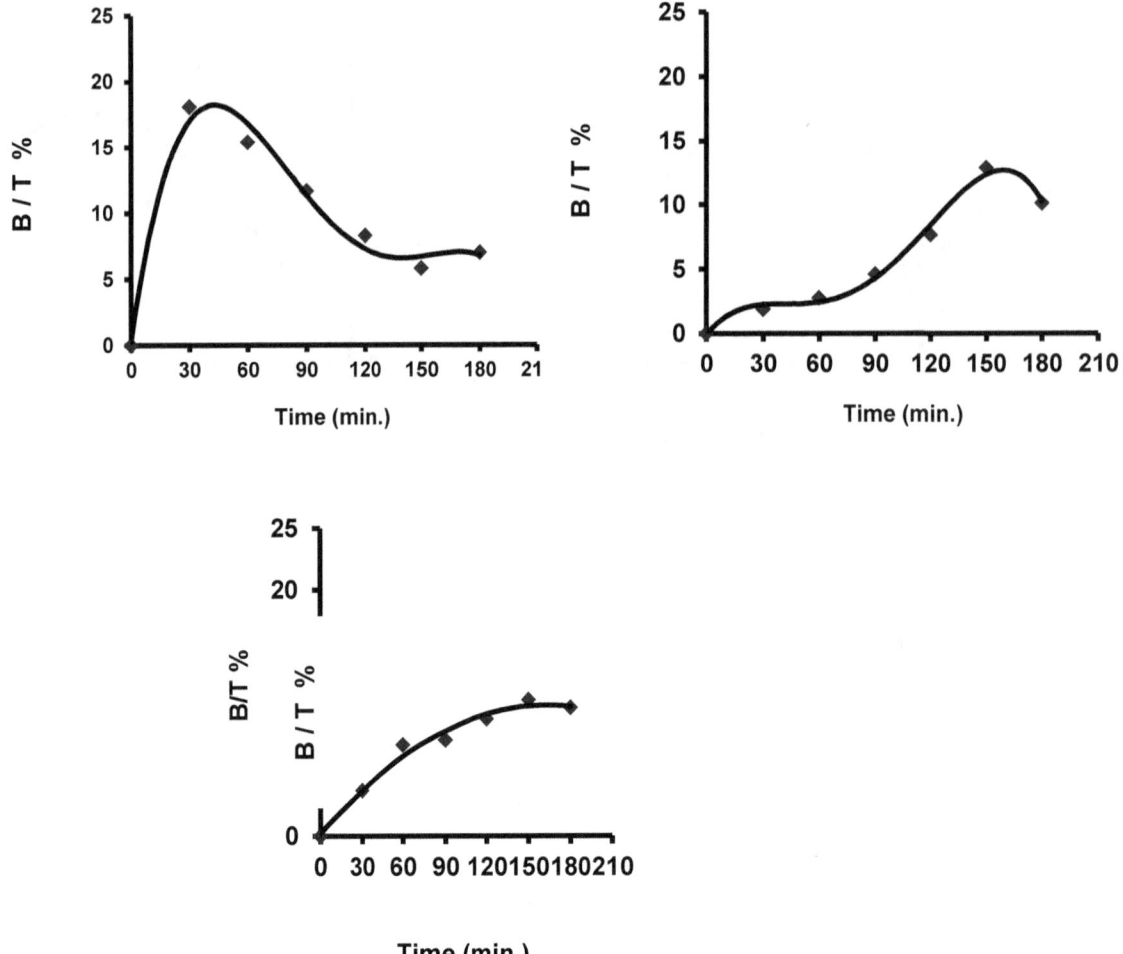

Figure (9): Time course of the binding of 125I-antiCA15-3 antibody with CA15-3 in postmenopausal malignant breast tumor. (All other details are explained in the text).

The Effect of Different Halides on the Binding

Different sodium halides at 0.01 M concentration were investigated to study their action on the binding 125I-anti CA 15-3 antibody with CA 15-3 in the three groups (benign breast tumors, Pre-and postmenopausal malignant breast tumors), as shown in figure (10).

The presence of the sodium halides in the incubation medium tends to promote the binding of 125I-anti CA 15-3 antibody to CA 15-3 in these groups, the following sequence of effects have occurred.

1. Benign breast tumor tissue homogenate

NaI > NaBr > NaCl > NaF

2. Premenopausal breast cancer tissue homogenate

NaCl > NaI > NaBr > NaF

3. Postostmenopausal breast cancer tissue homogenate
NaCl > NaBr > NaF > NaI

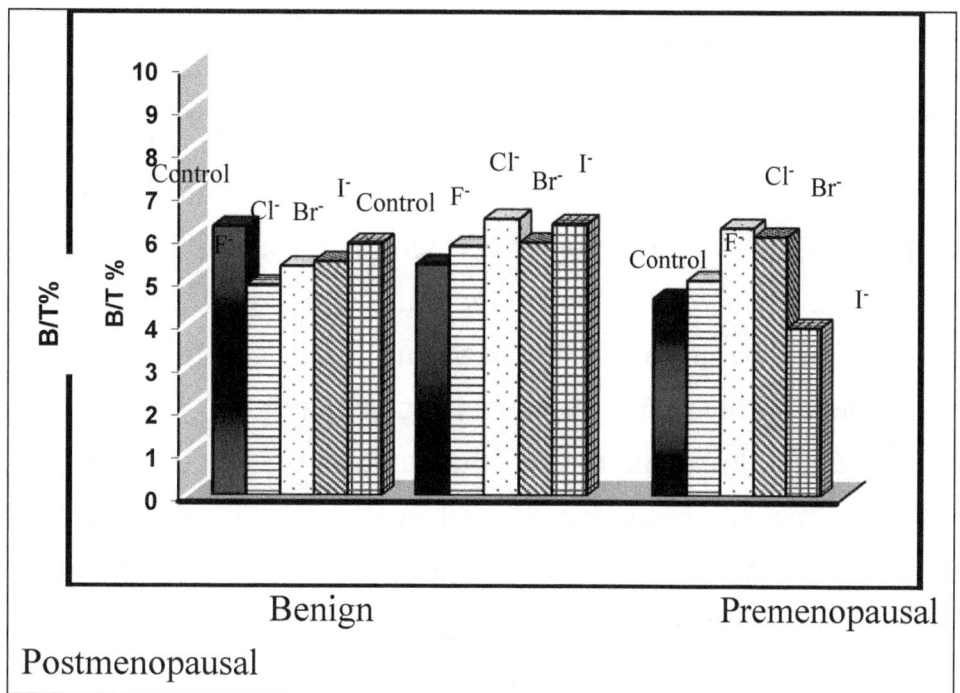

Figure (10): Effect of different halides on the binding of 125I-anti CA 15-3 antibody with CA 15-3. (All other details are explained in the test).

As shown in figure (10), the sodium halides inhibited the binding in benign breast tumors, according to the decreasing ionic radius and increasing radius of hydration. It seemed that fluoride ion causes lower binding, this could be due to higher electro negativity of fluoride ion that tend to interact with the positive residue in the binding site of the antibody and/or the antigen which lead to decrease the interaction between CA 15-3 and its antibody.

Melander and Horvath (1977) reported that the effect of halide salt type on hydrophobic interactions is quantified by its molar surface tension increment (MSTI) that is a measure of the increase in surface tension by the salt. On the other hand, figure (11) shows the effect of different halides salts at 0.01 M on the extent binding of 125I-anti CA 15-3 antibody to pre-and postmenopausal malignant breast tumors homogenate. It seems that halides salts increased the binding, especially NaCl, this could be due to that NaCl in lower concentration (0.15M) or in physiological concentration, increased the binding between CA 15-3 and its antibody.

The Effect of Monovalent and Divalent Cations on the Binding

The effect of different salts on the extent of binding of 125I-anti CA 15-3 antibody to CA 15-3 in benign and malignant breast tumors are shown in figure (11).

The results indicate that the binding process is sensitive to the presence of cation metal ions. $CuSO_4.5H_2O$ at concentration (25mM) was shown to increase the binding more than other divalent cations, while $ZnCl_2$ increased the binding less than other divalent cations. One hypothesis assumes that salts may alter the nature of the hydrophobic forces controlling stabilization of the complex formed and these vary depending on the nature of the interacting groups. From the results illustrated in figure (11), it is suggested that these salts maybe provide some conformational changes in the CA 15-3 and the charged groups of the binding domain of the antibody and antigen molecule, that hinder maximal binding are shielded. If the interaction is dominated by ionic strength, high salt concentration lowers the affinity.

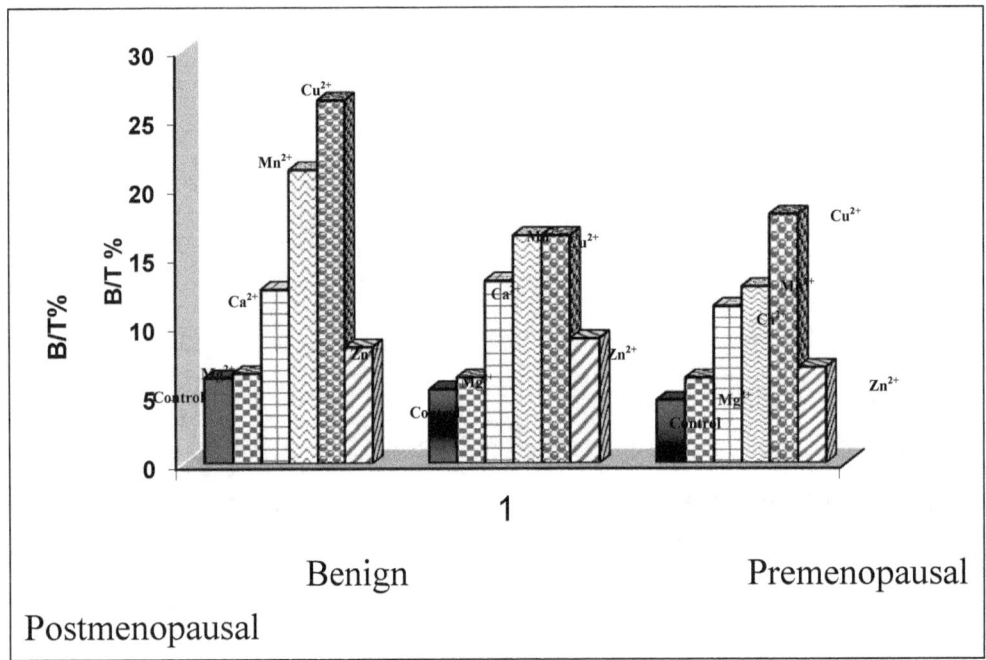

Figure (11): Effect of different divalent cations on the binding of 125I-anti CA 15-3 antibody with CA 15-3. (All other details are explained in the text)

Figure (12) shows the effect of monovalent cations (KCl and NH4Cl) on the extent of the binding of CA 15-3 to its antibody 125I-anti CA 15-3 in benign and malignant breast tumors. KCl at 25mM was shown to increase the binding in benign and premenopausal malignant breast tumors as compared with the control value, while KCl at the same concentration slightly inhibiting the binding in postmenopausal malignant breast tumors. These results may be due to conformational changes. NH4Cl at 25 mM was shown to inhibit the binding but to a lesser extent.

This result shows that NH4Cl effect on the binding is nearly unremarkable. Presumably, the lesser degree of hydration permits greater interaction of the salt with an anionic group located in the antibody-combining site and then inhibits the complex formation.

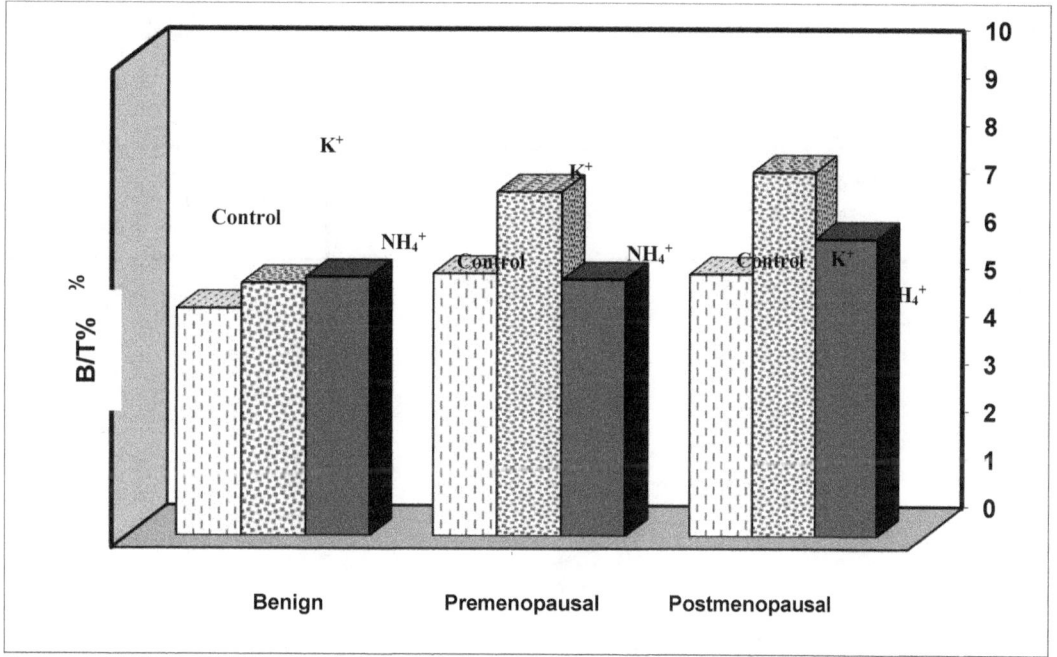

Figure (12): Effect of different monovalent cations on the binding of 125I-anti CA 15-3 antibody with CA 15-3. (All details are explained in the text).

Recovery of CA 15-3

This method was used to estimate the percent recovery of CA 15-3 in supernatant fractions of benign and malignant breast tumors homogenates. The results are summarized in table (12) and indicate that the CA 15-3 extracted from benign breast tumors, and CA15-3 extracted from malignant breast tissues homogenate were recovered more than CA 15-3 extracted from postmenopausal malignant breast tumors homogenates were recovered more than CA 15-3 extracted from premenopausal malignant breast tumors homogenates. Also the results indicate that total CA 15-3 can be determined through the developed method of immunoradiometric assay, as well as the percent of recovery indicates the precision of the used method.

Table (12): Recovery of CA 15-3. (All other details are explained in the text).

Type of CA 15-3	Measured B/T	Expected B/T	Recovery%
			Measured / Expected
Benign (Fibroadenoma)	16.23	24.84	65.34
Premenopausal (IDC)	20.04	27.64	72.50
Postmenopausal (IDC)	24.20	26.80	90.30

Purification of CA-15-3 Of Breast Tumors

CA15-3 is high molecular weight glycoprotein (>400 KD) identified at the apical side of alevoli and duct of mamary glands. Several authors have isolated, purified and characterized CA15-3 from different sources;either by the isolation of CA15-3 from a breast cancer patient's sera, using affinity chromatography, gel filtration, and then characterized by SDS-PAGE ,or by purifing a high molecular weight glycoprotein from human milk and breast carcinoma by using gel filtration, affinity chromatography and then PAGE In the present study , benign breast tumors and premenopausal malignant breast cancer were used as a source for partial purification of CA15-3 and then determination its yield. The factors effect the binding of partial purified CA15-3 to its antibody 125I-anti CA15-3 antibody were also studied.

Materials and Methods

Patients

The same patients tissues mentioned in the text were used in the following experiments. Benign breast tumor and premenopausal malignant breast cancer homogenates that showed maximal binding in the preliminary test in were used for the purification of CA15-3.

. Isolation of CA15-3 by Sepharose CL-4B Column

Preparation of the Column

The dimensions of the column were chosen according to the following equation .

$$\text{Diameter} = \sqrt{\frac{m}{10}}$$

Where:
m = amount of protein in mg.
L = 30 x diameter
Where:
L : length of the column

Preparation of the Gel

The gel was prepared by allowing the pre-swollen gel to swell again in PBS buffer (0.05 M) pH 7.0, then left to settle and the excess of buffer was decanted. The step was repeated several times. Suction was then used to degas the gel and slurry was left for 24 hrs to equilibrate with buffer.

The swollen gel was suspended and carefully poured into a vertical glass-column (0.7 x 30 cm) down the wall using a glass rod. After the gel has settled, the column was equilibrated with PBS for 24 hrs.

Void Volume Determination

The void volume of the column was determined by using blue dextran 2000 at concentration of 2 mg.mL-1 dissolved in PBS buffer pH 7.0 , then the elution was carried out with the same buffer at a flow rate of 20 mL.hrs-.

Fractions of 2 mL were collected and their absorbance was measured at 600 nm. Figure (13) shows the elution profile of blue dextran 2000. The volume of the buffer required to elute the blue dextran which represents the void volume was (6 mL).

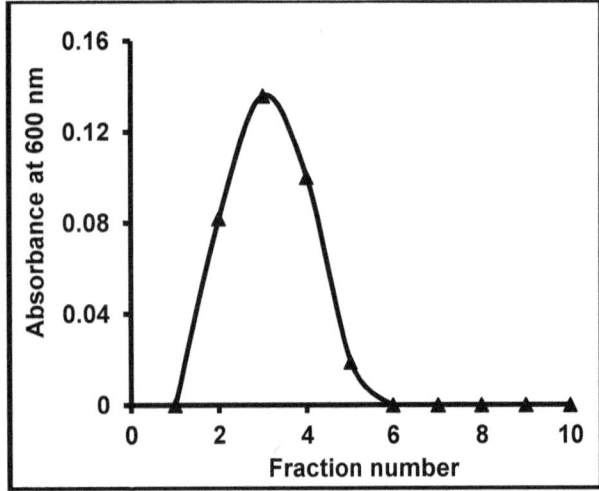

Figure (1): The elution profile of blue dextran 2000. (All other details are explained in the text).

.Column-Calibration

The column was calibrated by gel filtration kit, purchased from pharmacia fine chemicals which contained standard proteins. Standard protein solutions were prepared according to the manufacturers instructions, then applied through two 0.5 mL portions, proteins 1 and 3 in the first portion, protien 2 and 4 in the second portion. Elution was carried out with PBS buffer at a flow rate of 20 ml.hrs-1. the absorbance of the fractions collected was measured at 280 nm to evaluated the elution volume (Ve) of the standard protein.

Standard Proteins

Pharmacia calibration kit for determination of M.wt by gel filtration was used. The kit comprises the highly purified proteins and their high M.wt are detailed in table (13).

Table (13): Standard proteins and their molecular weights (All other details are explain in text).

Protein	M.wt (KD)	Conc. mg.mL-1
Thyroglobulin	669	4.0
Ferritin	440	1.0
Catalase	322	6.0
Aldolase	158	6.0

Calculations

The Kav values of the proteins eluted were determined using the following equation:

$$K_{av} = \frac{V_e - V_o}{V_t - V_o}$$

Where:
Vo= Void volume
Ve=Elution volume of each protein
Vt=Total gel - bed volume.

The calibration curve of Kav values vs. log M.wt. of the proteins were plotted.

Separation Procedure

Procedure

The sample of tissue homogenate (0.5 mL) containing approximately 3.43 mg protein was applied to the surface of gel, equilibrated with 0.15 M PBS buffer pH 7.2 for benign and premenopausal malignant breast tumor respectively. The sample was eluted by using the same buffer pH (7.0 and 7.6) for (benign and premenopausal malignant breast tumors respectively) with a flow rate of 20 mL.hrs-1 and fractions volume 2 mL were collected, gel filtration was carried out at 10 oC. The protein content of each fraction was determined using Lowry.et.al method.

The fractions contained CA15-3 were identified by the assay method. The binding of each fraction was calculated and plotted against the elution volume. The degree of purification (folds) of CA15-3 was calculated from the following formula.

$$\text{Purification fold of CA15-3} = \frac{\text{Specific binding of purified CA15-3}}{\text{Specific binding of crude CA15-3}}$$

Then yield % was determined as follows:

$$\text{Yield \%} = \frac{\text{Total protein content of purified CA15-3}}{\text{Total protein content of crude CA15-3}} \times 100$$

Dialysis for Concentration

After preparing dialysis tube, the fractions that contained high levels of the binding activity were pooled and concentrated by dialyzing against sucrose at 4 oC for 2hrs to get the required concentration to be used in the next experiments.

The Choice of the Optimum Conditions for the Binding of the Partially Purified CA15-3 to 125I-Anti CA15-3 Antibody

Optimum Protein Concentration

One hundred microliters of increasing amount (50,100,150,200 and 250) µg.mL-1 protein of the dialyzable fractions of the partially purified CA15-3 from benign breast tumor was incubated with 50 µL of 125I-anti CA15-3 antibody (0.35 mg.mL-1) and completed to a final volume of 500 μL with 0.15 M PBS pH 7.0. The assay tubes were incubated for 90 min. at 45 oC. Two additional tubes, containing 50 μL (0.35 mg.mL-1) of 125I-anti CA15-3 antibody only, for total radioactivity computation, were set a side until counting.

Steps 4,5 and 6 of the experiment were repeated. The same experiment was repeated on premenopausal malignant breast tissues homogenates (100 μg.mL-1 protein) with PBS buffer pH 7.6 and incubation time for 90 min at 15 oC.

Calculations

The (B/T) % was calculated as mentioned in experiment and plotted against increasing amounts of protein concentration.

Influence of 125I-Anti CA15-3 Antibody on the Binding

Procedure

Fifty microliters of increasing concentration (0.070, 0.140, 0.175, 0.210, 0.245, 0.280 mg.mL-1) of 125I-anti CA15-3 antibody were added to 100 μL (150 μg.mL-1 protein) of partially purified CA15-3 from benign breast tumors. The reaction was completed to 500 μL with PBS pH 7.0. The assay tubes were incubated for 90 min at 45 oC. Two additional tubes containing increased concentration of 125I-anti CA15-3 antibody only, for total counts were counted. Steps 4,5 and 6 of the experiment were repeated. The same experiment was repeated on premenopausal malignant breast tissues homogenate (100 μg.mL-1 protein) with PBS pH 7.6 and incubation time for 90 min at 15 oC.

Calculations

The (B/T) % was calculated as mentioned in experiment and plotted against increasing concentration of 125I-anti CA15-3 antibody.

Optimum pH

Procedure

To determine the optimum pH, 100 μL of a dialyzable fractions of partially purified CA15-3 from benign breast tumors (150 μg.mL-1 protein) were added to 20 μL of 125I-anti CA15-3 antibody (0.140 mg.mL-1). The volume of each fraction was completed to 500 μL with 0.15 M PBS of different pH (6.8 , 7.0 ,7.2 , 7.4 , 7.6 , 7.8 , 8.0). The assay tubes were incubated for 90 min at 45 oC. Two additional tubes, containing 20 μL (0.140 mg.mL-1) of 125I-anti CA15-3 antibody only , for total count , were set aside until counting. Steps 4,5 and 6 of experiment were repeated. The same experiment was repeated on premenopausal malignant breast tissues homogenates (100 μg,mL-1 protein) and 25 μL (0.175 mg.mL-1) of 125I-anti CA15-3 antibody was incubated for 90 min at 15 oC.

Calculations

The (B/T) % was calculated as mentioned and plotted against their corresponding pH values.

Optimum Temperature

Procedure

Twenty microliters (0.140 mg.mL-1) of 125I-anti CA15-3 antibody was added to 100 μL dialyzable fractions of partially purified CA15-3 from benign breast tumors (150 μg.mL-1 protein).

The volume of reaction was completed to 500 μL with 0.15 M PBS buffer pH 7.0. The assay tubes were incubated for 90 min at 45 oC. The same steps were repeated at (37, 25, 15, 5oC). Two additional tubes containing 20 μL (0.140 mg.mL-1) of 125I-anti CA15-3 antibody only, for total count, were set aside until counting. Steps 4,5 and 6 of experiment were repeated.

The same experiment was repeated on the premenopausal malignant breast tissues homogenates (100 μg.mL-1 protein) and 25 μL (0.175 mg.mL-1) of 125I-anti CA15-3 antibody in PBS buffer pH 7.0, with incubation time 90 min at 15 oC. The experiment was repeated at different temperatures (45, 37, 25 and 5 oC).

Calculations

The (B/T) % was calculated as mentioned in experiment nd plotted versus temperatures of incubation.

The Effect of Incubation Time

Twenty microliters (0.140 mg.mL-1) of 125I-anti CA15-3 antibody were added to 100 μL of dialyzable fractions of partially purified CA15-3 from benign breast tumors containing (150 μg.mL-1 protein). The reaction volume was completed to 500 μL with 0.15 M PBS buffer pH 7.0 , then incubated at 37 oC for (30, 60, 90, 120, 150, 180 min). Two additional tubes counting 20 μL (0.140 mg.mL-1) of 125I-anti CA15-3 antibody for total counts , were set aside until counting. Steps 4,5 and 6 of the experiment were repeated. The same experiment was repeated on the premenopausal malignant breast tissues homogenates (100 μg.mL-1 protein) and 25 μL (0.175 mg.mL-1) of 125I-anti CA15-3 antibody with 0.15 M PBS buffer pH 7.0 and incubated at 15 oC for (30, 60, 90, 120, 150 and 180 min).

Calculations

The (B/T) % was calculated as metioned in experiment and plotted versus the time of incubation for each group.

. Stability of CA15-3 at –20 oC

Procedure

Crude and purified CA15-3 were stored at –20 oC for several time intervals. The frozen specimen was thawed at the end of each interval and the binding activity was measured at optimum conditions as described in the text. The remaining activity was calculated and plotted against storage periods.

Calculations

The (B/T) % was calculated as mentioned in experiment and plotted versus time storage for each group.

Results and Discussion

Partial Purification of CA15-3

Isolation of CA15-3 was performed by gel exclusion chromatography technique. CA15-3 was found to be separated from aggregates and other protein having smaller molecular weight by sepharose CL-4B. Figure (14, A & B) shows the elution profile of CA15-3 from benign breast tumors and premenopausal malignant breast cancer homogenates. The homogenate was loaded on the column as described in section . The void volume (Vo) of column was (6 mL) as predicted from the elution profile of the blue dextran. The elution was performed with PBS buffer. The resultant fractions containing the binding activity of CA15-3 were collected, pooled and concentrated, then subjected to protein determination .

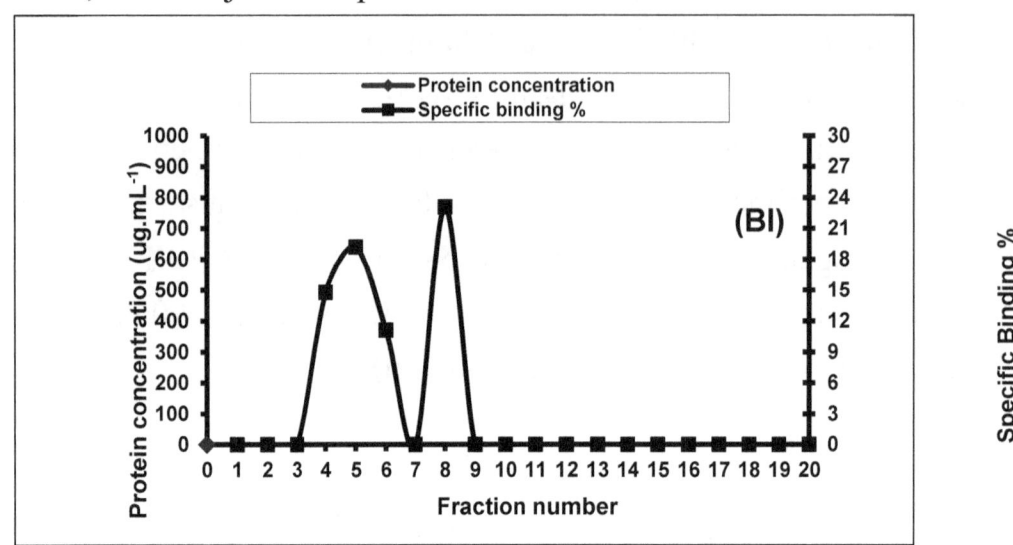

Figure (14A): The elution profile of human CA15-3 from benign breast tumors (BI). (All other details are explained in the text).

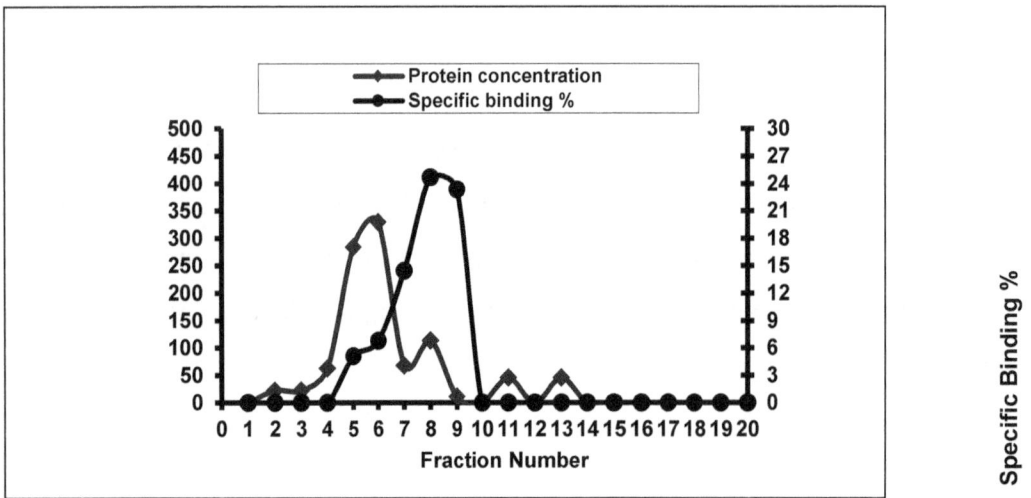

Figure(14B): The elution profile of human CA15-3 from premenopausal malignant breast cancer (MI). (All other details are explained in the text).

The elution volume Ve and then Kav values for the two peaks of CA15-3 (BI & MI) from benign breast tumors and malignant breast cancer respectively were calculated. The molecular weight of the partially purified

CA15-3 obtained from figure (15) was 440 KD for peak (BI) and peak (MI) in two cases.

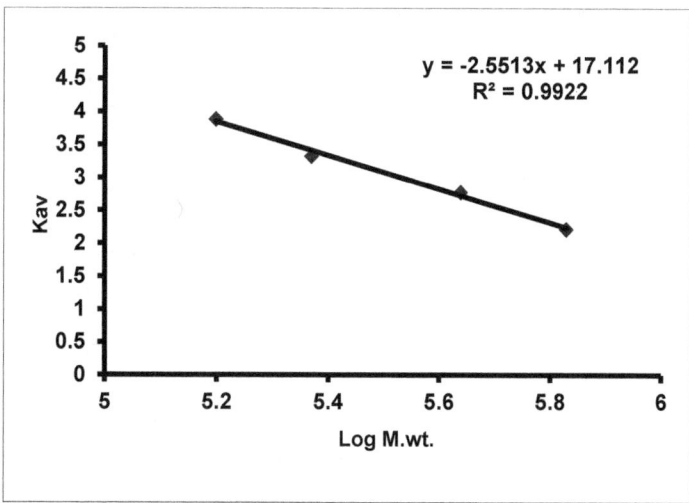

Figure (15): Calibration curve for determination of M.wt by gel filtration chromatography. (All other details are explained in the text).

The values ranged from 300-450 KD. Peaks of partially purified CA15-3 may be heavily aggregated, CA15-3 was obtained near the void volume of the column under separation conditions. From these results it was concluded that these components are capable of binding to the 125I-anti CA15-3 antibody with different affinities and in general CA15-3 type (BI) have lower binding affinities than CA15-3 type (MI), the isolation of CA15-3 from benign breast tumors on gel filtration column showed 3.02 folds of purification for peak (BI), while the isolation of CA15-3 from premenopausal malignant breast cancer showed 5.0 folds of purification. Table (14) illustrates the purification parameters for the different purified CA15-3 forms isolated by gel exclusion chromatography technique. The glycosylation of the protein backbone may differ in carcinoma cells from normal epithelial cells causing a wide range of molecular weight for this mucin.

Table (14): Partial purification of CA15-3 by gel filtration. (All other details are explained in the text).

CA15-3 Source	Total protein	Specifically bound 125I-anti CA15-3	Specifically binding 125I-anti CA15-3/mg protein	Yield %	Purification fold

		mg .mL-1				
Benign	Crude extract	3.43	10.17	2.97	100	1.00
	Gel filtration on sepharose CL-4B	2.91	30.70	10.55	84.84	3.02
Malignant	Crude extract	3.43	8.18	2.39	100	1.00
	Gel filtration on sepharose CL-4B	2.21	40.93	18.52	64.43	5.00

The Choice of Optimum Conditions for the Binding of Partially Purified CA15-3 with 125I-Anti CA15-3 Antibody

Optimum Protein Concentration

Figure (16) shows the effect of increasing amounts of partially purified CA15-3 to a fixed amount of 125I-anti CA15-3 antibody to produce (125I-anti CA15-3 antibody/CA15-3) complex, that grow in size until they formed a precipitate. Above this zone an equivalence between CA15-3 and its antibody concentration is obtained, and amount of complex shows no further increases. A further addition of CA15-3 give rise to a solubilization of complex. The results revealed that 150 µg protein was the most appropriate concentration for the binding of (BI) and 100 µg protein for (MI). From these results, it could be concluded that the binding of 125I-anti CA15-3 antibody with its partially purified CA15-3 (BI) needed a higher amount of protein concentration than partially purified CA15-3 (MI). This is may be due to lower concentration of CA15-3 in benign breast tumor as compared with malignant breast tumors. According to these results, in all subsequent experiments, (150 µg.mL-1 protein) in benign breast tumors and (100 µg.mL-

1 protein) in malignant breast tumors were used, since they give the highest binding.

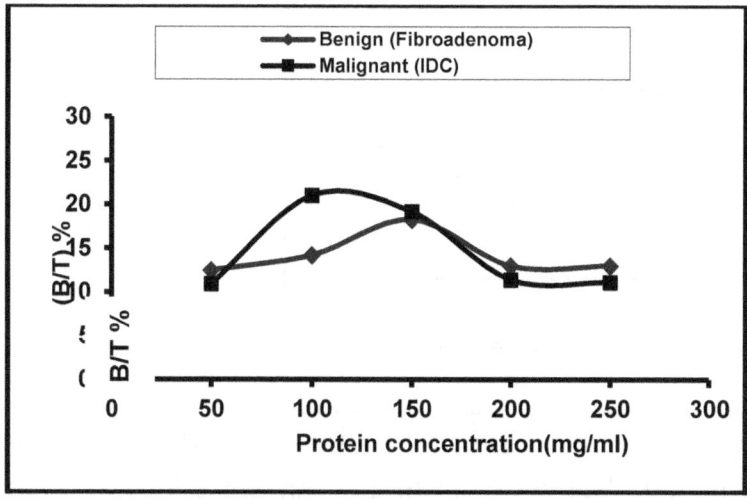

Figure (16): Influence of protein concentration on the binding of 125I-anti CA15-3 antibody with partially purified CA15-3 from breast tumors. (All other details are explained in the text).

Influence of 125I-anti CA15-3 Antibody on the Binding

Figure (17) illustrate the effect of 125I-anti CA15-3 antibody concentration on the binding with partial purified CA15-3 from benign breast tumors and premenopausal malignant breast cancer.

The maximum binding obtained when 0.140 mg.mL-1 of antibody in benign breast tumors and 0.175 mg.mL-1 of antibody in malignant breast tumors were used. From these results, it was found that (BI) purified fraction was saturated with small concentration of 125I-anti CA15-3 antibody than those required for (MI). This is may be due to the increasement of the epitope (is the part of an antigen molecule that binds to any single antigen-combining site) in partially purified CA15-3 in malignant breast tumors as compared to benign breast tumors.

According to these results, in all subsequent experiments (0.140 mg.mL-1) and (0.175 mg.mL-1) of 125I-anti CA15-3 antibody in benign and malignant breast tumors were used, since they give the highest binding.

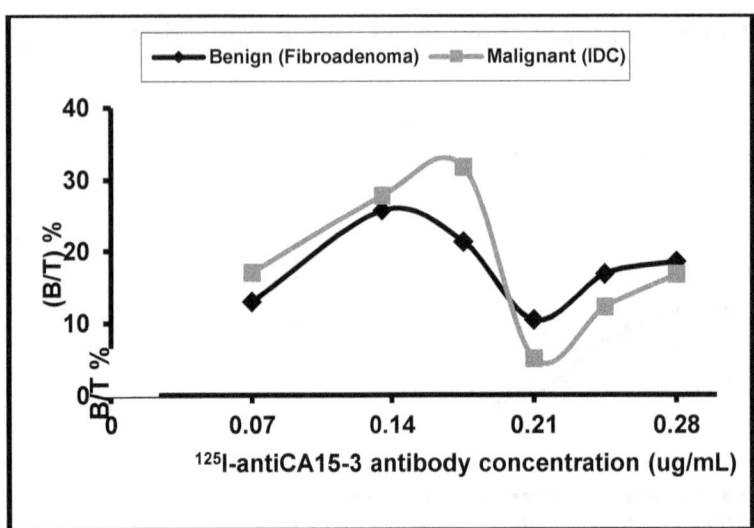

Figure (17): Effect of 125I-anti CA15-3 antibody concentration on its binding with partially purified CA15-3 from breast tumors. (All other details are explained in the text).

Optimum pH

Figure (18) shows the effect of increasing pH on the binding of 125I-anti CA15-3 antibody to its purified antigen. The results revealed that the optimum pH for (BI) and (MI) purified fractions for the binding with its antibody was 7.0. These results indicate that the binding was pH dependent.

The similarity in pH (7.0) suggests that the CA15-3 isolated from different sources of tissues either benign or malignant breast tissues homogenates possesses the same epitopes in both cases. That means the induction of protonation-deprotonation process occurs within the same changed polar groups on the amino acid residues present in the binding domain. According to the results obtained, the pH of the buffer used in all subsequent experiments was adjusted to pH 7.0.

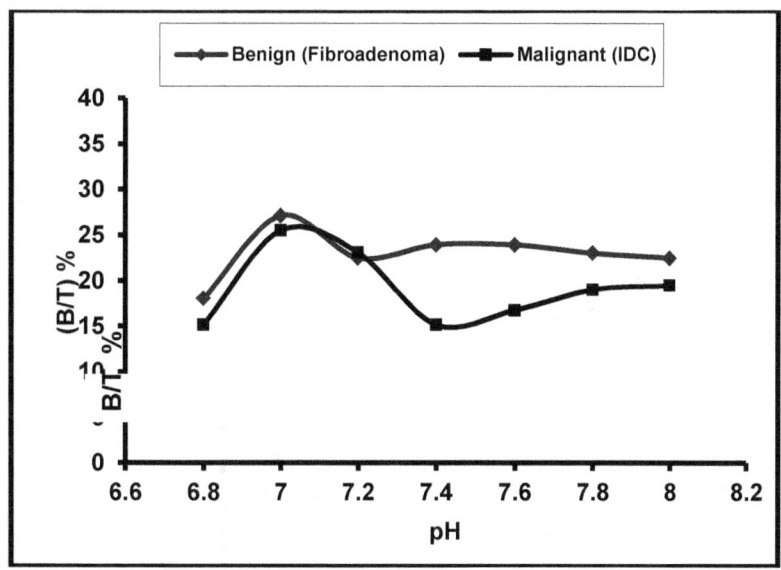

Figure (18): pH effect on the binding of 125I-anti CA15-3 antibody with partially purified CA15-3 from breast tumors. (All other details are explained in the text)

Optimum Temperature

The temperature dependency of the isolated CA15-3 binding to its antibody 125I-anti CA15-3 was investigated.

Figure (19) show the optimum temperatures on the binding of 125I-anti CA15-3 antibody was 37oC with partially purified CA15-3 (BI) and 15oC with partially purified CA15-3 (MI).

The difference of the temperature between crude and purified CA15-3 occurs in benign breast tumors, i.e. the optimum temperature was 45oC of the binding of 125I-anti CA15-3 antibody to crude CA15-3 while in the purified CA15-3 (BI) was 37oC. On the other hand, the optimum temperature in both crude and partially purified CA15-3 from premenopausal malignant breast tumors was 15oC.

The temperature dependency of the binding suggests that the whole process is controlled by diffusion of the interacting of 125I-anti CA15-3 antibody to CA15-3 in benign and malignant breast tumors.

In view of these results, the temperatures (37 & 15 oC) for both benign and malignant breast tumors were used in all subsequent experiments.

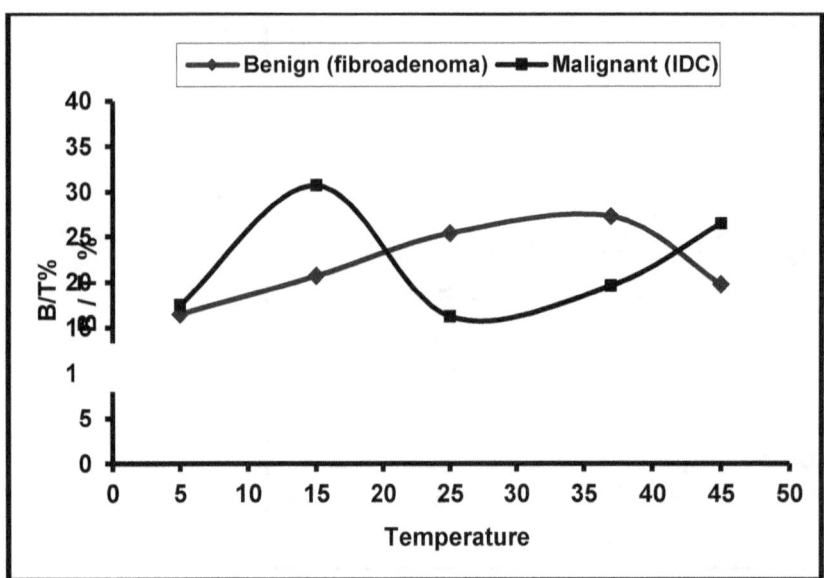

Figure (19): Effect of the temperature on the binding of 125I-anti CA15-3 antibody with partially purified CA15-3 from breast tumors. (All other are explained in the text)

The Effect of Incubation Time

Figure (20) shows the time required for the highest binding of 125I-anti CA15-3 antibody to partially purified CA15-3 in (BI) and (MI) was 90 min at 37 and 15 oC respectively.

In view of these results, the incubation time used in all subsequent experiments was 90 min.

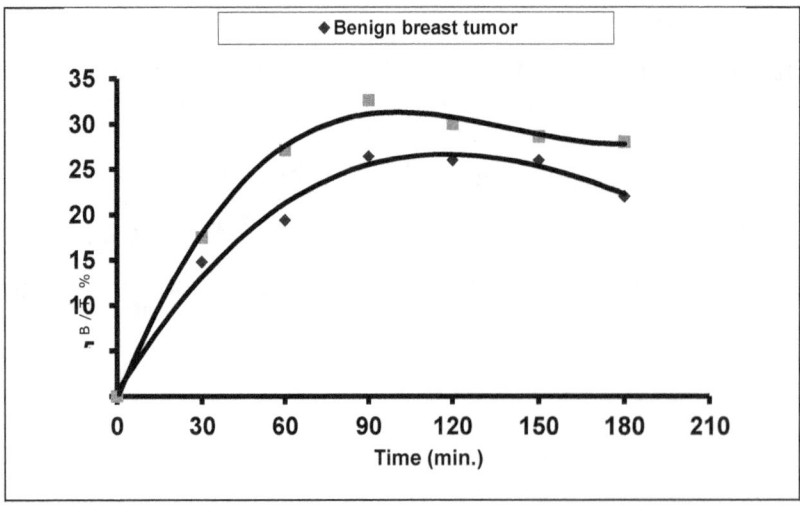

Figure (20): Time dependence of 125I-anti CA15-3 antibody binding with partially purified CA15-3 from breast tumors. (All other details are explained in the text)

Stability of CA15-3 at –20 oC

The crude and isolated fractions of CA15-3 from malignant breast tumors were stored at –20 oC during the experiments. It was carried out in order to study the stability of CA15-3 and check their efficiencies of the binding through out the storage period. The results showed that CA15-3 of crude fraction was more stable than the isolated fractions as shows in figure (21). This result is in agreement with Al-Atrakchi observations.

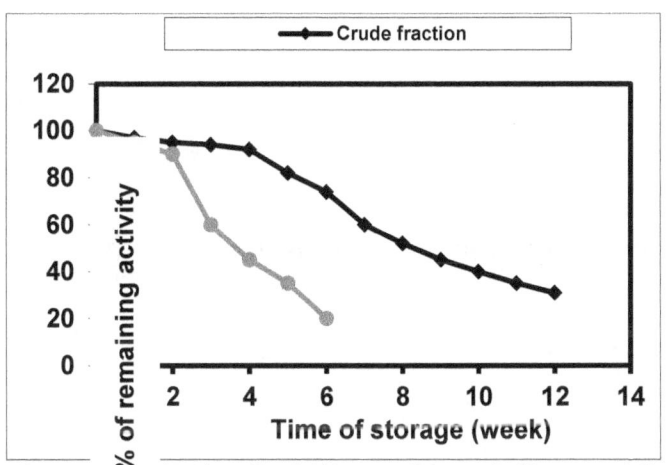

Figure (21): Stability of partially purified and crude CA15-3 upon storage at Abstract

Kinetic and thermodynamic parameter associated with the binding of 125I-anti CA15-3 antibody to partially purified CA15-3 in both cases, benign and malignant breast tumors were investigated.

It was shown that the reaction in all studied cases follow pseudo-first order reaction kinetics. The maximum binding (Bmax) of partially purified CA15-3 in benign breast tumors (Fibroadenoma) was 10.48x10-3 mg.mL-1 after 90 minutes incubation at 37oC, while the (Bmax) of partially purified CA15-3 in malignant breast tumors (IDC) was 13.38x10-3 mg.mL-1. The (Bmax) was decreased with increasing temperature. The values of affinity constant (Ka) were dependent on the temperature, Ka increased from 14.18

mg-1.mL at 5oC to 31.65 mg-1.mL at 45oC in benign breast tumors (Fibroadenoma), while Ka was increased from 13.87 mg-1.mL at 5oC to 23.81 mg-1.mL at 45oC in premenopausal malignant breast tumors (IDC). The association constant K_{+1} increased with temperature in benign breast tumors (Fibroadenoma). On the other hand, K_{+1} was independent of temperatures in premenopausal malignant breast tumors (IDC). The Van't Hoff plot demonstrated a linear relationship between Ka and 1/T, using the partially purified CA15-3 in benign and malignant tumor homogenate. Arrhenius plot indicate that there was a linear-relationship between log K_{+1} and 1/T. The transition state thermodynamic parameters (Ea, ΔH*, ΔG*, ΔS*) for the formation of (125I-antiCA15-3 antibody /CA15-3) were determined.

Kinetic and thermodynamics of Binding CA-15-3 to its antibody

The specific reaction between an antibody (Ab) and an antigen (Ag) is usually driven by electrostatic forces between oppositely charged amino acids, hydrogen bonding, and hydrophobic interactions. The equilibrium reaction, termed "biospecific interaction", is characterized by the affinity of reactants to form Ag-Ab complex.

Kinetic studies supplement the information for differences between the initial, final states of each reactant and an intermediate activated complex, (i.e, the pathway taken by the reactants reach the final product). On the other hand, thermodynamics of the binding describes the system in its initial, final states. Using kinetic and equilibrium data also determined thermodynamic formation constant.

Al-Mudhuffar et.al, have many studies on the kinetic and thermodynamic of protein-protein interaction in human breast tissue, like kinetic and thermodynamic of purified steroid receptor of malignant breast tumors with hormone, also kinetic and thermodynamic studies on the binding of lectin in human malignant breast to glycoprotein.

In this chapter, the basic mathematical analysis was described and used to explain the mechanism through kinetics of binding of CA15-3 from both breast tumor homogenates (fibroadenoma and Infiltrating ductalcarcinoma) to its antibody to form (125I-anti CA15-3 antibody / CA15-3) complex in partially purified fraction.

Materials and Methods

Kinetic Studies

The Time-Course of the Binding of 125I-anti CA15-3 Antibody with CA15-3 in Breast Tumor Homogenate

1. One hundred microliters of partially purified CA15-3 from benign breast tumor (fibroadenoma) and premenopausal malignant breast tumor (Infitrating ductal carcinoma, IDC) containing (150 and 100 μg.mL-1 protein) respectively, were added to (20 and 25 μL) of 125I-anti CA15-3 antibody containing (0.140 and 0.175 mg.mL-1) respectively.

2. The volume of reaction were completed to 500 μL with PBS buffer pH 7.0.

3. All tubes were incubated at 37oC at different time intervals (30, 60, 90, 120, 150, 180) min.

4. Steps 3, 4, 5 and 6 of experiment were repeated.

5. To determine the time-course of partially purified CA15-3 binding to 125I-anti CA15-3 antibody at different temperatures, step 1,2,3 and 4 in the same experiment were repeated at different temperatures 5, 15, 25 and 45Co.

Calculation

The values of (B/T)% were calculated as described and plotted against incubation time at each temperature for both types of homogenates.

Determination of Kinetic Parameters of 125I-Anti CA 15-3 Antibody Binding with Partially Purified CA in Benign and Malignant Breast Tumors

Determination of the affinity constant (Ka) and the maximal binding capacity (Bmax) of:

A. Partially Purified CA15-3 in Benign Breast Tumor Homogenate Binding with 125I-Anti CA15-3 Antibody

1. One hundred microliters of partially purified CA15-3 from benign breast tumor (Fibroadenoma) containing (150 µg.mL-1 protein) were added to increasing volumes (4, 8, 12, 16, 20 and 24 µL) of 125I-anti CA15-3 antibody containing (0.0280, 0.0560, 0.0841, 0.1121, 0.1402 and 0.1684 mg.mL-1) to each assay tube. The final volume of each assay tube was completed to 500 µL with PBS buffer pH 7.0.

2. All tubes were incubated for 90 min at 37oC.

3. Steps 3, 4, 5 and 6 in experiment were repeated at different temperatures (5, 15, 25 and 45oC).

4. The time of incubation required to reach the equilibrium state are reported in table (15) according to the following:

Table (15): The time of incubation for benign and malignant breast tumor homogenate at different temperatures

Temp. oC	Time (min.)	
	Benign breast tumor homogenate (Fibroadenoma)	Malignant breast tumor homogenate (IDC)

5	180	180
15	60	90
25	90	150
37	90	90
45	180	90

Calculations

1- The B/F ratio was computed for each tube, where:

B: is the bound radioactivity (mean counts in c.p.m), which represent the formation of (125I-anti CA15-3 /CA15-3) complex.

F: is the free radioactivity (mean counts in c.p.m.), which represents the (unbound or unreacted), 125I-anti CA15-3 antibody.

T: is the total activity (mean counts in c.p.m.)

F = T (total counts) - B (bound radioactivity)

2- The concentration of (125I-anti CA15-3/CA15-3) complex in mg.mL-1 which found after time (t) was calculated from the following equation:

$$B(mg.mL^{-1}) = \frac{B(c.p.m)}{T(c.p.m)} \times \text{Concentration of }^{125}I-\text{anti CA15}-3 \text{ antibody in the incubation medium in mg.mL}^{-1}$$

3- The affinity constant and maximal binding capacity were determined according to Scatchard equation.

$$\frac{B}{F} = \frac{1}{K_d} \times (B_{max} - B)$$

$$K_a = \frac{1}{K_d} = \frac{K_{+1}}{K_{-1}}$$

Where: Ka = affinity constant

Kd = dissociation constant

Bmax = maximal binding capacity

The value of affinity constant of the binding Ka at each temperature can be calculated from the slop of the straight line, while the value of the total concentration of CA15-3 (Bmax) in breast tumor homogenate for each group was calculated from the intercept of the x-axis.

B. Partially Purified CA15-3 in Human Malignant Breast Tumor Homogenate Binding with 125I-Anti CA15-3 Antibody

1. One hundred microliters of partially purified CA15-3 from premenopausal malignant breast tumor (IDC) containing (100 µg.mL-1 protein) were added to increasing volumes (5, 10, 15, 20, 25 and 30 µL) of 125I-anti CA15-3 antibody containing (0.035, 0.070, 0.105, 0.140, 0.175 and 0.210 mg.mL-1) to each assay tube. The final volume of each assay tube was completed to 500 µL with PBS buffer pH 7.0.

2. All tubes were incubated for 90 min at 15oC

3. Steps 3, 4, 5 and 6 in experiment were repeated at different temperatures (5, 25, 37 and 45 oC).

4. The times of incubation required to reach the equilibrium state are reported in table (16).

Calculations

The method outlined in experiment was followed exactly to obtain the values of Ka and Bmax at each temperature as shown in figure (3).

The Thermodynamic Studies of 125I-Anti CA15-3 Antibody Binding to Partially Purified CA15-3 in Benign and Malignant Breast Tumors

The same steps mentioned in section in the text were performed using the dialyzable protein fraction of benign and malignant breast tumor homogenate from fibroadenoma and (IDC) as the partially purified CA15-3 source.

Calculation

1. The thermodynamic parameters of standard state were obtained from Van't Hoff plot, the values of the natural logarithm of equilibrium constant (affinity constant Ka) obtained at different temperatures were plotted against the reciprocal values of the absolute temperature in Kelvin (1/T), according to the following equation:

$$\ln K_a = \frac{\Delta S^o}{R} - \frac{\Delta H^o}{RT}$$

Where:

ΔH_o = the enthalpy change of the standard state.

ΔS_o = the entropy change of the standard state.

R = the gas constant (8.314 J.K-1.mol-1).

ΔH_o value obtained from the slop of a linear relationship of the plot.

The change in Gibbs free energy of the standard state ΔG_o was obtained from the following equation:

ΔG_o = -RT Ln Ka

Where Ka is the affinity constant, while the standard state entropy change was obtained from :

$$\Delta S^\circ = \frac{\Delta H^\circ - \Delta G^\circ}{T}$$

2. The thermodynamic parameters of the transition state were obtained from Arrhenius plot of Ln K+1 values against (1/T) values, that given a linear relationship according to the following equation:

$$\text{Ln } K_{+1} = \text{Ln } A - \left(\frac{E_a}{RT}\right)$$

Where:

A: Arrhenius constant .

The values of activation energy (Ea) of the binding reaction can be determined from the slop of the straight line.

The enthalpy of transition state ΔH^* was obtained from:

ΔH^*=Ea-RT

Transition state of free energy change ΔG^* is calculated from the following equation:

$$\Delta G^* = -RT \text{ Ln} K_{+1} + RT \text{ Ln} \frac{KT}{h}$$

where K and h were Boltzmann and Plank's constant which equal (1.38x10-23 J.K-1), (6.62x10-34 J.sec-1) respectively.

The change in entropy of the transition state ΔS^* is calculated from the following equation:

$$\Delta S^* = \frac{\Delta H^* - \Delta G^*}{T}$$

Results and Discussion

Kinetic Studies

The Time-Course of the Binding of 125I-anti CA15-3 Antibody with CA15-3 in Breast Tumor Homogenate

Figure (24) shows the time – course of the formation of (125I-anti CA15-3 /CA15-3) complex at five different temperatures (5, 15, 25, 37 and 45 °C) of partially purified CA15-3 from benign and malignant breast tumors homogenates samples.

The concentration of (125I-anti CA15-3/CA15-3) complex formed after time (t) was calculated from the following equation:

$$[\text{125I-antiCA15-3/CA15-3}] \text{ in mg.mL}^{-1} \text{ after time (t)} = \frac{\text{Count (c.p.m.) of 125I-antiCA15-3 specifically bound after time (t)}}{\text{Total counts (c.p.m.) of 125I-anti CA15-3 used in the incubation}} \times \text{Concentration of 125I-antiCA15-3 in the incubation (mg.mL}^{-1}\text{)}$$

The results of time-course pattern at different temperatures indicated that the equilibrium binding studies is temperature and time dependent process. In case premenopausal malignant breast tumor (IDC) the maximum binding occurs at 15 °C (after incubation for 90 minutes), while in benign breast tumors (fibroadenoma) the maximum binding occurs at 37 °C at the same incubation time. This is may be due to the different source of CA15-3. Several authors studied the time – course of purified steroid receptors of malignant breast tumors, others studied the time – course on the binding of lectin in human malignant breast to glycoprotein, these studies revealed that the time-course must be done to find the maximum binding at different incubation time as a step to prepare the kinetic and thermodynamic studies.

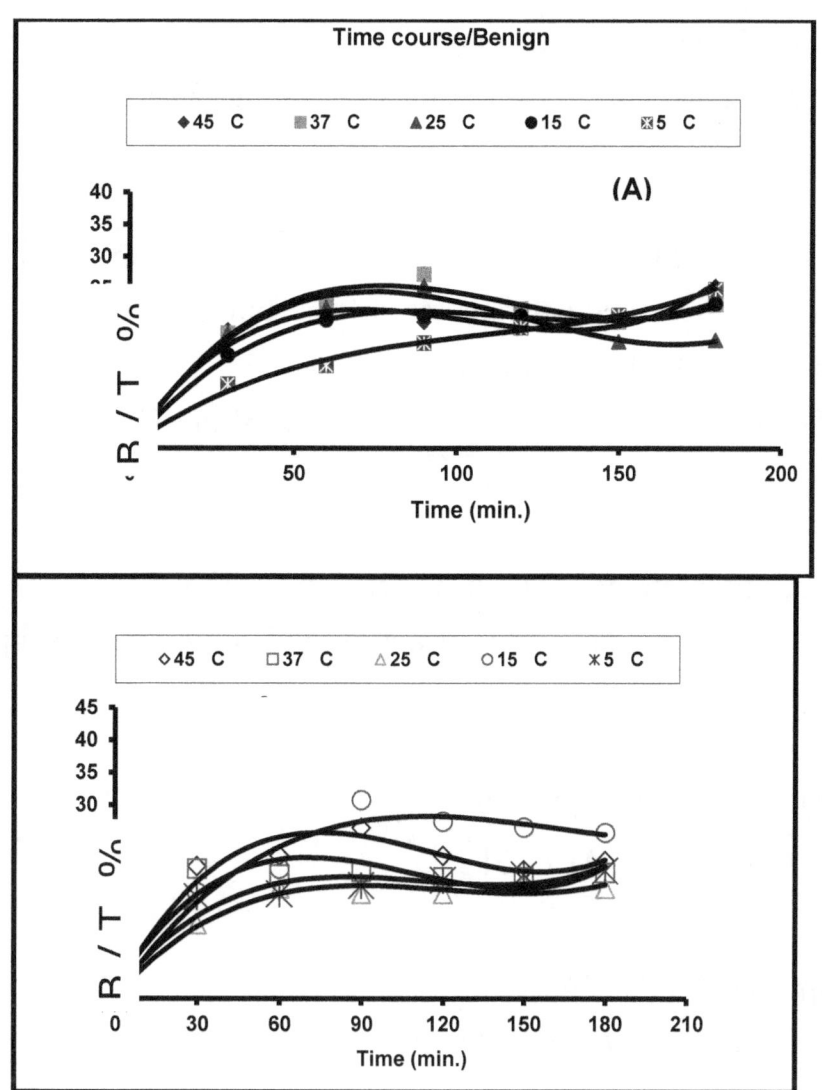

Figure (22): Time-Course of 125I-anti CA15-3 binding to partially purified CA15-3 in:

(A)   Benign tumor (Fibroadenoma) tissue homogenate.
(B)   Malignant tumor (IDC) tissue homogenate.
(All other details are explained in the text).

Determination of Kinetic Parameters of 125I-Anti CA15-3 Antibody Binding with Partially Purified CA15-3 from Benign and Malignant Breast Tumors

The time course of (125I-anti CA15-3/CA15-3) complex formation was carried out to describe the kinetic parameters of the binding. The simplest proposed model representing this interaction is:

$$125I\text{-antiCA15-3} + CA15\text{-}3 \underset{K_{-1}}{\overset{K_{+1}}{\rightleftharpoons}} [125I\text{-antiCA15-3/CA15-3}]$$

Where:

K+1: is the association rate of 125I-anti CA15-3 to /or CA15-3.

K-1: is the dissociation rate of (125I-anti CA15-3/CA15-3) complex formed.

At equilibrium:

$$K_a = \frac{[^{125}I - antiCA15 - 3/CA15 - 3]}{[^{125}I - antiCA15 - 3][CA15 - 3]} \quad \ldots\ldots\ldots(2)$$

$$K_d = \frac{[^{125}I - antiCA15 - 3][CA15 - 3]}{[^{125}I - antiCA15 - 3/CA15 - 3]} \quad \ldots\ldots\ldots(3)$$

Thus:

$$K_a = \frac{1}{K_d} = \frac{K_{+1}}{K_{-1}} \quad \ldots\ldots\ldots\ldots\ldots\ldots\ldots\ldots(4)$$

Where:

The value Ka and maximal binding capacity (Bmax). Were calculated from Scatchard plot at five different temperatures at incubation time of 90 minutes, figure (23) and (24).

It is clear from table (16), that the affinity constant (Ka) is depended on the type of the tumor (i.e., benign or malignant) and on the temperature. Ka increased with increased temperature for the same tumor (Fibroadenoma), Ka increased from 14.18 mg-1.mL at 5oC to 31.65 mg-1.mL at 45oC. Whereas the values of dissociation constant (Kd) was calculated by using equation (4), which show that the lowest Kd value of (125I-anti CA15-3/CA15-3) complex occurs at 45oC at time of incubation 180 minutes.

The concentration of CA15-3 in partially purified fractions of (Fibroadenoma) was determined to be 10.48x10-3 mg.mL-1 and the maximum binding (Bmax) occurred after 90 minutes incubation at 37 oC. While in the same table the maximum Ka value for the binding 125I-anti CA15-3 antibody with CA15-3 present in partially purified fraction of (IDC) occurred at 15oC and it was increased with temperature in the following order: 5 >15 >25 >37 > 45 oC.

The lowest Kd value of (125I-anti-CA15-3 /CA15-3) complex occurs at 45 oC at the time of incubation.

Scatchard plot analysis gave straight line as shown in figure (27) and (28) indicating that the (125I-anti CA15-3/CA15-3) complex is directed against the same epitopes on CA15-3 molecules. On the other hand, the maximum binding occurred at 15oC and was 13.38x10-3 mg.mL-1 also shows that the (Bmax) decreased with increasing temperatures of incubation.

Table (16): The Kinetic parameter of 125I-anti CA15-3 antibody binding to partially purified CA15-3 in breast tumor homogenate. (All other details are explained in the text).

Temp oC	Benign breast tumors (Fibroadenoma)			Malignant breast tumors (IDC)		
	Binding Capacity Bmaxx10-3 (mg.mL-1)	Ka (mg-1.mL)	Kdx10-2 (mg.mL-1)	Binding Capacity Bmax x10-3 (mg.mL-1)	Ka (mg-1.mL)	Kdx10-2 (mg.mL-1)
5	9.22	14.18	7.05	10.82	13.87	7.21
15	8.05	16.73	5.98	13.38	20.78	4.81
25	9.02	16.38	6.10	12.57	20.84	4.79
37	10.48	18.66	5.36	9.63	22.22	4.50
45	6.67	31.65	3.16	11.67	23.81	4.20

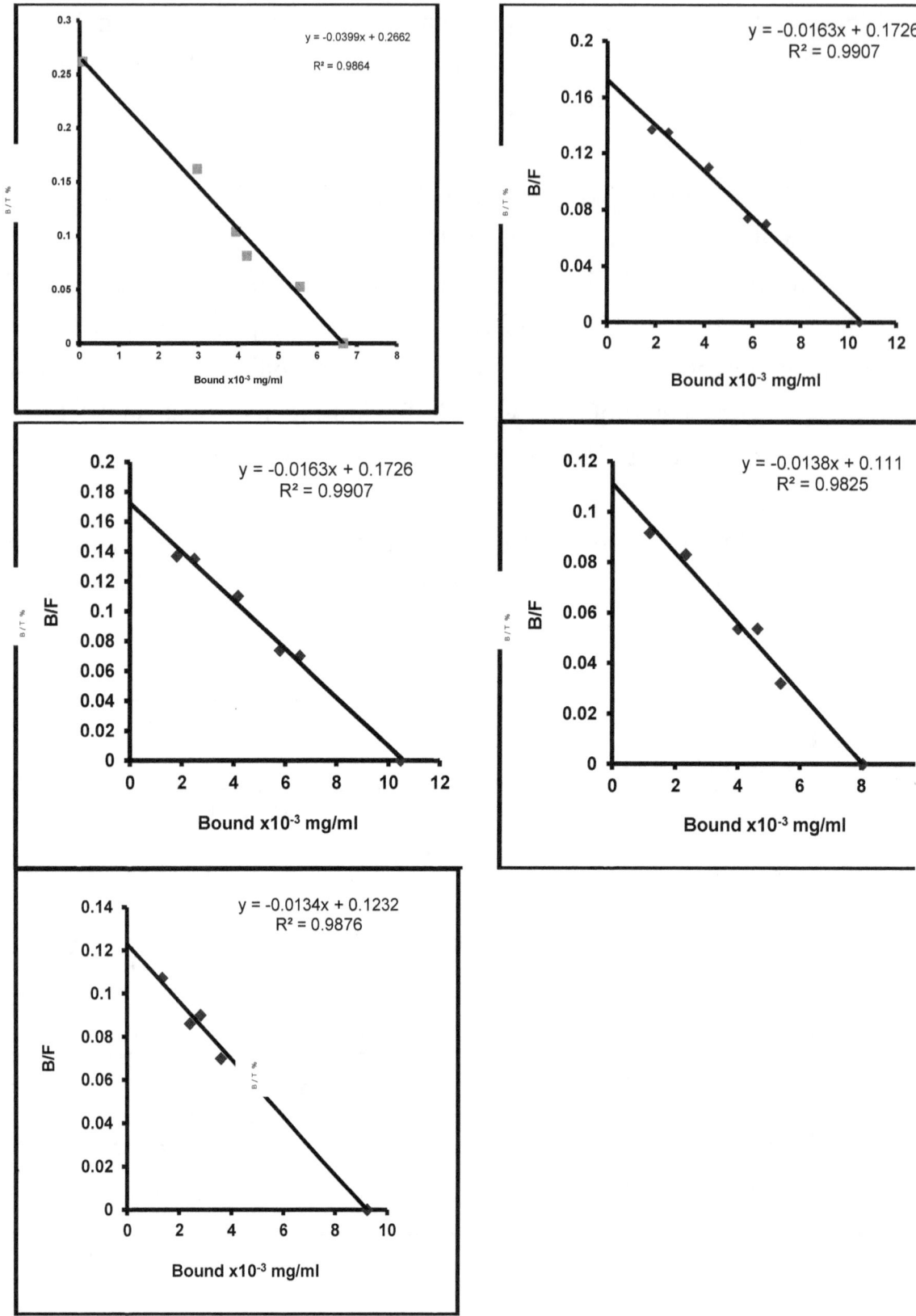

Figure (23): Scatchard plot of 125I-anti CA15-3 antibody binding to the partially purified CA15-3 in benign breast tumors (Fibroadenoma) at five different temperatures. All details are explained in the text.

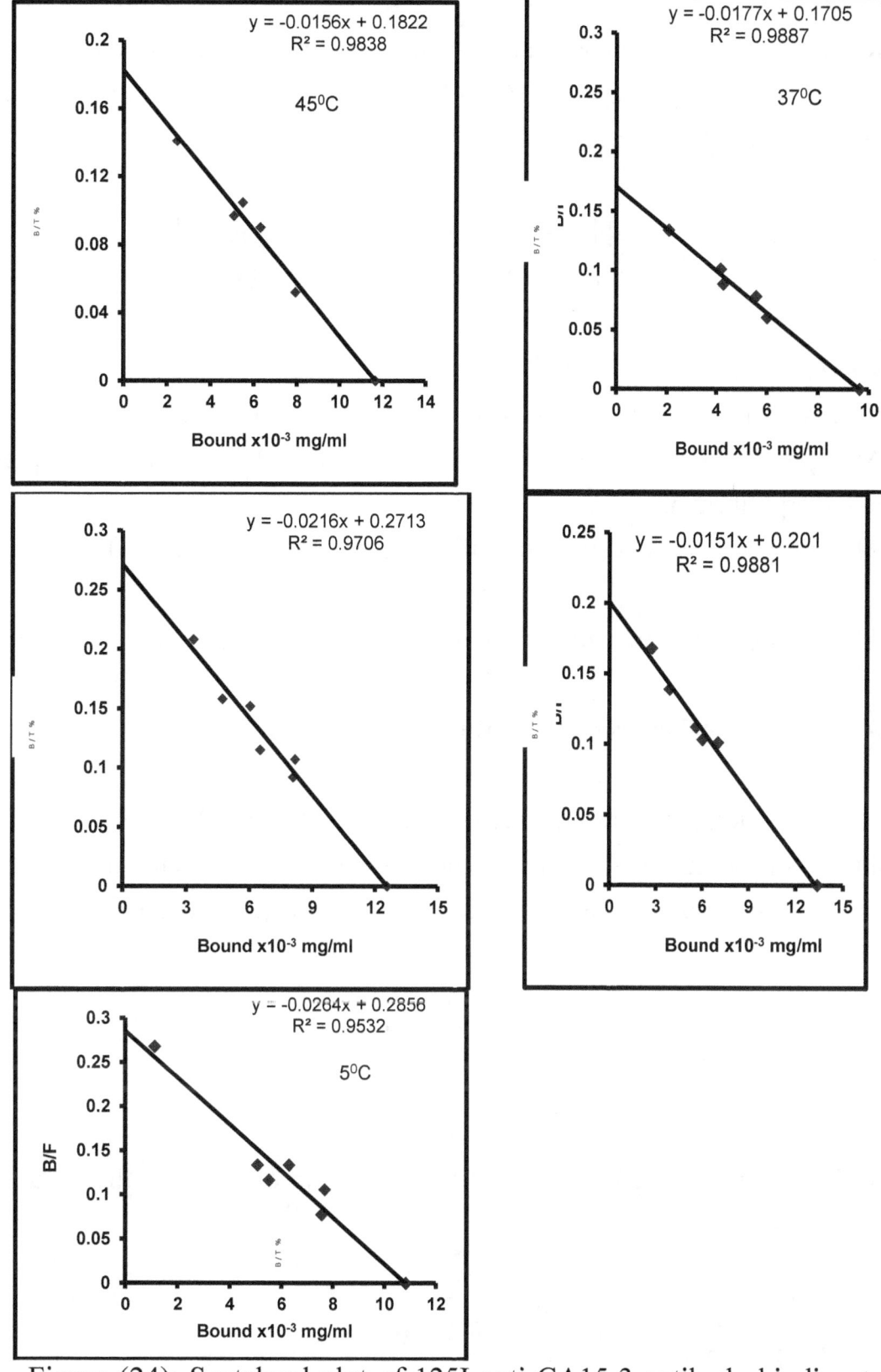

Figure (24): Scatchard plot of 125I-anti CA15-3 antibody binding to the partially purified CA15-3 in Malignant breast tumors (IDC) at five different temperatures. All details are explained in the text.

However, the time-course data shown in figure (25) could be used to determine the reaction order of CA15-3 binding to its specifically 125I-anti CA15-3 using the following equation:

$$Ln[AbAg]_e \left[\frac{[Ab]_t - [AbAg]_t [AbAg]_e / [Ag]_t}{[Ab]_t [AbAg]_e - [AbAg]_e} \right] = K_{+1} t \left[\frac{[Ab] + [Ag]_t - [AbAg]_e}{[AbAg]_e} \right] \quad \text{......(5)}$$

Where:
k+1 : is the kinetic association constant in mg-1. min-1. mL.
$[AbAg]_e$: is the concentration of (125I-antiCA15-3/CA15-3)complex formed at equilibrium.
$[AbAg]_t$: is the concentration of (125I-antiCA15-3/CA15-3) complex after time (t).
$[Ab]_t$: is the total concentration of 125I-anti CA15-3 antibody in mg. mL-1.
$[Ag]_t$: is the total concentration of CA15-3 in mg. mL-1.

Equation (5) represents the second order kinetics, but the percent of binding was in some cases, small and most labeled antibody remains free and only small fraction binds even at equilibrium, i.e , $[Ab]_t \gg [AbAg]_e$

Thus :

$$[Ab]_t \gg \frac{[AbAg]_t [AbAg]_e}{[Ag]_t}$$

So that the following equation could be used in order to fit the pseudo-first order kinetics:

$$Ln \frac{[AbAg]_e}{[AbAg]_e - [AbAg]_t} = K_{+1} t \frac{[Ab]_t [Ag]_t}{[AbAg]_e} \quad \text{..........(6)}$$

On the other hand, figure (25) and (26) show the plot of $\ln \frac{[AbAg]_e}{[AbAg]_e - [AbAg]_t}$ Against time (t) in both benign and malignant breast tumors, which give a straight line with a slope equal to the observed value of first rate constant Kbos in min-1. The rate constant (k+1) in mg-1. mL. min was calculated at five different temperatures by using the following equation

$$K_{obs} = K_{+1} \frac{[^{125}I - antiCA15-3]_t [CA15-3]_t}{[^{125}I - antiCA15-3/CA15-3]_e} \quad \ldots\ldots\ldots(7)$$

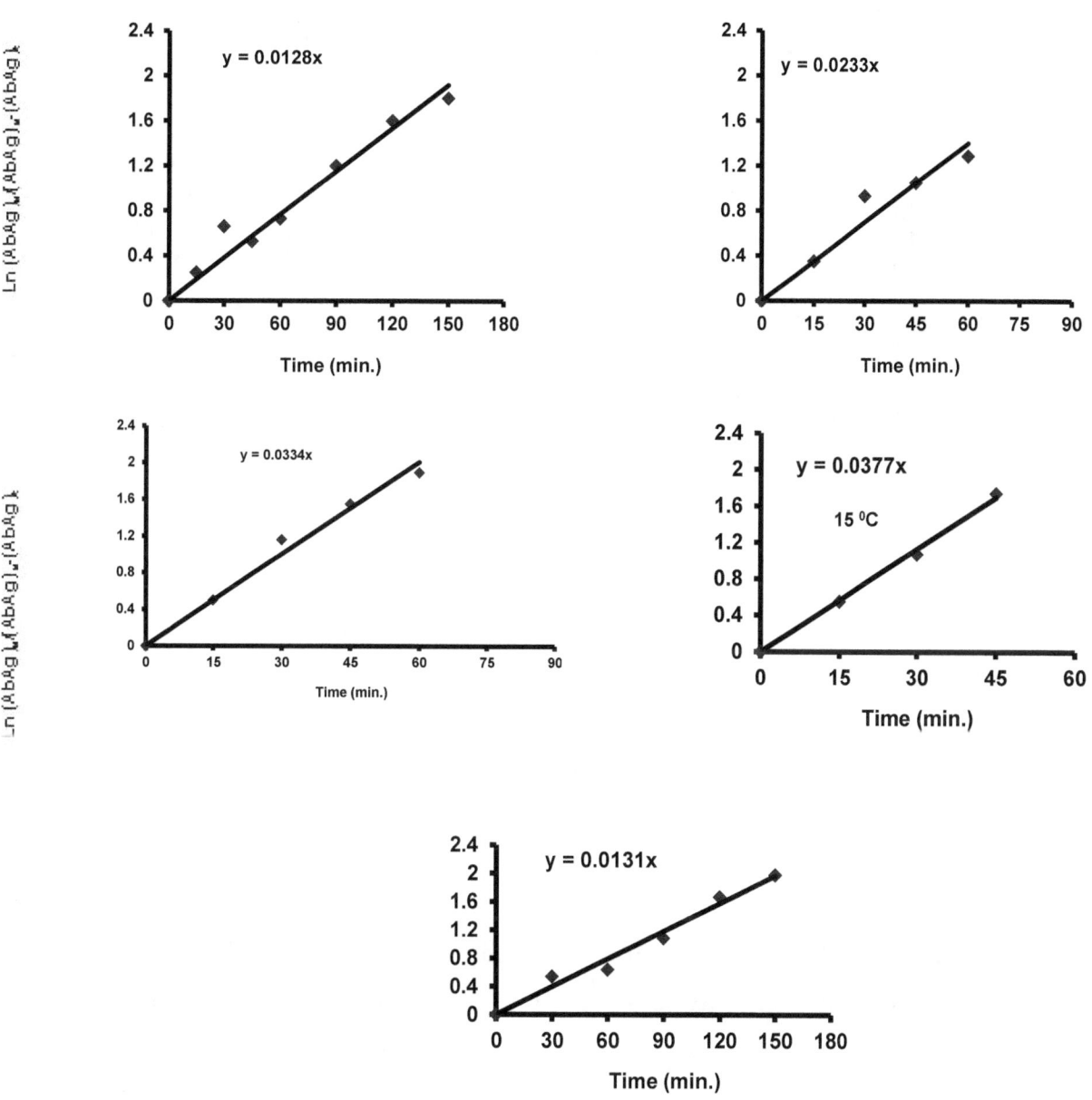

Figure (25): Kinetics of 125I-anti CA15-3 antibody binding to partially purified CA15-3 in benign breast tumors (Fibroadenoma). All details are explained in the text.

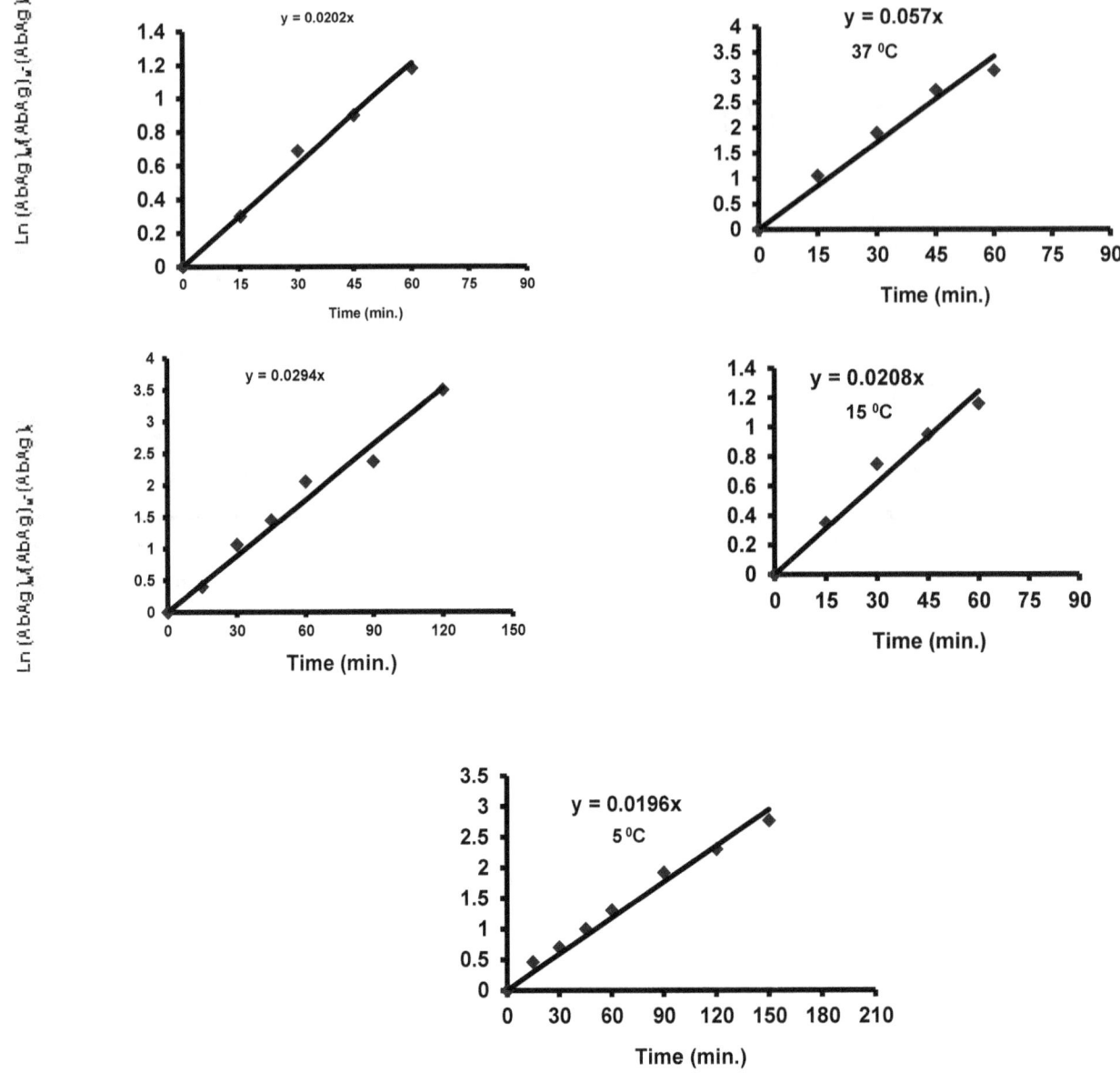

Figure (26): Kinetics of 125I-anti CA15-3 binding to partially purified CA15-3 in malignant breast tumors (IDC). (All details are explained in the text).

The value of k-1 at five temperatures was calculated by using equation (4). Whereas, the half-life time of association (t ½)ass. , Which represented the time needed for the formation of half amount of the complex at equilibrium was determined from the concentration of the complex at equilibrium and the time-course curve. The half-life time of dissociation (t ½) diss. , was calculated from the following relation:

$$(t_{1/2})_{diss.} = \frac{\ln 2}{k_{-1}} = \frac{0.693}{k_{-1}}$$

$$(t_{1/2})_{ass.} = \frac{\ln 2}{k_{obs}} = \frac{0.693}{k_{+1}}$$

The value of kobs. , k+1, k-1, (t ½)ass. ,(t ½)diss. at five different temperatures are summarized in table (17). Data analysis in this table shows that highest rate for the association reaction k+1 , in benign breast tumors (Fibroadenoma) and malignant breast tumors (IDC) occurs at 37°C and 15°C respectively , while the lowest rate occurs at 45°C. This means the dependence of reaction rate on temperature (Table 3) that also shows the values of the rate constant for the reverse reaction k-1 calculated from equation (4). Results show that the rate of dissociation of 125I-anti CA15-3 antibody, from its CA15-3 is temperature independent.

(17): The effect of temperature on the kinetic parameters of 125I-anti CA15-3 binding to partially purified CA15-3 in benign and malignant breast tumors at five different temperature.

Temp. °C	$k_{obs} \times 10^{-3}$ (min^{-1})		K_{+1} mg^{-1}.ml.min^{-1}		$k_{-1} \times 10^{-1}$ (min^{-1})		$(t_{1/2})_{ass}$ (min)		$(t_{1/2})_{diss}$ (min)	
	Benign (Fibroadenom)	Malignant (IDC)	Benign (Fibroadenoma)	Malignant (IDC)	Benign (Fibroadenoma)	Malignant (IDC)	Benign (Fibroadenoma)	Malignant (IDC)	Benign (Fibroadenoma)	Malignant (IDC)
5	12.8	20.20	48.69	45.81	15.38	34.68	54	34	45	20
15	23.3	57.00	60.65	116.16	32.50	73.75	30	12	21	9
25	33.4	24.9	93.98	35.48	57.37	18.07	21	28	12	38
37	37.7	20.30	103.54	46.61	61.89	22.43	18	34	11	31
45	13.1	19.60	35.40	21.34	24.96	10.58	53	35	28	66

The Thermodynamic Studies of 125I-Anti CA15-3 Antibody to the Partially Purified CA15-3 in Benign and Malignant Tumors

Thermodynamic Parameters of Standard State

Figure (27) and (28) show Van't Hoff plot of the binding of 125I-anti CA15-3 antibody to the partially purified CA15-3 in benign breast tumors (Fibroadenoma) and malignant breast tumors (IDC) respectively, at different temperatures (5 , 15 , 25 , 37 and 45 °C).

These figures revealed that the equilibrium binding constant (affinity constant) for CA15-3 to its antibody is a temperature dependent. The results indicated that ΔH°, in general, had small values and their positive sign ascertains that the reaction was nearly endothermic. The ΔH° value in the case of the binding of 125I-anti CA15-3 antibody to partially purified CA15-3 in benign breast tumors 12.71 KJ.mol-1 was higher than that in case of binding in malignant breast tumors (IDC) 6.7 KJ.mol-1, so more energy is needed in case of benign breast tumor for the reaction (binding) to occur. The small positive value of ΔH° may indicate a favorable interaction between 125I-anti CA15-3 antibody with partially purified CA15-3 in both cases.

These include the non-covalent interaction, which are fundamentally electrostatic in nature such as charge-charge, charge-dipole, dipole-dipole, charge-induced dipole, dipole-induced dipole interactions, and hydrogen bonds. The sum of these types of interactions can yield some stabilization to the folded structure of the complex .

The other values of thermodynamic parameters of standard state at five temperatures, such as ΔG° values and ΔS° values are summarized in table (18) and (19).

Table (18): Thermodynamic parameters at standard state of 125I-anti CA15-3 to the partially CA15-3 in benign breast tumors (Fibroadenoma). (All other details are explained in the text).

Temp. °C	ΔH° KJ .moL-1	ΔG° KJ .moL-1	ΔS° J .mol-1.K-1
5	12.71	-36.87	137.20
15	12.71	-38.59	138.42
25	12.71	-39.88	138.10

| 37 | 12.71 | -41.82 | 139.01 |
| 45 | 12.71 | -44.30 | 143.30 |

Table (19): Thermodynamic parameters at standard state of 125I-anti CA15-3 to the partially purified CA15-3 in malignant breast tumors (IDC). (All other details are explained in the text).

Temp. °C	$\Delta H°$ KJ .moL-1	$\Delta G°$ KJ .moL-1	$\Delta S°$ J .mol-1.K-1
5	6.70	-36.82	156.55
15	6.70	-39.11	159.06
25	6.70	-40.48	158.32
37	6.70	-42.27	157.97
45	6.70	-43.54	157.99

The negative values of $\Delta G°$ reflects the stability of the complex hence. The high affinity of the reactants. The high negative values of $\Delta G°$ for the binding reaction are controlled by high positive $\Delta S°$ values of the complex formed. So, our system is characterized by the sole contribution of $\Delta S°$ to the stability of the complex formed, which $\Delta H°$ has little or no effect. Whereas, the negative values of $\Delta G°$ indicates that the reaction is spontaneous at the standard condition. On the other hand, the high positive of $\Delta S°$ suggest that the binding was entropically driven. Entropy has a driven force for the occurrence of the binding reaction, this indicates that the hydrophobic interactions played an important role in the stability of complex formation.

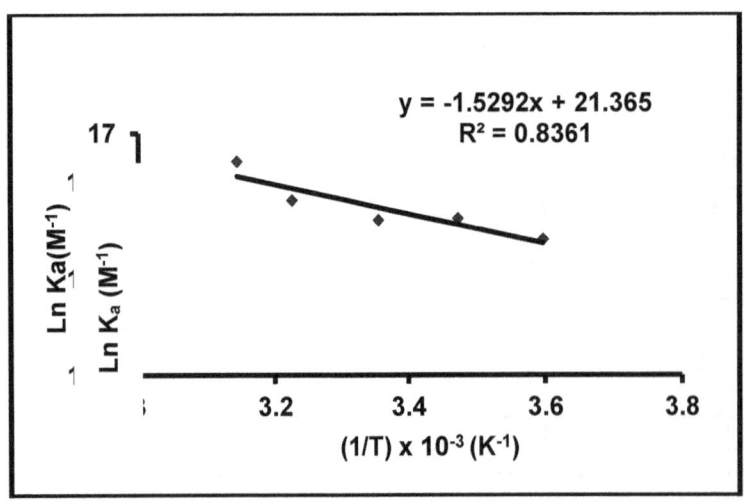

Figure (27): Van't Hoff plot for the binding of 125I-anti CA15-3 antibody to the partially purified CA15-3 in benign breast tumors (Fibroadenoma). All details are explained in the text.

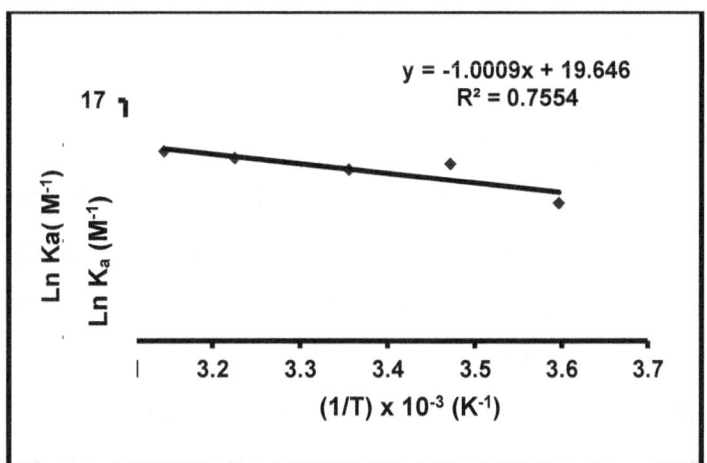

Figure (28): Van't Hoff plot for the binding of 125I-anti CA15-3 antibody to the partially purified CA15-3 in malignant breast tumors (IDC). All details are explained in the text.

B. Thermodynamic Parameters of Transition State

Transition state theory postulated that the interaction of two substances to form the final product proceeds through the formation of an activated complex (transition state).

Consequently, the association of 125I-anti CA15-3 antibody with its CA15-3 can be represented as follows:

$$^{125}I - antiCA15-3 + CA15-3 \rightarrow [^{125}I - antiCA15-3/CA15-3]^{\ddagger} \rightarrow [^{125}I - antiCA15-3/CA15-3]$$

State(A)   An Activated Complex   Final Product
       Transition State   State(B)

Thermodynamic parameters (ΔH*, ΔG* and ΔS*) of the transition state were determined from the application of Arrhenius equation to the kinetic data. Figure (29) and (30) show Arrhenius plots for the binding of CA15-3 to its antibody, the slope of the line represents the activation energy (Ea) of the binding reaction, the linear relationship indicates the dependency of the association rate constant of the binding of CA15-3 to its antibody for benign and malignant breast tumors homogenate on temperature.

Table (20) and (21) show the values of thermodynamic parameters of the transition state (Ea, ΔH^*, ΔG^* and ΔS^*).

The high values of activation energy 9.96 KJ.mol-1 and 41.76 KJ.mol-1 of CA15-3 partially purified from benign and malignant breast tumors respectively, represents the required energy to overcome the energy barrier of the transition state for the formation of (125I-anti CA15-3 antibody / CA15-3) complex. Also the value of activation energy is in accordance with the high positive values of ΔG^*, which indicates that the formation of the activated complex is a non-spontaneous process and requires a lot of energy (equal to Ea) to overcome the transition state energy barrier and giving the final product, whereas the high negative ΔS^* revealed that the activated complex had a more order structure than the reactants.

From the result obtained of the thermodynamic parameters in the transition state, it can be concluded that the positive values of ΔH^* and high positive values of ΔG^* are favorable to overcome the energy barrier of the transition state, the high negative values of ΔG^* is mainly attributed to the decrease in entropy of the transition state ($\Delta S^* < 0$).

In addition the positive values of ΔH^* show that the heat content of the activated complex is more than that in isolated species .

It is proposed that the formation of a complex occurs in the two steps. The first is the stabilization of the complex by hydrophobic interactions and second is the stabilization by short range interactions , such as electrostatic interaction, hydrogen bonding and Van der Waals interactions .

Hydrophobic interactions contribute to the complex stability via high positive entropy change ($\Delta S^* > 0$), while electrostatic interactions, hydrogen bonding and Van der Waals interactions contribute to the stability of the complex via negative entropy change ($\Delta S^* > 0$) .

The thermodynamic data indicate that the binding of 125I-anti CA15-3 antibody to partially purified CA15-3 are entropy driven and in agreement with the concept that hydrophobic interaction play an important rote in the formation of (125I-anti CA15-3 antibody / CA15-3) complex.

Table (20): Thermodynamic parameters at transition state of 125I-anti CA15-3 antibody to the partially purified CA15-3 in benign breast tumors (Fibroadenoma). (All other details are explained in the text).

Temp. °C	Ea KJ . mol-1	ΔH^* KJ . mol-1	ΔG^* KJ . mol-1	ΔS^* J . mol-1. K-1
5	9.96	7.65	58.94	-184.50
15	9.96	7.57	60.62	-184.20
25	9.96	7.48	61.72	-182.01
37	9.96	7.38	64.06	-182.84
45	9.96	7.32	68.62	-192.77

Table (21): Thermodynamic parameters at transition state of 125I-anti CA15-3 antibody to the partially purified CA15-3 in malignant breast tumors (IDC). (All other details are explained in the text).

Temp. °C	Ea KJ . mol-1	ΔH^* KJ . mol-1	ΔG^* KJ . mol-1	ΔS^* J . mol-1. K-1
5	41.76	39.45	59.08	-70.61
15	41.76	39.37	59.09	-68.47
25	41.76	39.28	64.14	-83.42
37	41.76	39.18	66.12	-86.90
45	41.76	39.12	70.00	-97.11

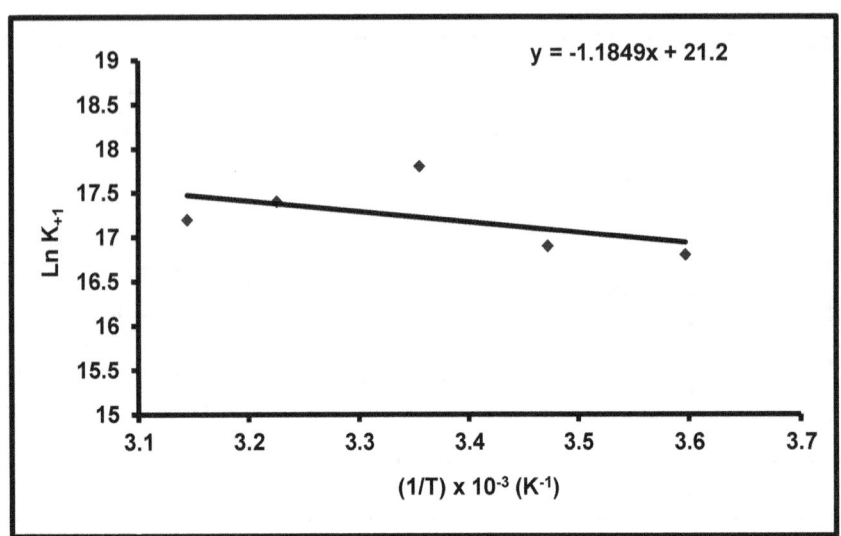

Figure (27): Arrhenius plot for the binding of 125I-anti CA15-3 to the partially purified CA15-3 in benign breast tumor (Fibroadnoma). All details are explained in the text.

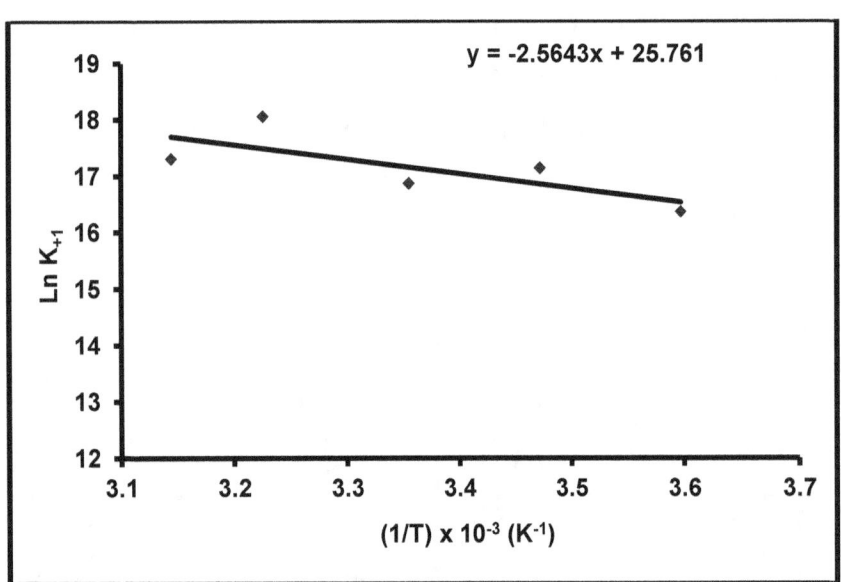

Figure (28): Arrhenius plot for the binding of 125I-anti CA15-3 to the partially purified CA15-3 in malignant breast tumor (IDC). All details are explained in the text. Gel filtration technique was used to separate 125I-anti CA 15-3 antibody bound to partially purified CA 15-3 using benign (Fibroadenoma) and malignant (IDC) breast tissue homogenate (as CA 15-3 source) from unbound (Free) 125I-anti CA 15-3 antibody.

characterization
of complexes of CA 15-3

Introduction

The characterization of the complexes (125I-anti CA 15-3 antibody/ CA15-3) from both benign and malignant breast tumors was carried out through the ultraviolet spectroscopic studies. Factors affecting the absorption properties of the two types of complexes such as pH, solvent polarity (solvent perturbation technique), spectrophotometric pH titration, and thermal stability in the presence of different concentrations of sodium chloride have been studied. pH titration of the two types of the complexes show that about (41.43%) and (44.29%) of histydyl residues are located on the surface of the two types of protein complexes (benign and malignant) respectively, while (40%) and (50%) of tyrosyl residues are buried interiorly in the complexes of (benign and malignant) respectively.

Molecules absorb light; the efficiency of absorption depend on both the structure and environment of the molecule making absorption spectroscopy a useful tool for characterizing both small and large molecule.

The ultraviolet absorption spectra of protein solutions in the region 250 to 310 nm are contributed from phenylalanyl, tyrosyl and tryptophanyl residues. But at the shorter wavelengths the contributions come from other groups such as histidyl residues and the peptide bond . Changes in the environment of these chromophores can lead to alteration in the absorption spectrum, and the conformational changes of a protein may also involve environmental changes of its chromophoric groups . A variety of environmental changes (e.g. pH, temperature) can affect the absorption spectrum if the interaction of chromophore and perturbing agent affects the ground and excited states, the altered spectrum of the chromophore can be shifted to longer (red shift) or shorter (blue shift) wavelengths. The shift may or may not be accompanied by a change in intensity of the spectrum . Saif-Alla, P.H., studied the UV spectra of h-PRL-antibody complex and CA15-3 molecule .

Interaction of h-CA 15-3 partially purified from benign (fibroadenoma) and malignant (IDC) tissues homogenate with its antibody is an example of protein-protein association. Although several new immunochemical techniques were developed to study such interactions , UV spectral remain as one of the most important methods in immunology because it provides a sensitive and quantitative measurements for the study of antibody structure and its specific ligand binding .

Very limited work concerning the physical properties of CA 15-3 specially those related to UV spectroscopy has been done, also the UV studies on CA 15-3 antibody interaction are not wide spread. Hence, this work is planned to study the association of the partially purified h-CA 15-3 and its antibody at different conditions.

Materials and Methods

Methods

Gel Filtration Technique for Separation of Free and Bound 125I -Anti CA 15-3 Antibody

Preparation of the Column

The dimensions of the column were (1x30 cm) chosen according to the equation .

Preparation of the Gel and Determination of Void Volume

The sepharose CL-4B was used to separate free and bound 125I - anti CA 15-3 antibody, and was prepared as mentioned , the void volume was determined and found to be 10 mL.

Separation Procedure of (125I-Anti CA 15-3 Antibody/CA15-3) Complex

A) Partially Purified CA15-3 from Benign Breast Tumor (Fibroadenoma) and its Antibody 125I -Anti CA 15-3

1- Partially purified CA 15-3 (475µL) containing (0.665 mg. mL-1) was incubated with 120 µL of 125I-anti CA 15-3 antibody (0.8412mg. mL-1) and complete the reaction to a final volume of 700 µL with PBS buffer 0.15 M pH 7.0. The tubes were incubated for 90 min. at 37oC.

2- At the end of incubation, the mixture was applied to the surface of a sepharose CL-4B (1x30 cm) with a bed volume (23.5 cm3) equilibrated with PBS buffer 0.15M, pH 7.0. Elution was carried out using the same buffer to separate CA 15-3 bound to 125I-anti CA 15-3 antibody from unbound (Free) CA 15-3 and 125I-anti CA 15-3 antibody with a flow rate (1 mL per 7 min), and fraction volumes of 1 mL were collected.

3- The radioactivity of each fraction was counted by gamma counter for one minute.

4- Protein concentration was measured at 280 nm.

5- One hundred and twenty microliters of 125I-anti CA 15-3 antibody (0.84 mg. µL-1) was completed to 700 µL with PBS buffer (0.15M, pH7.0), then this volume was injected to the column as mentioned in step2, then steps 2,3 and 4 were repeated.

Calculations

1. Radioactivity (c.p.m) of each eluted fraction was plotted against the fraction number.

2. The absorbance of each eluted fractions was measured at 280nm, and the absorbance was plotted against the fraction number.

3. The percent radioactivity was calculated by dividing the sum of the radioactivity of the fractions under each peak by the sum of radioactivity of all peaks appeared in the profile:

$$\text{Percent radioactivity of each peak} = \frac{\text{Radioactivity per peak (c.p.m)}}{\text{Sum of radioactivity of all peaks (c.p.m.)}} \times 100$$

B) Partially Purified CA15-3 from Premenopausal Malignant Breast Tumors (IDC) and Its Antibody 125I-anti CA 15-3

Reagents

Buffer PBS 0.15 M, pH 7.0 containing 0.02% sodium azid was prepared as described previously in section (2.1.1.3).

Procedure

1. Four hundred and twenty four microliters of partially purified CA 15-3 (0.147 mg. mL-1 protein) and incubated with 106 µL of 125I-anti CA 15-3 antibody (0.743 mg.mL-1) in a final volume 700 mL with PBS buffer 0.15M pH 7.0. The tubes were then incubated for 150 min at 15oC.

2. Steps 2,3,4 and 5 in section (5.2.1.3 A) were repeated.

Calculation

The same calculation that mentioned in section (5.2.1.3 A) was used to calculate the radioactivity; protein was measured at 280nm and the percent of radioactivity of each peak was determined.

The UV Spectrum of (125I-Anti CA 15-3 Antibody/CA15-3) Complex from Benign and Malignant Breast Tumors

The gel filtration profile gave two peaks. The fractions under each peak were pooled and the absorption spectrum was scanned in UV Region against the appropriate blank in the reference beam.

The UV. Spectrum of 125I-Anti CA 15-3 Antibody

Half milliliter of 125I-anti CA 15-3 antibody was placed in a 0.25 cm cuvette in the sample beam and the absorption spectrum was measured immediately against an appropriate blank in the reference beam.

The UV Spectrum of Partially Purified CA 15-3

Half milliliter of partially purified CA 15-3 from benign (Fibroadenoma) and malignant (IDC) breast tumors was placed in a 0.25 cm curette in the sample beam and the absorption spectrum was measured immediately against an appropriate blank in the reference beam.

Factors Affecting the Absorption Properties of (125I-Anti CA 15-3 Antibody/CA 15-3) Complex from Benign and Malignant Breast Tumors

The pH Effect on the Complex

Procedure

Two hundred and fifty microliters of pooled fractions under the first peak that represent (125I-anti CA 15-3 antibody/CA 15-3) complex, was completed to 500µl with different buffers at different pH values (4 to 11), then each sample beam and the buffer at the adjusted pH in the reference beam. The absorption spectrum was scanned.

Calculations

The molar absorption coefficient (ε) for (125I-anti CA 15-3 antibody/CA 15-3) complex at 278 nm was calculated from Lambert-Beer's law.

Effect of Solvent Polarity on UV Spectra of the Complex

The effect of 20% ethanol, and the same amount for ethylene glycol, glycerol, sucrose, urea, dimethyl sulphoxide, dioxane, and polyethylene glycol; on the complex. Two hundred and fifty microliters of complex from benign and malignant breast tumors of pooled fractions under the first peak were completed to 500 µL with phosphate buffer containing any of the following solvent at pH 7.4 in the test cell and the 20% ethanol, ethylene glycol, glycerol, sucrose, urea, dimethyl sulphoxide, dioxane, and polyethylene glycol was adjusted and placed in the reference cell using 0.25 cm cuvette (i.e., the experiment was repeated by using solvents individually).

Calculations

The absorption spectrum of each sample was scanned immediately in the area of (200-350 nm).

Spectrophotometric pH Titration on the Complex

A series of complex from benign (Fibroadenoma) and Malignant (IDC) breast tumors (250 μL) were completed to 500 μL with buffer at pH ranging from 8 to 11. The maximum absorbance of each sample was measured at 295 nm; the absorbance of λ max at each pH value was plotted versus the corresponding pH. Other series of complexes isolated from benign (Fibroadenoma) and malignant (IDC) breast tumors (250 μL) were completed to 500 μL with buffer at pH ranging 4 to 8. The maximum absorbance of each sample was measured at 211nm. The absorbance of λmax at each pH value was plotted against the corresponding pH.

The Effect of NaCl Concentration on the Thermal Stability of the Complex by UV Spectral Studies

Reagents

Twenty percent ethylene glycol buffur was prepared by dissolving 20mL of ethylene glycol in 80mL of phosphate buffer. NaCl (0.01M) in 20% ethylene glycol was prepared by dissolving 0.05844 gm of NaCl in 100mL of 20% ethylene glycol buffur, while NaCl (0.1M) in 20% ethylene glycol was prepared by dissolving 0.5844 gm of NaCl in 100mL of 20% ethylene glycol buffer.

Procedure

Two hundred and fifty microliters of complex from benign (Fibroadenoma) or malignant (IDC) breast tumors were completed to a final volume 500 μL with 20% ethylene glycol buffer pH7.4 containing 0.01 M NaCl Each mixture was placed in 0.25 cm cuvette in the sample beam and the buffer at the adjusted pH in the reference beam. The absorbtion was measured at the wavelength of (292 and 295 nm) at different temperatures 20, 30, 40, 50, 60, 70oC. The experiment was repeated for each complex with another solution (20% ethylene glycol 0.1 M NaCl), at 295 nm.

Calculations

The absorbance of each complex was plotted against the different temperatures at two wavelengths (292 and 295 nm).

Effect of Urea, KCl and (Urea, KCl) Mixture on the Spectrum of the Complex

Two hundred and fifty microliters of complex isolated from benign (fibroadenoma) and malignant (IDC) were pipetted in a set of three tubes. The volume was completed to 500 μL with PBS buffer at pH 7.4 contains (0.03 KCl, 8 M urea and mixture 1:1 of both 0.03 KCl and 8M Urea) respectively, then each sample was placed in 0.25cm cuvette in the sample beam and the buffer at the same pH in the presence of the same salt in the reference beam.

Calculations

The absorption spectrum of each sample was scanned immediately in the area of (200-350 nm).

Results and Discussion

Protein UV light maximum absorption is at approximately 280nm, caused by tryptophan, tyrosine and (to a lesser extent) phenylalanine residues, and at lower wavelength (215-230 nm) due to polypeptide chain backbone. Absorbance at 280 nm varies for each protein. The absorbance at lower wavelengths is directly related to the amount of polypeptide material and is usually considerably more sensitive than at 280nm. however, many buffers and other molecules also absorb at these lower wavelengths (phosphate and tris buffers are acceptable but the preservative sodium azide absorbs strongly).

Absorbance at 215-230 nm is useful for monitoring peptides that may not contain tryptophan or tyrosine.

Gel Filtration Technique for Separation of Free and Bound 125I-Anti CA15-3 Antibody

Figure (29) and (30) show the results of gel filtration technique to separate 125I-anti CA 15-3 antibody bound to partially purified CA 15-3 from benign (Fibroadenoma) and malignant (IDC) breast tumors respectively. The profile of separation revealed two peaks. The first peak represents (125I-anti CA 15-3 antibody/CA 15-3) complex, the second peak represents the unbound (Free) 125I-anti CA 15-3 antibody. Figure (31) show the gel filtration profile of 125I-anti CA 15-3 antibody, the results revealed only one peak in the same position of the second peak of figures (29) and (30), which represent the unbound 125I-anti CA 15-3 antibody. The percent of 125I-anti CA 15-3 antibody/ CA 15-3) complex was 49.74% in benign (Fibroadenoma) breast tumors patients, while the percent of complex was 56.20% in malignant breast tumors patients (IDC). On the other hand the percent of 125I-anti CA 15-3 antibody was 34.40% in benign breast tumors (Fibroadenoma) and 31.42% in malignant breast tumors (IDC). This is because the epitope of CA 15-3 in malignant breast tumors was higher than in benign breast tumors.

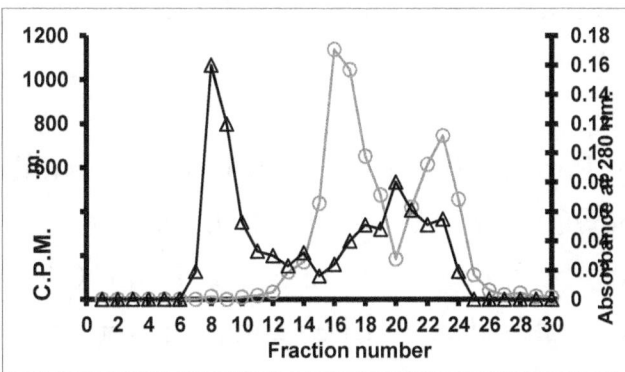

Figure (29): The elution profile of the isolated complex (125I-antiCA15-3 antibody/CA15-3) and free antibody in benign breast tumors on Sepharose

CL-4B. (○) radioactivity, (△) protein. (All other details are explained in the text).

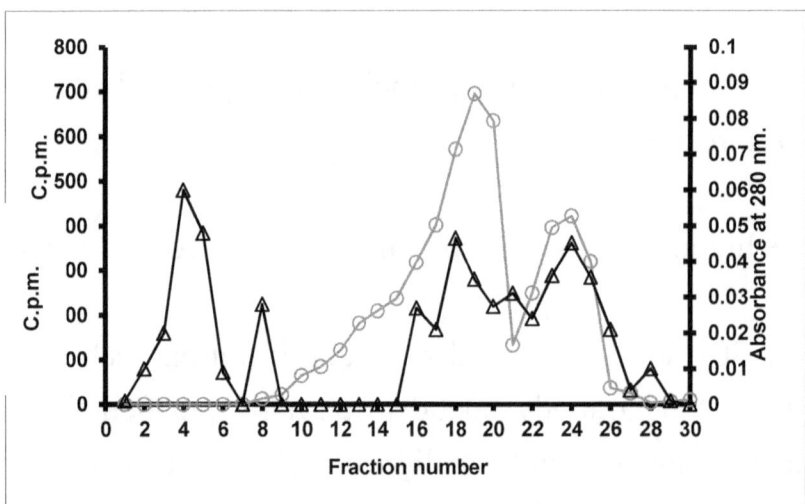

Figure (30): The elution profile of the isolated complex (125I-antiCA15-3 antibody/CA15-3) and free antibody in malignant breast tumors (IDC) on Sepharose CL-4B, (○) radioactivity, (△) protein. (All other details are explained in the text).

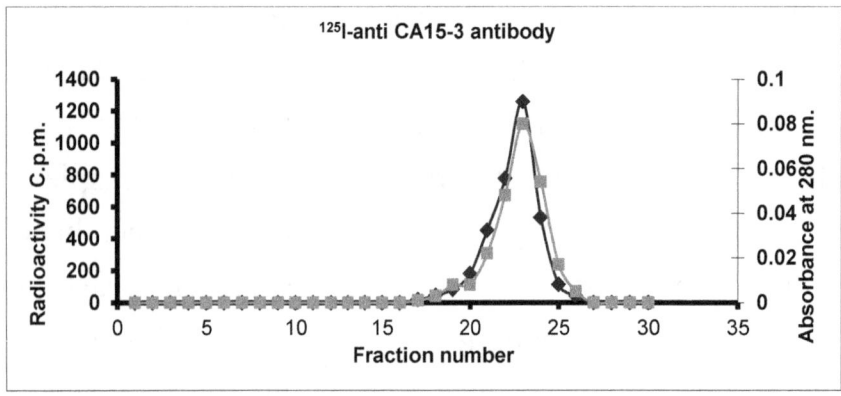

Figure (31): The elution profile of the 125I-antiCA15-3 antibody on Sepharose CL-4B, (■) radioactivity, (♦) protein. (All other details are explained in the text).

The UV Spectra of Partially Purified CA 15-3, Anti CA 15-3 Antibody and (125I-Anti CA 15-3 Antibody/CA 15-3) Complex Molecules

The UV spectra of partially purified h-CA 15-3, 125I-anti CA 15-3 antibody and (125I-anti CA 15-3 antibody/CA 15-3) complex were scanned from 200-350 nm to determine the absorption spectra, and the alternation in the UV spectra as a results of their interaction.

The UV Spectrum of Partially Purified CA 15-3

The UV spectra of partially purified h-CA15-3 in benign tumors (Fibroadenoma) and malignant tumors (IDC) at neutral pH shows that the λmax for purified CA15-3 from benign (Fibroadenoma) consisted of two peaks; a large one at 208nm and smaller one at 270nm, while the UV spectra of purified CA 15-3 from malignant tumors (IDC) shows two peaks at 205 and 270nm as shown in table (1). Therefore it seemed that each human CA 15-3 has a characteristic spectrum and can be identified by its peaks, the first peak (at 208nm or 205nm) such results could be due to the amide group in polypeptide bond of h-CA 15-3 molecule with contribution of the histidyl residues, while the second peak (at 270) is assigned to the side chain chromophore of phenylalanine or tryptophyl residues.

The UV spectrum of 125I-Anti CA 15-3 Antibody

The UV spectrum of 125I-anti CA 15-3 antibody at neutral pH shows that the λ max consisted one peak at 203.6nm, which is assigned to the amide groups in the polypeptide bond, with contribution of hisidyl residues as shown in table (22).

The UV spectrum of (125I-Anti CA 15-3 Antibody/CA 15-3) Complex

The UV spectra of partially purified CA 15-3 extracted from benign (Fibroadenoma) and malignant (IDC) bound to 125I-anti CA 15-3 antibody at neutral pH show that the λ max is consisted of two peaks at (203.4nm and 274nm) in benign complex, while the λ max is consisted of two peaks at (204.2 nm and 278nm) in malignant complex as shown in table (1). The first peak at (274nm or 278nm) is assigned to tyrosyl residues, it is very weak band and it seems that the tyrosyl residues in the benign or malignant complexes is located on the surface of protein complex.

The strong absorption of the second peaks (at 203.4 or 204.2nm) arises form electronic transition in the peptide backbone itself and is therefore sensitive to backbone conformation.

Table (22): The λ max valves of (125I-anti CA 15-3 antibody/CA 15-3) complex, partially purified CA 15-3 and unbound (Free) 125I-anti CA 15-3 antibody in both cases benign and malignant breast tumors. (All other details are explained in the test).

No.	Fractions	Benign λmax (nm)	Malignant λmax (nm)
1	CA 15-3 partially purified	208, 270	205, 270
2	125I-anti CA 15-3 antibody	203.6	203.6
3	125I-anti CA15-3 antibody/ CA15-3) complex	203.4, 274	204.2, 278

Factors Affecting the Absorption Properties of (125I-Anti CA15-3 Antibody/ CA15-3) Complex from Benign and Malignant Breast Tumors

The Effect of pH on the Complex

The pH of the solvent determines the ionization state of ionizable chromophores in the protein molecule . The UV spectrum of isolated (125I-anti CA 15-3 antibody/CA15-3) complex from benign (Fibroadenoma) and malignant (IDC) breast tumors was determined at different pH (2, 4, 6, 7, 7.4, 8, 9, 10, and 11). Table (2) shows the effect of different pH on both complexes. At an acidic pH 2 and neutral pH (7,7.4) the both complexes benign (Fibroadenoma) and malignant (IDC) have one maximum wavelength near 200 nm as compare to UV spectrum of h-CA 15-3 and the 125I anti CA15-3 antibody.

The λ max of CA 15-3 (270 and 208 nm) in benign (Fibroadenoma) and its antibody λ max (203.6) disappeared. The λ max of CA 15-3 (270 and 205 nm) in malignant (IDC) and its antibody λmax (203.6) also disappeared. The absorption near 200 nm is characteristic of the amide group in the polypeptide bond of the complex . The blue shift is due to the increasing of hydrogen bond formed in the presence of highly positively charged state . The disappearance of λmax 280 nm of tyrosine and phenylalanine due to conformational changes and chromophore in native complex were buried in the interior of their complexes. Protein shows a strong absorption in range

(180-225 nm), absorption at such wavelength arises from electronic transition in the polypeptide backbone itself and is therefore sensitive to back bone conformation.

At pH (4, 6, 9, 10, and 11) no band was observed and all peaks disappeared. The disappearance of the λ max at these pH's may be due to conformational changes of the protein complex.

Table (23): The effect of different pH on λ max values of (125I-anti CA 15-3 antibody/CA 15-3) complex. (All other details are explained in the text).

pH	λ max (nm)	
	(125I-anti CA 15-3 antibody/CA15-3) benign complex	(125I-anti CA 15-3 antibody/CA 15-3) malignant complex
2	200	200
4	-	-
6	-	-
7	200	200
7.4	200	200
8	200	200
9	-	-
10	-	-
11	-	-

Effect of Solvent Polarity on UV Spectra of the Complex

The immediate environment of a chromophore affects its absorption. The determination of whether an amino acid is internal or external by measuring the spectra of protein in a polar and non-polar solvent is called the solvent perturbation method . In fact, proteins are rarely studied in completely non-polar solvents because most proteins are either insoluble or denatured in these solvents. However, significant solvent effects can be induced by use of a mixture of water and substance of a reduced polarity such as ethanol, ethylene glycol, polyethelyne glycol, sucrose, dioxane and dimethyl sulfoxide (DMSO)(208). Several spectra changes were obtained in the precence of these perturbants, like the alteration of λmax positions and intensities of protein spectrum and the appearance of new chromophores on the surface of the complex. These chromophores on the region of the protein disappeared in the absence of the solvent. One of the main assumptions of the solvent perturbation technique is that solvent alters the peak positions and intensities by altering the energy and probably of electronic transitions. Other considerations include the following:

a. Polarization effect

b. Change in permanent dipole moment during excitation, which will tend to produce either a short wave or a long wave shift depending on the nature of the electronic transition and wheather the solute is a hydrogen donor or hydrogen acceptor.

The effects of different solvents on the (125I-anti CA 15-3 antibody /CA 15-3) complex from benign (Fibroadenoma) and malignant (IDC) breast tumors at pH 7.4 were investigated. The data obtained are illustrated in table (24). It was found that one λ max specific for the amide groups of polypeptide bond at pH 7.4, this shift toward the shorter wavelength is due to the n-π* transitions in the presence of 20% ethanol, ethylene glycol and glycerol. In the presence of polyethylene glycol there was a significant red shift in the λ max (204 nm) in benign (Fibroadenoma) complex and λ max (205nm) in malignant (IDC) complex. When 20% Dioxane was used there were a significant red shift in the λ max (220nm) of the amide bond at pH 7.4, which assigned to tyrosyl residue. The value of λ max is for n-π* transitions which occur at longer wavelength because the nonbonded electrons in the anion are available for interaction with the π electron system of the ring, while in the presence of 20% sucrose the complex has a slight blue shift and show λ max at 202 nm and 201nm in both benign and malignant complexes. Finally the effect of 20% DMSO on the complex, show that the amide bands at pH 7.4 were disappeared, this may be due to the denaturation of protein complex in presence of 20% DMSO.

The application of spectrophotometric solvent perturbation on the complex is to determine the location of tyrosyl residues, whether they are buried and inaccessible or exposed and accessible to the solvent approach.

Laskowski has listed the major assumptions of solvent perturbation experiments. There are: (1) buried chromophors are unperturbed, that is only the groups located on the surface or near the surface of the protein should experience the perturbing effects of the solvent; groups buried in the interior of the protein, not accessible to the solvent; which should not be affected and consequently could not contribute to the overall spectral shift observed. (2) No conformational changes take place upon addition of perturbant, and (3) the solvation layer around the chromophore contains the same concentration of perturbation experiments when employed at convenient concentrations (often 20%), do not appear to produce conformational changes in most protein studied under reasonable conditions of pH, ionic strength, and temperature. This concentration is large enough to cause measurable shifts in the spectra of chromophoric residues. Conformational changes can be expected if perturbation is carried out under conditions in which the protein structure has marginal stability (low-or high pH for many protein). Chromophore may not completely bury. It has been, distinguished between chromophores in crevices and chromophores that are partially buried . The former are observed to be fully perturbed by perturbant solvent smaller than a certain critical size (e.g ethanol), but not by larger perturbants (e.g polyethylene glycol). The degree of exposure is thus determined only by the size of perturbant molecule (solvent) or by the size of the crevice in which the chromophore is located . Partially buried chromophores on the other hand, show a degree of exposure that depends on the nature of the perturbant, rather than on its size. The observed degree of exposure decrease in the order:

Sucrose ≥ Glycerol ≥ Ethyleneglycol ≥ Methanol, Ethanol > Polyethylene glycol ≥ Dimethyl sulfoxide.

The first perturbant in this series modify the solvent nonspecifically, while the later ones in the series may specifically interact with chromophore.

When comparing the effects of the six solvents used, ethanol, ethleneglycol, glycerol, dioxane, polyethylene glycol and dimethylsulfoxide at pH 7.4 on the UV spectrum, especially on the shift of λ max, which is due to tyrosyl residues. It seems that the maximum effect was observed in the presence of 20% dioxane as perturbant solvent, where there was a shift in the λ max about 16nm, while minimum effect was observed in the presence of 20% polyethylene glycol where the λ max remained unchanged. Since the change in the λ max of tyrosyl residues does not depend on the size of the perturbant solvent, the tyrosyl residues showing the changes in λ max ; absorbance must be partially buried.

Table (24): The effect of 20% of ethanol, ethyleneglycol, glycerol, polyethylene glycol, sucrose, dioxane, DMSO and on the λ max of (125I-Anti CA 15-3 Antibody/ CA15-3) complex at pH 7.4. (All other details are explained in the text).

Solvent of 20% of	λ max (nm)	
	(125I-anti CA 15-3 antibody/ CA15-3) benign complex	(125I-anti CA 15-3 antibody/ CA 15-3) malignant complex
Ethanol	Near 200	200
Ethylene glycol	Near 200	Near 200
Glycerol	Near 200	Near 200
Polyethylene glycol	204	205
Sucrose	202	201
Dioxane	220	220
DMSO	-	-

Spectrophotometric pH Titration of the Complex from Benign and Malignant Breast Tumors

To study (125I- anti CA 15-3 antibody/ CA15-3) complex structure, this requires the determination of pka values for proton dissiociation from ionizable amino acid side chains, because these values give an indication of the location of amino acid in the protein. This can often be done spectrophotmetrically because dissociation often changes the spectrum of one of the chromopores (tyrosyl). For proteins this usually amounts to the titration of the phenolic groups of tyrosine residues. By the measurement of the absorption at 295 nm (λ max for the ionized form of tyrosine), or observation of histidine dissociation by measurment at 211nm.

The titration curves of (125I- anti CA 15-3 antibody/ CA15-3) complex from benign (Fibroadenoma) and malignant (IDC) for both histidyl and tyrosyl residues are illustrated in figure (5-4 A&B) respectively. Figure (32) shows that the pka for histidine is (6.69) for (125I- anti CA 15-3 antibody/ CA15-3) complex from benign breast tumors, while the pka for histidine is (6.65) for (125I- Anti CA 15-3 Antibody/ CA15-3) complex from malignant (IDC) breast tumors. From the same curve it could be concluded that about (41.43%) histidyl residues are located on the surface of the protein complex of benign (Fibroadenoma), while about (44.29%) histidyl residues are located on the surface of the protein complex from malignant (IDC). The other residues are buried interior the benign and malignant complex. Figure (32 B) shows that the pka value of the benign complex of tyrosyl residues is (8.9) and it's about (40%) at tyrosine residues are internal and a large arise in the absorbance at very high pH was observed. While in the malignant (IDC) complex the pka value of tyrosyl is (8.4) and it's about (50%) this indicates that the internal tyrosines have become exposed to the solvent, which is the protein complexes in folded (become denatured).

The two curves also illustrated the low content of histidine compared to the high content of tyrosine in the benign and malignant complex.

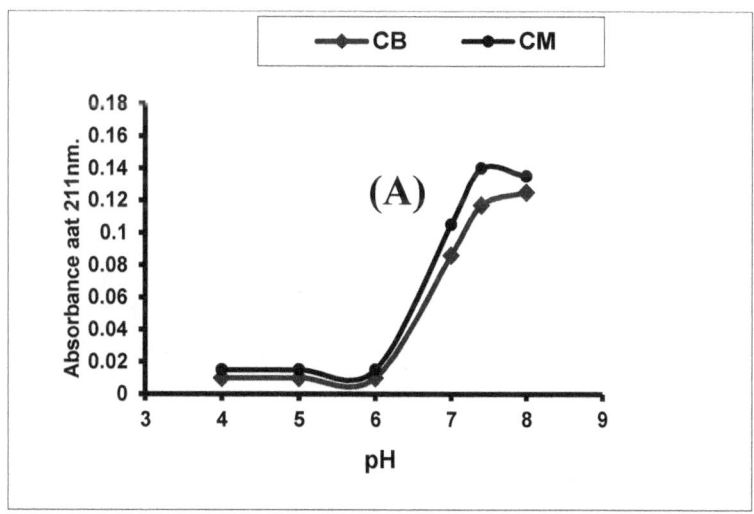

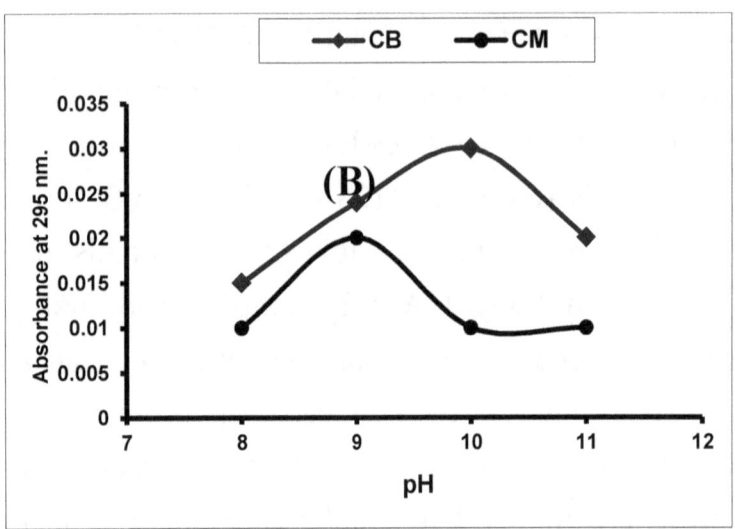

Figure (32): Spectrophotometric pH titration of 125I-anti CA15-3 antibody/CA15-3 complex from benign and malignant breast tumors:

(A) for histidine, (B) for tyrosine.

(CB): Complex of benign breast tumors, (CM): Complex of malignant breast tumors. (All other details are explained in the text).

The Effect of NaCl Concentration on the Thermal Stability of the Complex by UV Spectral Studies

The effect of different concentrations of NaCl on the thermal stability of the protein complex isolated from benign and malignant breast tumors was examined in this experiment. The values of absorbance at λ max (292, 295nm) for tryptophyl and tyrosyl residues respectively, in two different concentrations of NaCl 0.01 M and 0.1 M in 20% ethylene glycol buffer are shown in figure (5 A&B) and (6 A&B). The λ max was used to examine if the protein contains internal tryptophans and tyrosines .

As shown in figure (33 A&B), the absorbance of both tryptophane and tyrosine reach higher absorbance at 60oC, in the presence of 0.01 M NaCl in benign and malignant complex. The increment in the absorbance of both tryptophyl and tyrosyl residues with increasing temperature could be due to that buried chromophores becomes exposed to the solvent during thermal denaturation .

Figure (34 A) shown the absorbance of tyrosin reach higher absorbance at 70oC in the presence of 0.1 M NaCl in benign and malignant complex. On the other hand figure (34 B) shown the absorbance of tryptophane reach higher absorbance at 60oC and 30oC in benign and malignant breast tumors complexes in presence of 0.1 M NaCl respectively. Which means that the complexes were very stable at 70oC in presence of higher concentration of NaCl, 70oC was needed for unfolding benign and malignant complex at λ max 292 nm and benign complex was more stable at 60oC in presence of 0.1 M NaCl,while the temperature is decreased to 30 oC in the presence of 0.1M NaCl at λ max 295. This is due to conformational changes required more energy 70oC in presence of 0.1 M NaCl than in 0.01 M NaCl.

The decreased absorbance in presence of 0.1 M NaCl as compared with that in 0.01 M NaCl could be due to salt concentration. Each protein in solution containing salts will collect around it a counter ion atmosphere entriched in oppositely charged small ion (chloride ion, sodium ion) and such a cloud of ions will tend to screen the protein, the more effective electrostatic screening will be, and decrement in the absorption intensity will be observed .

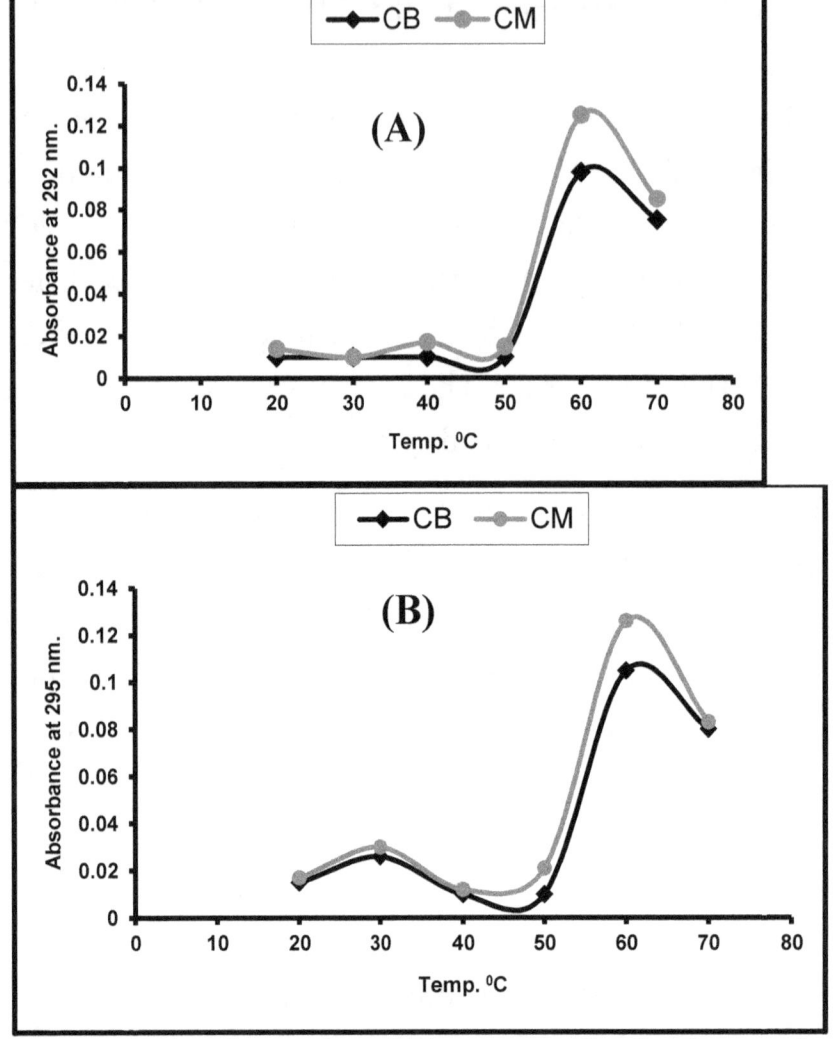

Figure (33): Thermal stability curve for benign and malignant: (A) at λmax 292 in the presence of 0.01 M NaCl, (B) at λmax 295 in the presence of 0.01 M NaCl. (CB): Complex of benign breast tumors, (CM): Complex of malignant breast tumors. (All other details are explained in the text).

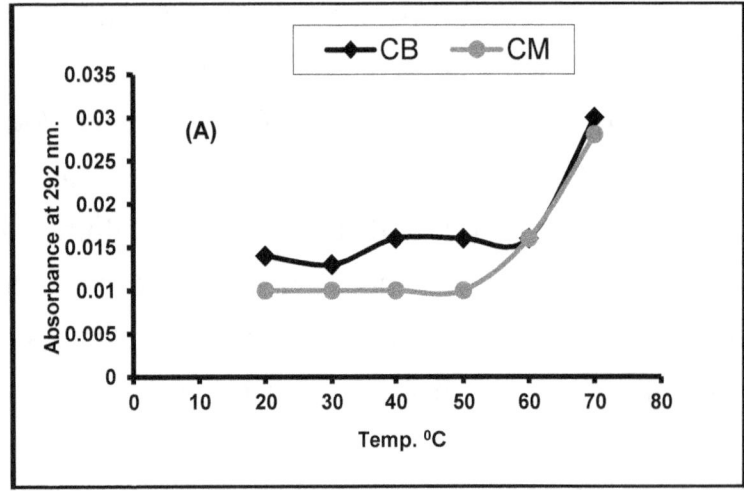

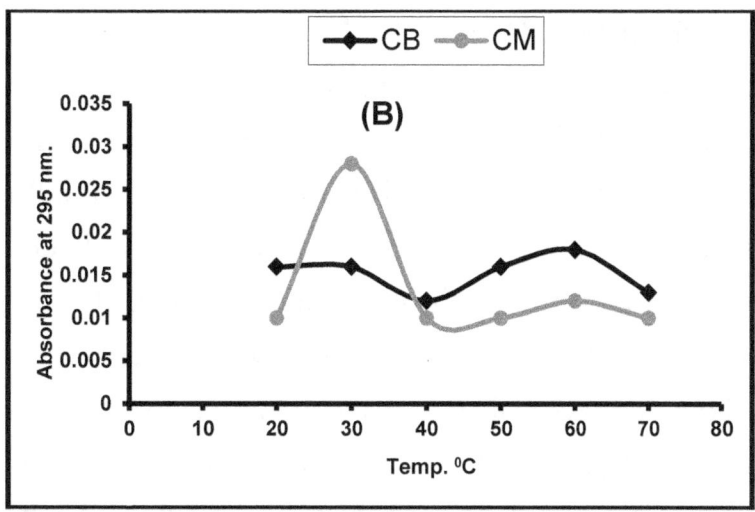

Figure (34): Thermal stability curve for benign and malignant: (A) at λmax 292 in the presence of 0.1 M NaCl, (B) at λmax 295 in the presence of 0.1 M NaCl. (CB): Complex of benign breast tumors, (CM): Complex of malignant breast tumors. (All other details are explained in the text).

Effect of Urea, KCl and (Urea, KCl) Mixture on the Spectrum of the Complex

The effect of 8 M urea, 0.03 M KCl and a mix of 1:1 of 8 M urea and 0.03 M KCl on the λ max of the benign (fibroadenoma) and malignant (IDC) complexes, were examined. The values of λ max are illustrated in table (4). When table (25) is compared with table (1), it seems that the presence of 8 M urea at pH 7.4, there was a red shift of the λmax1 of polypeptide bond from 200 to 227.4 nm in benign complex and a red shift of λmax1 from 200 to 226 nm in malignant complex respectively. While λmax2 of aromatic amino acid i.e., tyrosine residues in both complexes was disappeared. The red shift is due to intramolecular hydrogen bonding between the oxygen of the amide group and the solvent.

When 0.03 M KCl was used, there was no alternation in the position of the λ max2 of the tyrosyl at pH 7.4 in both benign and malignant complexes. There was a slight blue shift (3-4nm) in the λ max1 of the polypeptide bond in the benign and malignant complex spectra respectively. On the other hand the λ max of the aromatic ring of tyrosyl residues at (274 or 278nm) disappeared. Such blue shift can arise by introducing positive (K+) or negative (Cl-) charges near the chromophore (the amid group), which might interact with π-electron system of the amide group.

When 8 M urea was mixed with 0.03 M KCl there was significant red shift in λ max (203.4 and 204.2nm) to λ max (221.4 and 219.4nm) in both benign and malignant complexes. The same shift was observed when 8 M urea was used alone with each benign and malignant complexes, this mean that the red shift due to the effect of urea, but not to 0.03 M KCl. On the other hand, there was no alternation in positions of the λ max of the tyrosyl residues near 278nm. As was seen, the changes in absorption were near 230 nm and near 280 nm. This was also observed by Glazer who that solvent perturbation or denaturation of protein poduces may changes in absorption near 230 nm and 280 nm. Some of this change in absorption may be produced by change in the n-□□□absorption of poly peptide bond in protein either because of a change in their geometrical arrangement, or because of an environment changes.

Table (25): The effect of 8M urea, 0.03M KCl and mixture (urea+KCl) on the λ max of the complex UV spectrum at pH 7.4. (All other details are explained in the text).

Solvent	λ max (nm)	
	(125I-anti CA 15-3 antibody/ CA 15-3) Benign Complex	(125I-anti CA 15-3 antibody/ CA 15-3) Malignant Complex
Urea 8M	227.4	226
KCl 0.03M	200	200
Urea+ KCl mixture 1:1	221.4 278.6	219.4 278

Chapter Four

Immunoradiometric assay determination of CA19-9

Introduction

A solid-phase Immunoradiomertric Assay sandwich technique (IRMA) was used for the determination of the carbohydrate antigen 19-9 (CA19-9) defined by a monoclonal antibody ^{125}I-anti CA19-9. The antibody ^{125}I-anti CA19-9 reacts with CA19-9 found at low concentrations in sera of healthy women but increased slightly in sera of patients with breast cancer.

The factors affecting the binding of ^{125}I-anti CA19-9 antibody with CA19-9 in the breast tumor homogenate (benign and pre-and post-menopausal malignant) were determined. The results revealed that 100, 75 and 75 µg protein was the most appropriate amount of protein used in each incubation at pH 7.8, 8.0 and 7.0 respectively, with 0.0565 mg. mL^{-1} of ^{125}I-anti CA19-9 antibody for 4,1 and 6 h incubation time at optimum temperatures 25, 37 and 45 °C respectively. The use of 0.01 M sodium halides and 0.025 M of divalent salts were shown to cause different effects on the binding in the three groups.

The recovery of the method was calculated and found to be 99%, 98% and 95% for binding CA19-9 present in (benign and pre-and post-menopausal malignant) breast tumor homogenates respectively.

CA 19-9 is a carbohydrate antigen identified as a glycolipid-that is, sialylated lacto-N-fucopentose II ganglioside, which is a sialylated derivative of the Lewis a blood group antigen and is denoted as Le a . CA19-9 is synthesized by normal human pancreatic and biliary ductular cells and by gastric, colonic, endometerial, kidny, salivary gland, sweat gland and present in ductal epithelium of breast . In serum it exists as a mucin, a high-molecular weight (200-1000 KD) glycoprotein complete . The monoclonal antibody against CA19-9 was developed from a human colon carcinoma cell line, SW-1116 by Koprowski and associates . The Carbohydrate antigen 19-9 (CA19-9) (Koprowski etal.,1979), is specific carbohydrate fraction of a circulating antigen found in sera of normal adults (Koprowski etal.,1981) , has sialyl Lewis[a] structure and is present in individually expressing the Lewis[a] and /or Lewis[b] blood group antigen . CA19-9 is identified as a glycolipid- that is , sialylated lacto-N-fucopentose II ganglioside . In serum, it exists as a mucin , a high molecular mass (200-1000 KD) glycoprotein complex. In Normal tissues, sialyl Lewis[a] antigen is present in ductal epithelium of breast, kidney, salivary gland, and sweatglands. CA19-9 is measured with a double monoclonal immuno-radiometric assay .

Another techniques used for the detection of CA19-9 in tissues and sera were performed by an immunoperoxidase assay and by radioimmunoassay of samples from patients, and enzyme immunoassay for quantitative determination of CA19-9 in human serum. The upper limit of normal value 37.0 U.mL^{-1}. The abnormal expression of the sialyl Lewis [a] is closely correlated with various forms of cancer including pancreatic

Monoclonal antibody 19-9 derived from spleen cells of a mouse immunized with human colon adenocarcinoma cell line SW-1116. The epitope of this antibody is carbohydrate with the sugar sequence

NeuNAcα 2-3 Gal β 1-3 GlcNAc β 1-3 Gal...

$$\begin{array}{c} 4 \\ | \\ \textbf{Fuc α 1} \end{array}$$

Methodology

CA 19-9 is measured with a double monoclonal immunoradiometric assay, using monoclonal antibodies raised against the SW-1116 cell line. The antibody reacts with CA19-9 found at low concentrations in sera from healthy individuals, but frequently increased in sera from patients with adenocarcinomas. The upper limit of normal for healthy subjects has been defined by the cutoff value of 37.0 (U.mL-1). CA 19-9 has become an established marker for pancreatic cancer, but it must still be regarded as a research test for colorectal cancer.

Another methods to determinate CA 19-9 were enzyme-linked immunosorbent assay. Both the capture and the enzyme-conjugated antibody use the CA 19-9 monoclonal antibody. It should be noted that this antibody is useless for cancer diagnosis when a patient is lacking the enzyme for the synthesis of sialyl Le a. In Japanese, about 5-10% of the population lacks this enzyme. Determination carbohydrate antigen CA 19-9 levels in serum were also measured by radioimmunoassay (RIA). Immunohistochemical technique used for the distribution of CA19-9 in tissues using an immunoperoxidase assay. By this technique the CA 19-9 can be detected not only in cancerous tissues but also in non cancerous normal tissues.

Screening

Numerous studies have addressed the potential utility of CA 19-9 in adenocarcinoma of the colon and rectum.

The reported incidence of elevated serum CA 19-9 in colorectal cancer ranges from 20% to 40% . The incidence of elevated CA 19-9 in stage-related, with the highest sensitivity occurring in patients with metastases . However, the sensitivity of CA 19-9 was always less than that of the CEA test for all stages of disease . The false-positive rate (>37.0 U.mL-1) is 15% to 30% in patients with non-neoplastic diseases of the pancreas, liver and biliary tract . Consequently, CA 19-9 cannot be used for screening asymptomatic populations.

Monitoring Response to Treatment

Kouri et.al. compared CEA and CA 19-9 for predicting response to chemotherapy in 85 patients. Decreases in CEA more accurately reflect the response to therapy than did the decreases of CA 19-9. The pretreatment CA 19-9 value was, however, an important prognostic factor. Median survival was 30 months for patients with normal CA 19-9 values and 10.3 months for patients with elevated CA 19-9 values. CA 19-9 used to examined the serum levels and immunohistochemistry during the clinical course of female patient treatment with idiopathic interstitial pneumonia (IIp) that had elevated serum levels of CA 19-9 .

Clinical Application

Elevated levels (>37 U.mL-1) were seen in patients with pancreatic (80%), hepatobiliary (67%), gastric (40-50%), hepatocellular (30-50%), colorectal (30%), and breast (15%) cancer. Pancreatits and other benign gastrointestinal diseases show a 10 to 20% elevation; however, the levels are usually lower than 120 (U.mL-1). CA 19-9 levels correlate with pancreatic cancer staging (54). CA19-9 is useful in monitoring pancreatic and colorectal cancer. Elevated levels can indicate the recurrence before clinical finding by 1 to 7 months (134). Unfortunately, early detection of relapse may not be useful because of the lack of effective therapy for pancreatic cancer

A monoclonal antibody CA19-9 against sialyl Lewis [a] is a popular diagnostic agent for these tumors. The antibody is useless for cancer diagnosis when a patient is lacking the enzyme for the synthesis of sialyl Lewis [a]. In Japan, about 5-10% of the population lacks this enzyme leading to false negative results . CA19-9 represents the most important and basic carbohydrate tumor marker. The immunohistologic distribution of CA19-9 in tissues is consistent with the quantitative determination of higher CA19-9 concentrations in cancer than in normal of tissues . Recently reports indicates that serum CA19-9 level is frequently elevated in the serum subjects with pancreatic (80%), hepatobiliary (67%), gastric (40-50%), hepatocellular (30-50%), colorectal (30%) and breast (15%) cancer .

Research studies demonstrate that serum CA19-9 values may have utility in monitoring subjects with the above-mentioned diagnosed malignancies. A declining CA19-9 value may be indicative of a favorable prognosis and good response to treatment . Therefore, the development of immunoradiometric assay was planned to carry out the determination of the optimum conditions of ^{125}I-anti CA19-9 antibody.

Materials and Methods

Patients and Blood Samples

Thirty breast patients and specimens were used in this chapter, classified to three group of patients, one group with benign and two groups with malignant breast tumors. The fourth group is a healthy women used as control.

- **Group I:** Consisted of 10 patients with benign (Fibroadenoma) breast tumors.
- **Group II:** Consisted of 10 premenopausal patients with breast cancer (IDC).
- **Group III:** Consisted of 10 postmenpausal patients with breast cancer (IDC).
- **Group IV:** Consisted of 10 normal healthy subjects.

Blood samples were prepared while homogenization of breast tumor tissues was carried out. Statistical analysis was determined by student's t-test.

Methods

Determination of CA19-9 Levels in Sera of Patients with Benign and Malignant Breast Tumors

1. Anti CA19-9 monoclonal antibody coated on the ELSA fixed in the bottom of the tube.
2. Anti ^{125}I-CA19-9 monoclonal antibody, radioactivity content < 10 µCi (<370 KBq)
3. Six standard ready for use, Human serum, Human CA19-9 in sodium azide (0,14,30,66,130 and 255 $U.mL^{-1}$).
4. Diluent (0.0 $U.mL^{-1}$), human serum in sodium azide.
5. Control (35 $U.mL^{-1}$), human serum, human CA19-9 in sodium azide. Patients sera and control were used without dilution in this assay.

Procedure

The assay protocol is described in table (26).
Table (26): IRMA protocol of serum CA19-9 ($U.mL^{-1}$).

	CA19-9 (U.mL^{-1})						Control		Unknown Samples	
	0	14	30	66	130	255	Level I	Level II	1	2 etc.
Coated tube no.	1,2	3,4	5,6	7,8	9,10	11,12	13,14	15,16	17,18	19,20
Standards (μL)	←——————————————————→					100 μL				
Control serum or samples (μL)	←——————————————————→					100 μL				
Buffer (μL)	←——————————————————→					200 μL				
	Incubation for 3 h. at 37 °C in water bath									
	The solution was aspirated, and washed the tubes 3 times with 3 mL distilled water									
^{125}I-anti CA19-9 (μL)	←——————————————————→					300 μL				
	All tubes were mixed gently with vortex-type mixer and									
	Incubated for 3 hrs. at room temperature (18-25 °C)									
	The solution was aspirated, the tubes were washed 3 times with 3 mL distilled water									
	The remaining bound radioactivity was measured with gamma counter.									

Calculations

1. The mean net count for each group of tubes was counted in gamma counter for 1 min, represents the bound c.p.m.

2. The standard curve was constructed by plotting counts per min. (Y-axis) versus concentration of CA19-9 standard (X-axis) figure (6.1). Then the points were connected with straight-line segments.

Preliminary Test of the Binding of CA19-9 in Breast Tumor Tissues with 125I-Anti CA19-9 Antibody in Breast tumors Homogenates

The pellet and the cytosol fractions were obtained from the supernatant of breast homogenate were centrifuged at 4000 r.p.m. In order to detect CA19-9, 20 μL of crude cytosol fraction having 1100 μg protein were incubated with 60 μL (0.1356 mg.mL^{-1}) of ^{125}I-anti CA19-9 antibody. The volume of mixture was completed to 500 μL with PBS buffer pH 7.2, and then incubated at 37 °C for 3 hrs. The assay tubes were centrifuged at 4000 r.p.m. for 45 min. at 45 °C. The supernatant was discarded, the rims at tube were swabbed with cotton piece, then the complex formed was counted in gamma counter for 1 min. Pellet CA19-9 were determined by dissolving the sediment in PBS buffer pH 7.2 with ratio 1:5 (weight: volume), then 20 μL of supernatant fraction of pellet breast homogenate having 800 μg protein, was added to 60 μL (0.1356 mg.mL^{-1}) of ^{125}I-anti CA19-9 antibody. The same steps mentioned above were followed to determine the radioactivity of the complex formed. For total radioactivity two additional tubes with 60-μL of ^{125}I-anti CA19-9 antibody were counted in gamma counter.

Calculations

1. The counted radioactivity in each tube (expressed in c.p.m.) represents the bound fraction (B); (i.e., ^{125}I-anti CA19-9 antibody/CA19-9 complex).

2. The counted radioactivity in the tubes counting ^{125}I-anti CA19-9 antibody only represents the total radioactivity (T).

3. The (B/T) % ratio for each tube was calculated as follows:

$$(B/T)\% = \frac{\text{Sample counts (B)}}{\text{Total counts (T)}} \times 100$$

.Factors Effecting of 125I-Anti CA19-9 Antibody Binding to CA19-9 in Breast Tumors Homogenates

Effect of Protein Concentration on the Binding

Reagents

Sixty microliters (0.1356 mg.mL^{-1} protein) of ^{125}I-anti CA19-9 antibody were added to 20 μL of cytosolic fraction of benign (Fibroadenoma) and malignant (premenopausal IDC and postmenopausal IDC) breast tumors respectively, containing increasing amounts of protein (50, 75, 100, 150, 200 and 250 μg.mL^{-1}) and were completed to a final volume of 500 μL with 0.15 M PBS pH 7.2. The assay tubes were incubated for 3 hrs. at 37 °C. At the end of incubation, the assay tubes were centrifuged at 4000 r.p.m. for 45 min. at 4 °C. The supernatant was decanted; the rims at the tube were swabbed with cotton piece. The radioactivity of the complex formation was counted using gamma counter.

Calculations

1. The (B/T) % values were determined .
2. Values of (B/T) % were plotted against their corresponding amount of protein of the breast tumor homogenate.

Effect of 125I-Anti CA19-9 Antibody Concentration on the Binding

Procedure

Sixty microliters of increasing amounts (0.0226, 0.0452, 0.0565, 0.113, 0.1356, 0.226 mg.mL^{-1}) of ^{125}I-anti CA19-9 antibody were added to 20 μL of crude cytosolic fraction (100, 75 and 75 μg protein) for benign (fibroadenoma) and malignant (premenopausal IDC and postmenopausal IDC) respectively, completed to a final volume 500 μL with 0.15 M PBS pH 7.2. After incubation for 3 hrs at 37 °C the bound CA19-9 was determined as .

Calculations

1. **The (B/T) % values were determined .**
2. Values of (B/T) % were plotted versus the concentrations of ^{125}I-anti CA19-9 included.

Effect of pH on the Binding

Twenty microlites (100, 75 and 75 µg protein) of cytosolic fraction (fibroadenoma, premenopausal IDC and postmenopausal IDC respectively) were added to 25 µL (0.0565 mg.mL^{-1}) of ^{125}I-anti CA19-9 antibody respectively. The volume of the mixture was completed with PBS buffer of different pH (6.8, 7.0, 7.2, 7.4, 7.6, 7.8, 8 and 8.2) to a final volume 500 µL. After incubation for 3hrs at 37 °C, the bound CA19-9 was determined.

Calculations
1. **The (B/T) % values were determined.**
2. Values of (B/T) % were plotted versus the corresponding pH.

Effect of Temperature on the Binding

Procedure

Twenty microliters (100, 75 and 75 µg protein) of cytosolic fraction (Fibroadenoma, premenopausal IDC and postmenopausal IDC) were added to 25 µL (0.0565 mg.mL^{-1}) of ^{125}I-anti CA19-9 antibody respectively. The volume of mixture was completed to a final volume 500 µL with PBS buffer at pH 7.8 for fibroadenoma, pH 8.0 for premenopausal (IDC) and pH 7.0 for postmenopausal (IDC). The experiment was carried out at (5, 15, 25, 37 and 45°C) for 3hrs. After incubation the bound CA19-9 was determined.

Calculations
1. The (B/T) % values were determined
2. Values of (B/T) % were plotted versus the temperature.

Effect of Incubation Time on the Binding

Procedure

Twenty microliters (100, 75 and 75 µg protein) of cytosolic fraction (fibroadenoma, premenopausal IDC and postmenopausal IDC) were added to 25 µL (0.0565 mg.mL^{-1}) of ^{125}I-anti CA19-9 antibody respectively. The reaction mixture was completed to a final volume 500 μL with PBS buffer pH (7.8 , 8.0 and 7.0) respectively. The experiment was carried out at 25 °C , 37 °C and 45 °C for fibroadenoma , premenopausal (IDC) and postmenopausal (IDC) respectively. The incubation was carried out at different time intervals (1, 2, 3, 4, 5 and 6 hrs). The bound CA19-9 was estimated .

Calculations

1. The (B/T) % values were determined as in section (6.2.2).
2. Values of (B/T) % were plotted versus incubation time.

Effects of Different Halides on the Binding

Reagents

1. Halid reagents were prepared in concentration of 0.01M PB at pH (7.8, 8.0 and 7.0) individually, by dissolving each of 0.021gm of NaF, 0.0292gm of NaCl, 0.0515gm of NaBr, and 0.075gm of NaI in a final volume 50mL of PB and the pH was adjusted.

2. The breast tumors homogenates (fibroadenoma , premenopausal IDC and postmenopausal IDC) were prepared as described in section (2.1.7), except using PB-buffer instead of PBS at the same pH and same concentration was carried out the homogenization.

Procedure

using three groups of human breast homogenate (i.e., fibroadenoma, premenopausal IDC and postmenopausal IDC), by incubating 20 μL of the homogenate from each group containing (100, 75 and 75 μg protein) respectively with 25 μL (0.0565 mg.mL^{-1}) of ^{125}I-anti CA19-9 antibody. The reaction mixture was completed to a final volume 500 μL with PBS buffer pH (7.8, 8.0 and 7.0) containing 0.01 M of each of the following salts: NaF, NaCl, NaBr and NaI in each assay tube (A sample without the addition of any salt was used as a control). The assay tubes were incubated for (4,1 and 6 h) at 25 ,37 and 45°C for three group individually. The bound CA19-9 was estimated .

Calculations

1. **The (B/T) % values were determined**
2. Values of (B/T) % were plotted versus 0.01 M of NaX.

Effects of Monovalent and Divalent Cations on the Binding

1. Monovalent and divalent cations (0.025 M) were prepared in PB buffer, and then the pH was adjusted to 7.8, 8.0 and 7.0 individually by dissolving each of 0.0931 gm of KCl, 0.0668 gm of NH_4Cl, 0.2541 gm of $MgCl_2.6H_2O$, 0.1388 gm of $CaCl_2.2H_2O$, 0.2474 gm of $MnCl_2.4H_2O$, 0.3150 gm of $CuSO_4.5H_2O$, 0.1703 gm of $ZnCl_2$, in a final volume 50 ml of PB and the pH was adjusted.

Procedure

The experiment was carried out at optimum conditions using three groups of human breast homogenate (i.e., fibroadenoma, premenopausal IDC and postmenopausal IDC) respectively.

The same steps were followed to determine the effect of monovalent and divalent cations on the binding, except ; the buffer solution was PB (0.15 M) containing 0.025 M of the following salts: KCl, NH_4Cl, $MgCl_2.6H_2O$, $CaCl_2.2H_2O$, $MnCl_2.4H_2O$, $CuSO_4.5H_2O$ and $ZnCl_2$.

Calculations

1. The (B/T) % values were determined as in section (6.2.2).

2. Values of (B/T) % were plotted versus the 0.025 M of monovalent and divalent cations.

Recovery of CA19-9

Procedure

The experiment was carried out at optimum conditions. Known concentration of CA19-9 (255 U.mL^{-1}) was added to the three group of benign (fibroadenoma) and malignant (premenopausal IDC and postmenopausal IDC) breast tissues homogenates. The experiment was carried ou at optimum conditions that was obtained in the experiment.

Calculations

1. The bound (c.p.m.) of the reaction mixture added to tissue homogenate with ^{125}I-anti CA19-9 antibody, represent the measured value.

2. The bound (c.p.m.) of CA19-9 in tissue homogenate with ^{125}I-anti CA19-9 antibody only, represent the expected value.

3. The recovery % (yield) calculated as follows:

$$\text{Recovery \%} = \frac{\text{Measured values (c.p.m)}}{\text{Expected values (c.p.m)}} \times 100$$

Results and Discussions

Determination of CA19-9 levels in Sera of Patients with Benign and Malignant Breast Tumors

Serum CA19-9 levels were measured by a solid-phase "sandwich" Immunoradiometric Assay (IRMA), which is specifically recognized by the anti CA19-9 monoclonal antibody. The monoclonal antibody is coated on the solid phase, or radiolabeled with the iodine 125 and used as a tracer. The radioactivity of the bound is directly proportional to the amount of CA19-9 presents at the beginning of the assay.

CA19-9 levels in sera of patients with benign breast tumors (group I) and (pre-and post-menopausal) malignant breast tumors (group II and group III) were measured by immunoradiometric assay. Three groups were matched with one group of control subjects. Table (27) shows the results obtained from this study. CA19-9 concentration of specimens and control were determined directly from standard curve in figure (35). The level of serum CA19-9 in benign breast tumor patients was found to be 31.0 $U.mL^{-1}$ ($p<0.05$), where that of (pre-and post-menopausal) malignant breast tumor patients were found to be 33.1 $U.mL^{-1}$ ($p<0.05$) and 32.1 $U.mL^{-1}$ ($p<0.0005$) respectively. While in control, the level was found to be 28.8 $U.mL^{-1}$. Matching case and control subject proved to be important for controlling undesired variability. The mean CA19-9 was significantly high in postmenopausal patients ($p<0.0005$) while in premenopausal and benign breast tumors the mean of CA19-9 was significantly low ($p < 0.05$ Student's t-test).

Table (27): Sera CA19-9 levels (U.mL^{-1}) in patients with benign and malignant breast tumors. (All other details are explained in the text).

Group	Patients	No. of Cases	Age (year)	Serum CA19-9 U.mL^{-1} (mean ± SD)	P values
I	Benign breast tumors	10	18-35	31.0 ± 1.52	P<0.05
II	Premenopausal malignant breast tumors	10	35-43	33.1 ± 2.79	P<0.05
III	Postmenopausal malignant breast tumors	10	53-65	32.1 ± 0.13	P<0.0005
Control	Control	10	25-35	28.8 ± 0.631	

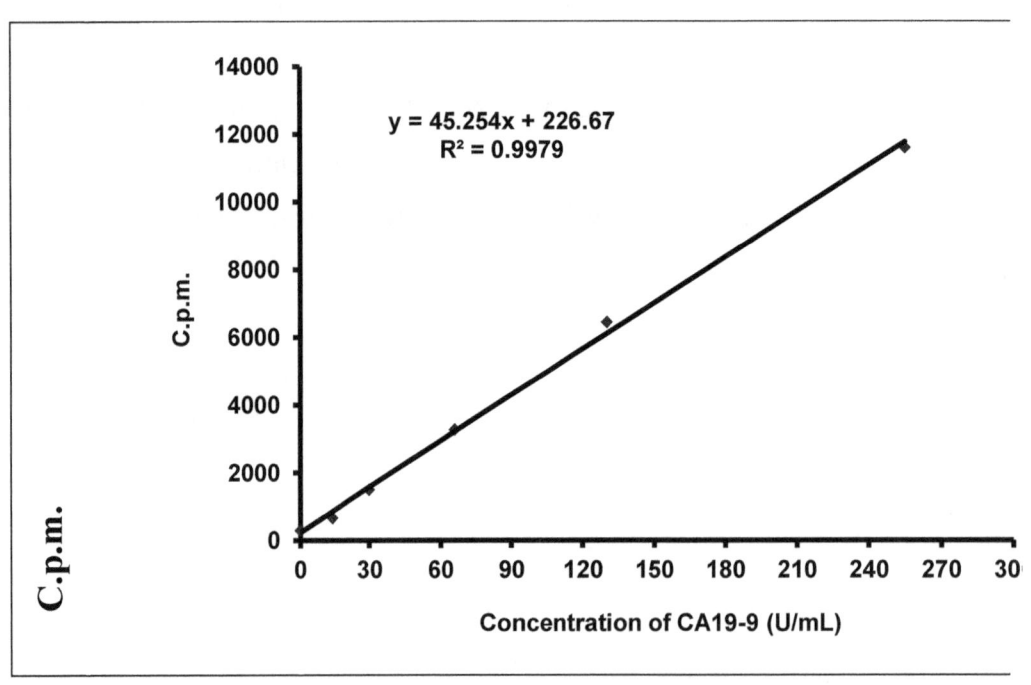

Figure (35): Standard curve of CA19-9. (All other details are explained in the text).

CA19-9 was at low concentration in sera of healthy individuals, these results are in agreement with several authors previously.

There were few studies to evaluate CA19-9 in breast tumors patients. Several investigators detected CA19-9 in bone metastasis in breast cancer patients and in patients without documented metastases and reported that CA19-9 level elevated in patients with metastases breast cancer.

When patients were analyzed with respect to the menopausal status, significant differences between the monastic and non monastic patients was detected.

Several studies proved the possibility of the role of carbohydrate antigen 19-9 as a tumor marker in colorectal cancer, pancreas, gastric, liver disease and esophageal cancer. Recently, European group proved that CA19-9 monitored in patients with tumors of gastrointestinal tract and endometrial cancer could be used as a tumor marker and can be helpful in monitoring patients with breast cancer. They observed significant increase of CA19-9 and CA15-3 in all patients.

Preliminary Test of the Binding of CA19-9 with 125I-Anti CA19-9 Antibody

Supernatant and pellet obtained at speed (4000 r.p.m) were investigated in the three groups of human breast tumor homogenate (fibroadenoma, premenopausal IDC and postmenopausal IDC). In each fraction, CA19-9 was detected through the incubation of ^{125}I-anti CA19-9 antibody with crude fraction supernatant and pellet individually for 3 h at 37°C in PBS buffer pH 7.2 as a medium to complete the reaction.

The separation of the bound antibody from unbound was carried out at 4000 r.p.m for 45 min. to precipitate the ^{125}I-anti CA19-9 antibody/CA19-9 complex formed.

Table (28): Incidence of CA19-9 in supernatant and pellet fractions in three different breast homogenate. (All other details are explain in the text).

Groups	Age(year)	B/T %	
		Supernatant fraction	Pellet fraction
Benign	34	5.32	1.43

Premenopausal (IDC)	43	5.48	2.03
Postmenopausal (IDC)	63	5.86	2.47

Table (28) shows the amount of binding B/T % values of pellet and supernatant fractions. The data revealed that CA19-9 in cytosolic fraction obtained from supernatant was higher in incidence than in pellet fraction, according to these results cytosolic fraction was collected. CA19-9 collected and the pellet was then discarded.

Factors Effecting of 125I-Anti CA19-9 Antibody Binding to CA19-9 in Breast Tumors Homogenates

Effect of Protein Concentration on the Binding

To obtain the optimum protein concentration of cytosolic fraction for the binding of CA19-9 with ^{125}I-anti CA19-9 antibody, cytosolic fraction containing increasing amount of soluble CA19-9 in the presence of fixed amount of ^{125}I-anti CA19-9 antibody was carried out . Figure (40) represent the formation of (^{125}I-anti CA19-9 antibody/CA19-9) complex in three cases (fibroadenoma, premenopausal IDC and postmenopausal IDC) and shows that (100, 75 and 75 µg protein) were the most appropriate concentration to give the maximum values of binding in crude fraction of three cases respectively. The decrease of the binding at high concentration of cytosolic fraction (in three cases) in the reaction mixture may be due to a conformational change in CA19-9 and ^{125}I-anti CA19-9 antibody rather than the formation of reversible inactive (^{125}I-anti CA19-9 antibody/CA19-9) complex and may be due to splitting antigen into large fragments with proteolytic enzymes .

In all subsequent experiments an amount of (100, 75 and 75 µg protein in three cases respectively), were used in the incubation mixture.

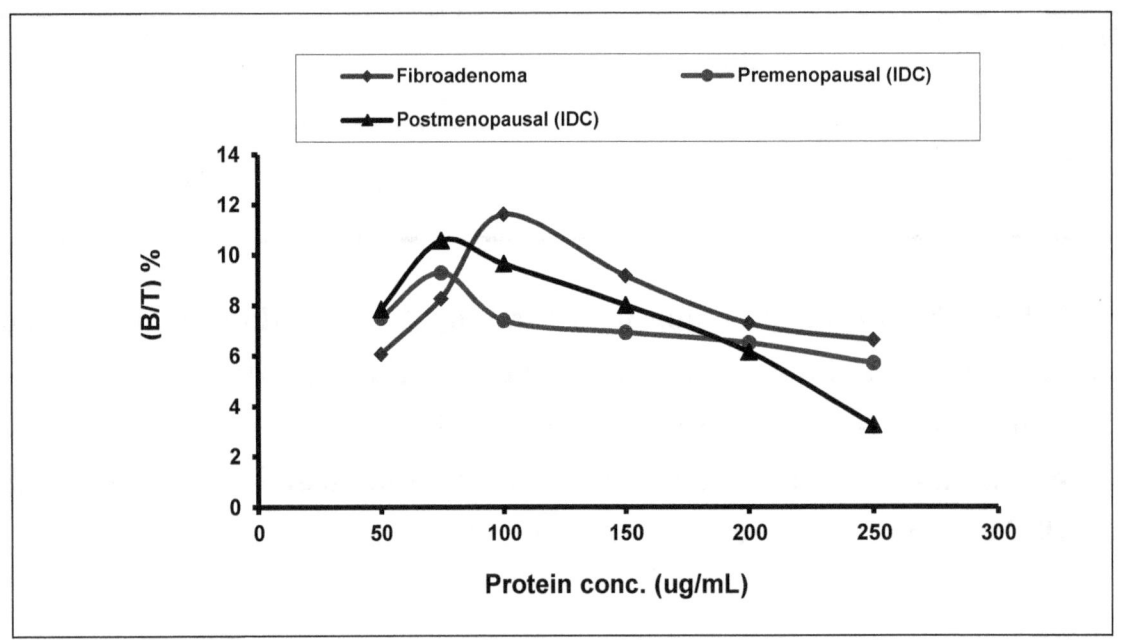

Figure (35): Influence of increasing protein concentrations on the binding of CA19-9 with ^{125}I-anti CA19-9 antibody. (All other details are explained in the text).

Effect of 125I-Anti CA19-9 Antibody concentration on the Binding

One of the most important factors that effect binding is the concentration of ^{125}I-anti CA19-9 antibody. To determine the suitable concentration of ^{125}I-anti CA19-9 antibody, cytosolic sample (100, 75 and 75 μg protein) in the three cases (fibroadenoma, premenopausal IDC and postmenopausal IDC) respectively were incubated with increasing concentration of ^{125}I-anti CA19-9 antibody, the incubation was carried out for 3 h at 37 °C. The results revealed that the optimum concentration of the ^{125}I-anti CA19-9 antibody to give the maximum binding in all three cases was (0.0565 mg.mL^{-1}). The results showed that an increase in the conc. of ^{125}I-anti CA19-9 antibody caused a decrease in the binding %. This is because the soluble complexes, and the excess of antibody cover all antigentic sites, which leads to complex formation inhibition. Accordingly in all subsequent experiments, 0.0565 mg.mL^{-1} of ^{125}I-anti CA19-9 was used as the optimum conc., which gives the highest binding %.

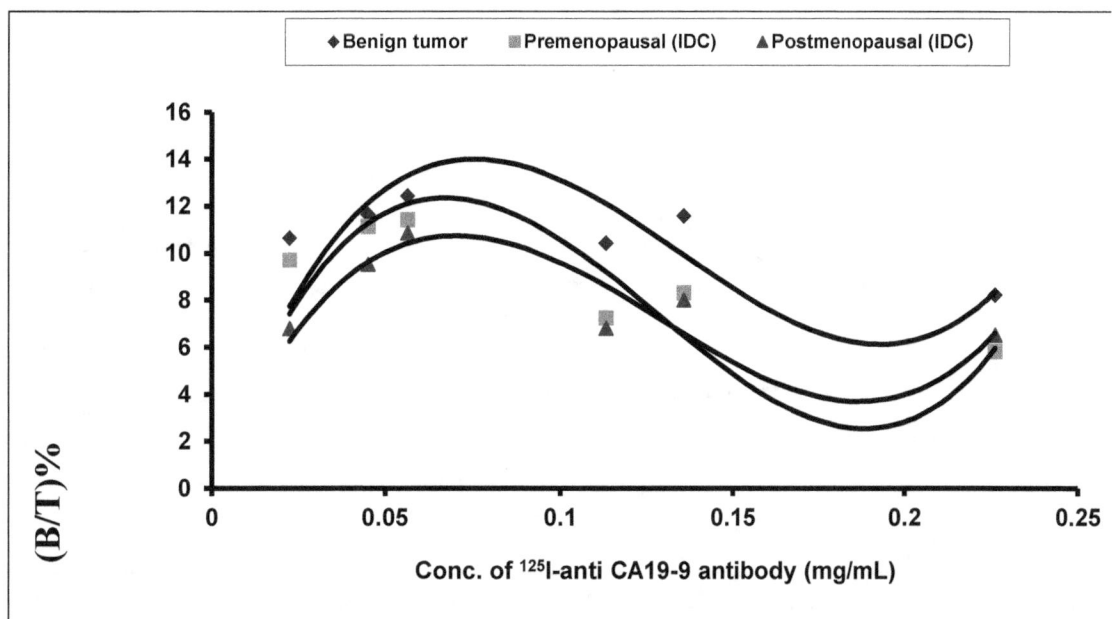

Figure (36): Effect of different concentration of ^{125}I-anti CA19-9 antibody on the binding with CA19-9. (All other details are explained in the text).

Effect of pH on the Binding

The effect of pH on the binding of radioactivity CA19-9 to its antigen CA19-9 was investigated. Figure (6-4) shows that the maximum binding of ^{125}I-anti CA19-9 antibody to its antigen CA19-9 was found to be (7.8, 8.0 and 7.0) in the three cases used (fibroadenoma , premenopausal IDC and postmenopausal IDC) respectively. The shift in pH of the environment may involve a protonation-deprotonation process occuring within the change of polar groups of the amino acids residues present in the binding domain .According to these results, the pH of the buffer used in all subsequent experiments were (7.8, 8.0 and 7.0) for the three cases respectively.

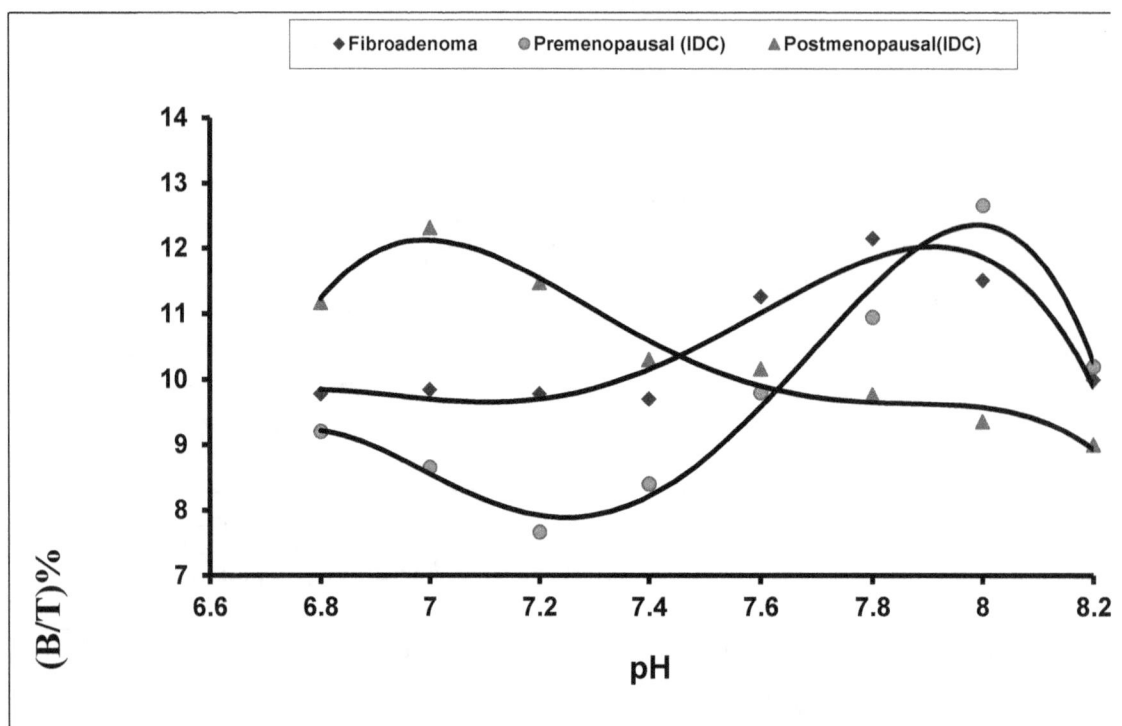

Figure (37): The effect of pH on the binding of CA19-9 with its antibody ^{125}I-anti CA19-9 antibody with CA19-9. (All other details are explained in the text).

Effect of Temperature on the Binding

Temperature dependency of the association of ^{125}I-anti CA19-9 antibody to its cytosolic fraction CA19-9 was investigated. Cytosol fraction of benign and malignant breast tumors was incubated for 3 hrs at different temperatures (5, 15, 25, 37 and 45 °C). Figure (38) reveals that the binding of ^{125}I-anti CA19-9 antibody to its cytosol fraction CA19-9 was increased when the temperature was raised from 5 to 25 °C in fibroadenoma and the maximal binding was obtained at 25 °C and from 5 to 37 °C in premenopausal (IDC) and the maximal binding was obtained at 37 °C. Finally from 5 to 45 °C in postmenopausal (IDC) and the maximal binding was obtained at 45 °C. The decrease in the binding at temperature higher than the optimum temperature is probably due to denaturation of CA19-9 molecules or due to proteolytic degradation of enzyme. According to these results (25 °C, 37 °C and 45 °C) respectively they will be used in all the subsequent experiments for the three cases used.

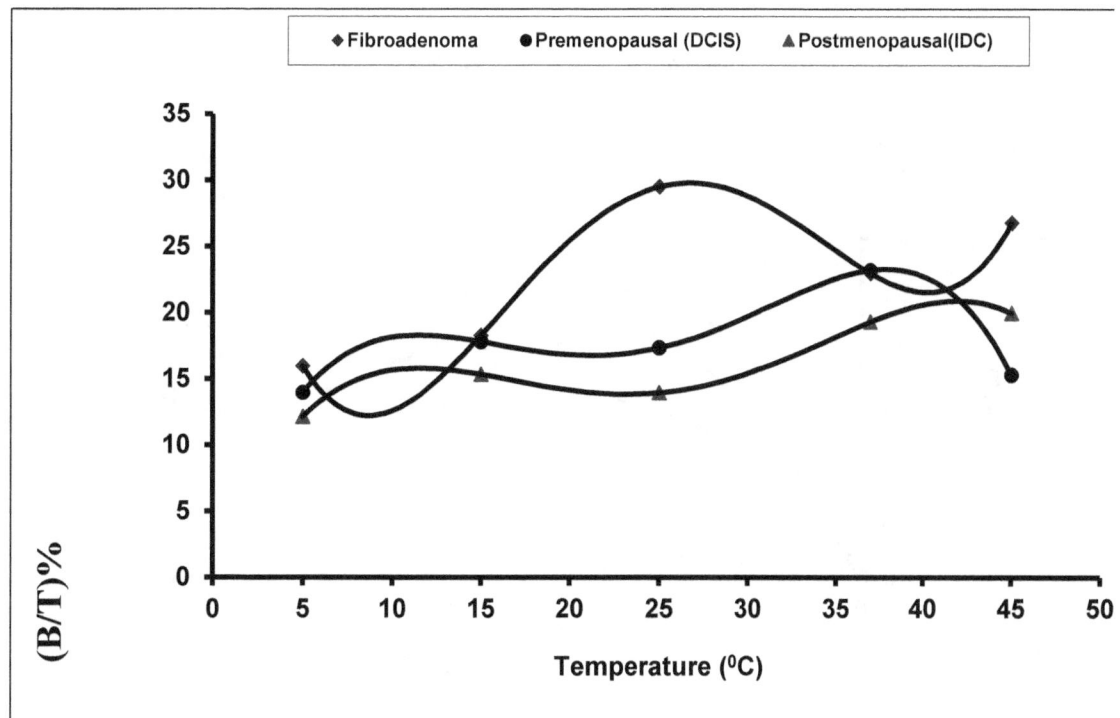

Figure 38): Effect of temperature on the binding of ^{125}I-anti $^{CA19-9}$ antibody with CA19-9. (All other details are explained in the text).

The Effect of Incubation Time on the Binding

To choose the most appropriate incubation time at (25, 37 and 45 °C) for the three cases used in this study (fibroadenoma, premenopausal IDC and postmenopausal IDC) respectively, the experiments were carried out at different time intervals. Figure (39) shows the results of this analysis. It seemed that the specific binding of ^{125}I-anti CA19-9 antibody to cytosolic fraction homogenate for the three cases were maximal at (4,1 and 6 hrs) respectively. In view of these results, the incubation time used in all subsequent experiments were (4,1 and 6 hrs) respectively.

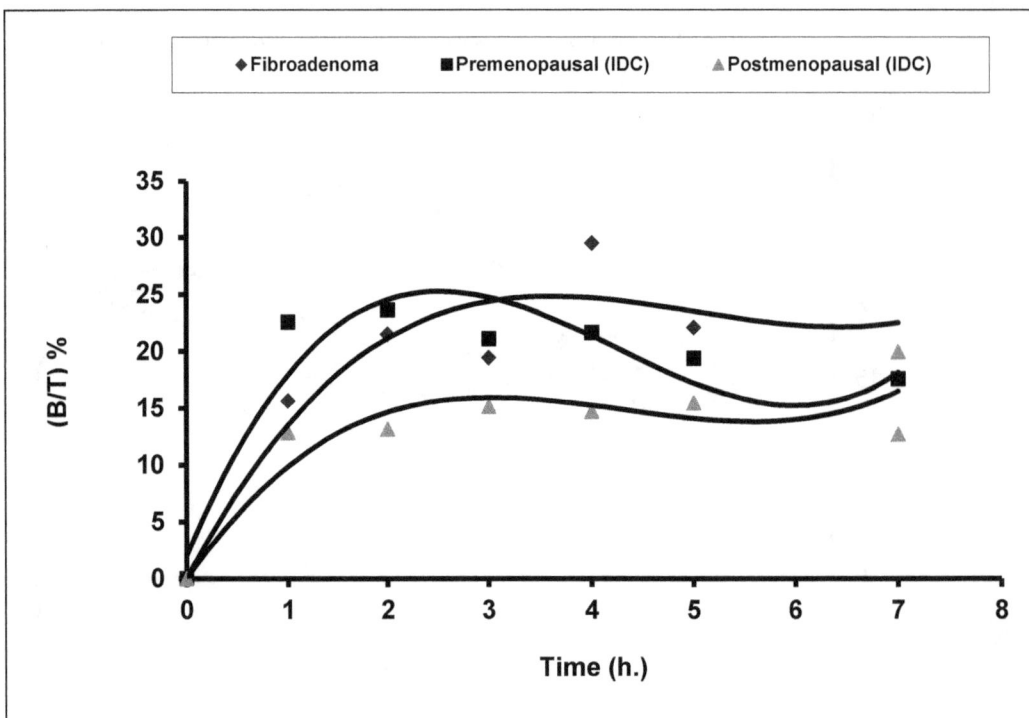

Figure (39): The effect of incubation time on the binding of ^{125}I-anti $^{CA19-9}$ antibody with CA19-9. (All other details are explained in the text).

Effect of Different Halides on the Binding

Figure (40) shows the effect of different halides salts (i.e., NaF, NaCl, NaBr and NaI) at 0.01 M concentration on the extent of ^{125}I-anti CA19-9 antibody binding to their cytosol fraction homogenate in benign and malignant breast tumors. The sodium halides (ion radius) in the incubation mixture of benign and postmenopausal malignant breast tumors induced inhibition of the percent of binding according to the following sequence:

NaI>NaF>NaCl>NaBr

While the sodium halides in the incubation mixture of premenopausal malignant breast tumor (IDC) induced activation of the percent of the binding in the order:

NaF<NaCl<NaI<NaBr

Melander and Horvath (1977) reported that the effect of halide salt type on hydrophabic interactions is quantified by its molar surface tension increment (MSTI) which is a measure of the increasing in a surface tension by the salt, also they found that parameter increases as the following sequence:

NaF>NaCl>NaI

The same researches found that halides with higher MSTI values will strengthen the hydrophabic interactions while halides with lower MSTI values reverse this effect. Thus the dependence of the extent of the binding in benign and malignant (pre-and post-menopausal) breast tumors on MSTI values of the corresponding halide further implicates the low involvement of hydrphobic forces in maintaining the stability of (^{125}I-anti CA19-9 antibody /CA19-9) complex formed.

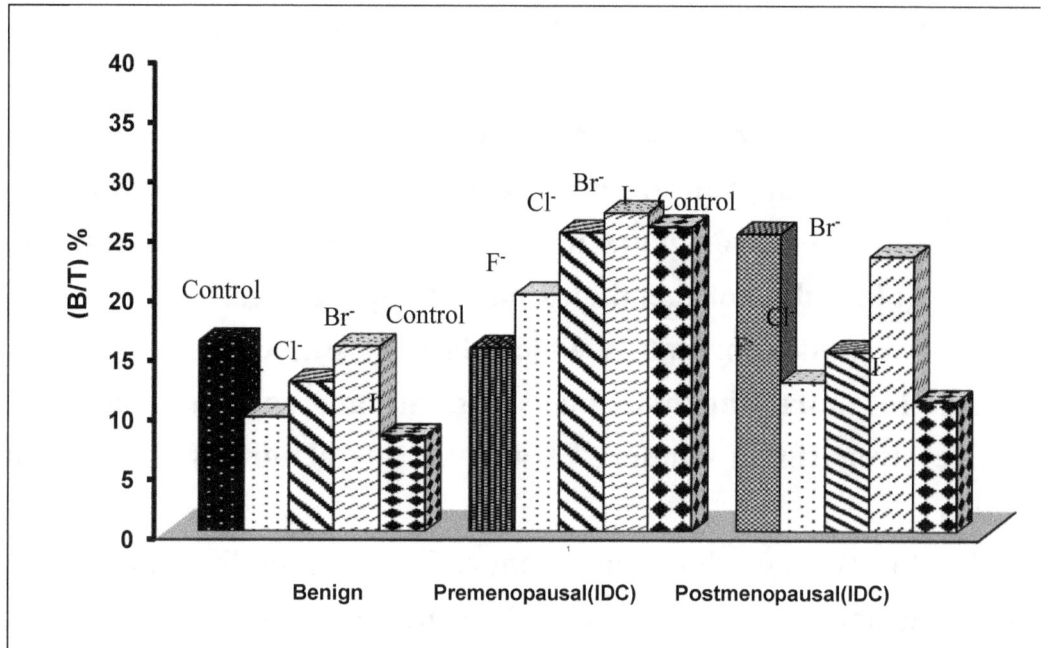

Figure (40): Effect of different halides on the binding of ^{125}I-anti $^{CA19-9}$ antibody with CA19-9. (All other details are explained in the text).

Effect of Monovalent and Divalent Cations on the binding

Figure (41) and (42) show the effect of different divalent and monovalent cations respectively on the binding value in benign and malignant breast tumors. The results indicate that the binding process is sensitive to the presence of cation metal ions. $CuSO_4.5H_2O$ at concentration 25 mM was showed to increase the binding two folds than the control as compared with other divalent cations.

$CaCl_2.2H_2O$ induced activation in the binding in benign (fibroadenoma) and malignant (premenopausal IDC), while induced inhibition in the binding in malignant (postmenopausal IDC). $ZnCl_2$ decreased the binding in two groups (fibroadenoma and premenopausal IDC), while $ZnCl_2$ increased the binding in malignant (postmenopausal IDC).

The frequency of the stimualtion of the binding of ^{125}I-anti CA19-9 antibody to its cytosolic fraction CA19-9 homogenate of the three groups by divalent cations is according to the following:

Postmenopausal breast cancer tissue homogenate (IDC)
$Cu^{+2} > Zn^{+2} > Mn^{+2} > Mg^{+2} > Ca^{+2}$

Premenopausal breast cancer tissue homogenate (IDC)
$Cu^{+2} > Ca^{+2} > Mn^{+2} > Mg^{+2} > Zn^{+2}$

Benign breast tumor tissue homogenate (Fibroadenoma)
$Cu^{+2} > Ca^{+2} > Mg^{+2} > Mn^{+2} > Zn^{+2}$

The binding of metal ions to proteins is a function of pH among the different classes of groups, such as carboxyl, amino, imidozol and tyrosyl (the unshared electron pairs for nitrogen, oxygen and sulfur atoms). The sites of binding of metal ions may range from elaborate chelate sites to simple complex formation which discrete single ligand groups in the protein. In short, chelation plays a dominant role in establishing the relative strengths of binding of a given metal ion by various sites in protein.

Figure (42) shows that monovalent cations inhibite the binding in benign and malignant premenopausal (IDC), while the monovalent cations induce activation of the binding in-group of malignant postmenopausal (IDC). The alternation of increased and decreased binding percent between these cations may be ascribed to the differences in tissues studied. The variation of results obtained between these divalent cations may be ascribed to the difference in tissue studied.

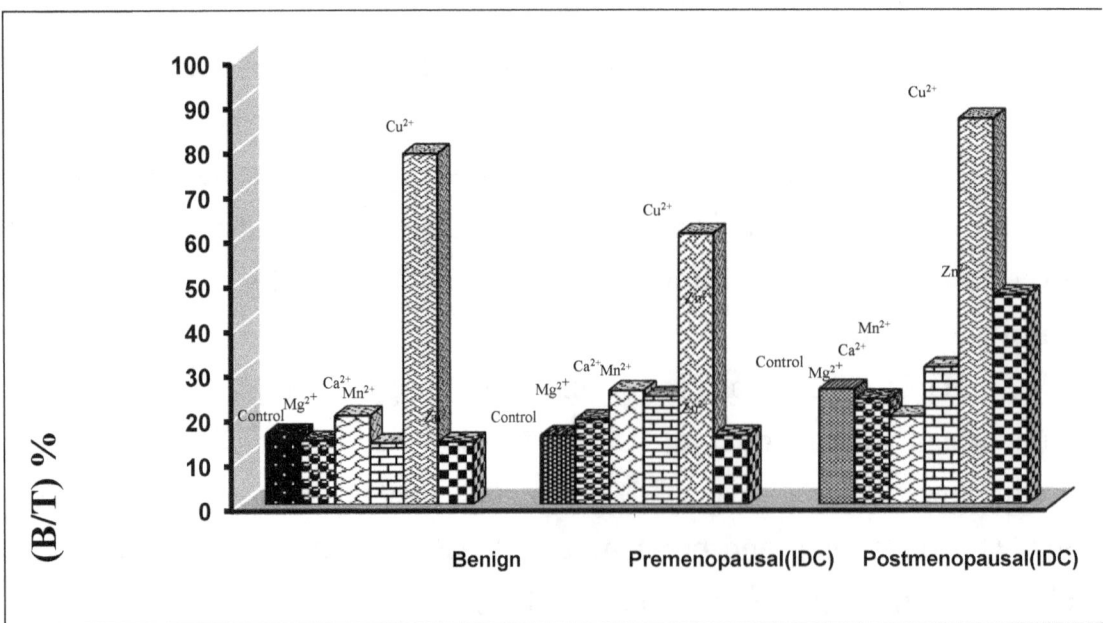

Figure (41): Effect of different cations on the binding of ^{125}I-anti $^{CA19-9}$ antibody with CA19-9 in different human breast tumor homogenate. (All other details are explained in the text).

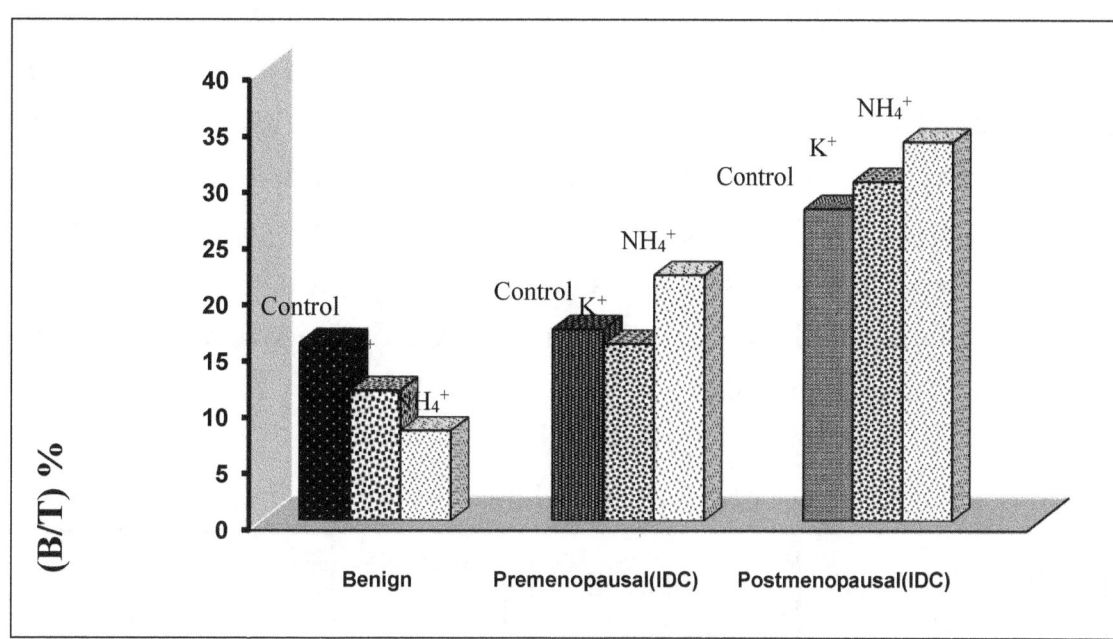

Figure (42): Effect of different monovaleat cations on the binding of ^{125}I-anti $^{CA19-9}$ antibody with CA19-9 in different human breast tumor homogenate. (All other details are explained in the text).

Recovery of CA19-9

The method used to estimate the percent recovery of cytosolic fractions of benign (Fibroadenoma) and malignant (pre-and post-menopausal IDC) breast tumors homogenates. The results summarized in table (29) indicate that CA19-9 extracted from malignant breast tumors homogenates was recovered less than CA19-9 extracted from benign breast tumor (fibroadenoma) and CA19-9 extracted from malignant premenopausal (IDC) malignant breast tumors homogenates was recovered more than CA19-9 extracted from postmenopausal (IDC) malignant breast tumors homogenates. Also the results indicate that total CA19-9 could determine through the developed method of immunoradiometric assay. The percent of recovery indicates the precision of the used method.

Table (29): Recovery of CA19-9 (AII details are explained in the text).

Type of CA19-9	Measured B/T %	Expected B/T %	Recovery % Measured/ Expected %
Benign	103	104	99
Premenopausal	192	195	98
Postmenopausal	166	175	95

Chapter Four
Biochemical Characterization of Human Chorionic Gonadotropin and It's Receptor in Sera and Tissues of Breast Tumors

Introduction

Human Chorionic Gonadotropin

Human chorionic gonadotropin (hCG) is the signature hormone of the placenta. It's a hetrodimeric glycoprotein hormone that also belongs to the cystine–knot growth factor family. hCG is a member of closely related pituitary glycoprotein hormones (TSH, FSH, and LH), present in mammals, which are important to the correct functioning of the reproductive system. In mammals the glycoprotein hormones consist of a common alpha subunit which in human is encoded by a single gene. Each hormone has different β subunit and this subunit is responsible for the different target specificity of each hormone.

Structure of hCG

The alpha subunit of hCG, which is identical in sequence to the alpha subunit of the pituitary glycoprotein hormones, is composed of 92 amino acids; while the target-receptor-specific β subunit of hCG has significant sequence homology, with 80% identity to β-LH subunit. The β-hCG subunit contains 145 amino acids with 30 additional amino acids at the carboxyterminus[4]. This modification is an essential facet of hCG biology allowing sera concentration in the pregnant mother to reach peak level of 50-100000mIU/ml (1-10mg/l). hCG also varies from LH by virtue of increased glycosylation, internal disulfide bond and sialic acid content.

Carbohydrates constitute approximately 30% by weight of each subunit[6]. Their moieties play a role in the secretion, stability, folding, subunit assembly of hormone and also it seems to be important to maintain the proper conformation of the hormone. Each carbohydrate moiety terminates in sialic acid, which accounts for 10% of the weight, of the molecule and confers a high degree of resistance to degradation and consequently a long plasma half- life. The solid crystal structure of hCG showed that each of the gonadotropin subunits is rich in disulfide bonds (the α subunit have 10 half-cystines residues that form five intrachain disulfide linkages and 12 half-cystines that form six conserved disulfide bridges. It also has structural homology to the disulfide-knot growth factor proteins. he three dimensional model of hCG, proposed a structure which predominantly composed of three helical segments ,two in the α and one in the β subunit forming antiparallel strands in each subunit. They are joined by three hairpin loops giving the hormone only a small hydrophobic core with a large interfacing area),as in figure (43).

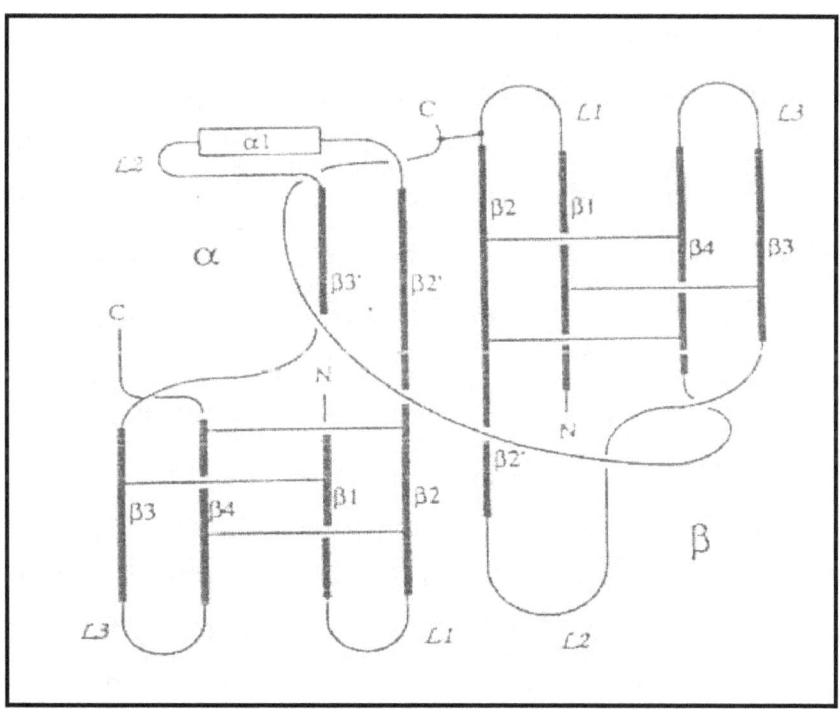

Figure (43): A schematic drawing of the hCG dimer Topology).

The β subunit, like the γ subunit of hCG, is composed of three loops. The β subunit contains six disulfide bridges which hold the molecule together when peptide bond cleavages take place in loop 2.

Human Chorionic Gonadotropin Forms

The combination of multiple subunits, multiple N-linked and O-linked oligosaccharide side chains causes significant heterogeneity in hCG structure).

hCG free subunits, degraded molecules, molecules with irregular N-and O-linked oligosaccharide side chains and fragments, which well informed, are present in sera, urine and other body fluids.

Blood and Urinary Forms of hCG

Although most measurements of hCG are made in serum, the source of hormone used for clinical treatment of infertility and other medical problems is usually a purified urinary fraction containing the biologically active forms of the hormone. All standard hCG reference preparations are all from the urine of normal pregnant women, that hCG serum measurements were based on.

Although a number of isoforms of hCG circulate in blood, most of these forms vary only by PI differences, due to variable content of the sialic acid residues. Only a small quantity of free subunit and various quantities of nicked hCG circulate in blood; in contrast, a much greater variety of molecular forms of hCG is present in urine due to proteolytic processing of hCG passing through the kidney. Many efforts to separate the several forms of hCG from different variety raw urine sources were evaluated.

- **Nicked hCG:**

The urinary metabolites of hCG and which may also exist to a lesser extent in blood are nicked form of hCG as well as nicked β subunit. Nicked hCG is simply a heterodimer with M.wt.(~36500D), it has peptide bond cleavages in loop 2 of the β subunit (between beta residues 44-49). Loop 2 is known to be exposed to solvent and is easily eliminated by protease . Dissociation of nicked hCG will result in free nicked β subunit, which is also found in a small extent in urine. The existence of nicked hCG was confirmed by isolation and gel electrophoresis. Reliable immunoassays for nicked forms of hCG were not available, due to diminished immunopotency after cleavages in β loop 2,until the development of a fairly specific nicked hCG immunometricassay and immunoassay systems. Anumber of studies suggested that nicked forms of hCG may have clinical significance as markers of certain cancers.

- **hCG βcf :**

Is the main urinary metabolite that produced by proteolysis of hCG or its β-subunit in the kidney since very little hCG βcf is detected in the blood. Core molecules are generally present in urine in much greater molar concentration (2-10 times) than dimeric gonadotrophins or free subunits. Other studies have shown that the molecule may be produced directly by placental cells in tissue culture and pituitary tissue. The structure of this fragment was reported since 1983 and was the first to be studied and discovered upon solution of the crystal structure of hCG. It is derived from hCG β subunit and is composed of residues 6-40 disulfide bridged to residues 55-92 and contaning tirmned carbohydrate groups with no sialic acid and the polypeptide chains are head together by disulfide bridges. Highly specific assays were developed to the hCG βcf and applyed in verious clinical situations). Extensive studies of its compartment distribution and concentration throughout pregnancy were conducted by de Medeirosetal, finding that it is increased in parallel to hCG throughout pregnancy, making its measurement a useful marker of ecto pregnancy. Furthermore, it has diagnostic applications in cancer tests including ovarian, lung, bladder and various gynecological cancers.

- **Intact hCG:**

It is the major hCG-related molecule present in sera that was separated as purified standard from different raw urine sources. It is composed of a heterodimer with intact polypeptide backbone. Composing of α subunit with 92 amino acids and β subunit with 145 amino acids residue polypeptide, mono and biantennary oligosaccharides and mostly O- linked trisaccharides. Detectable level of intact hCG in sera and urine has been reported, observing that its level in sera of nonpregnant women increase with age, and higher than in men[30,40]. It rises during the first tirmester (8-10 weeks) to maintain the steroid environment necessary for the pregnancy; following by a rapid decrease until 15 weeks. Intact hCG levels have been also measured in non trophoblastic tumor, observing a slight elevation in its levels as it is compared to a pregnant woman.

- **freeβ hCG:**

It is defined as mono subunit drived from dissociation of hCG. In this form, with M.Wt about (~22000D), only β subunit is present and no α- subunit; having biantennary N-linked oligosaccharides and mostly trisaccharides O-linked oligosaccharide. Using different techniques, the presence of βhCG in sera, urine, tissues, cyst fluid and cell line in different normal, benign and malignant tumor have been demonstrated. Free βhCG was found to be the most secreted form in the non trophoblastic tumors; while its level was found to be low or in normal range in the normal individuals.

- **free α hCG:**

The α-subunit of hCG was found to be synthesized and released by normal placental tissue as a small precursor form that convertes to a larger form prior to secrete; with finding that the free α form do not bind to purified βhCG subunit. It is derived from the dissociation of hCG having identical structure of the combined α subunit excepte its carbohydrate composition which prevents recombination. Many studies were undertaken for studying the presence of the free α subunit in different organs, its immunological and biological property and its anti α monoclonal antibodes. It was found to be in a small extent in urine and sera of healthy individuals. It has also been detected in sera, tissues and cytosol of the pregnant individuals, several endocrine and nonendocrine, and cell line tumors.

b. Pituitary Forms of hCG

Although hCG is not considered to be a pituitary hormone, there have been many reports of the presence of it in the urine and blood of healthy nonpregnant individuals. While the immunological studies indicated the likely presence of hCG in pituitary tissues, it was never isolated until 1996 by Birken.S,et.al. They have been able to isolate sufficient quantities of pituitary hCG and examine its primary structure, carbohydrate composition and in vitro bioactivity. The structure of pituitary hCG appears to be of an intermediate hormone between the structure of the placental hCG and that of the pituitary hLH, non essential hormone. Other forms of hCG from the pituitary were been studied. Western blotting techniques shown that the pituitary forms of hCG did not exhibit nicking. While immunological evidence reported that the hCGβcf is present in pituitaries at level about 1% that of hLH βcf ;such low levels make it difficult to be isolateed .

Production and Action of hCG
a. Normal Production and Action of hCG

Although trophoblast is the major source of hCG, a wide variety of normal tissues, including anterior pituitary, can produce hCG. Non trophobastic hCG is not glycosylated and its levels vary in the circulation due to rapid clearance. Some non trophoblasti tissues may not even release hCG, so that it will serve as a local ligand for hCG receptors.

- **Pregnancy:**

The appearance of hCG both in sera and urine soon after conception and its subsequent rise in concentration during early gestational growth make it an excellent marker for the early detection of pregnancy. Sera level of hCG rise rapidly, increasing with conspectus number, after implantation of the trophoblast and reach a peak at 8-10 weeks of gestation then decline and remain at a low level until term). hCG has multiple tropic effect during pregnancy; it has long been known to act on the ovaries to cause synthesis of estradiol and progesterone, which regulate reproductive tract function[(69,77,78)]. New evidence has revealed that hCG can also act directly on reproductive tract organs themselves and regulate their function.

- **Fetus:**

hCG levels are very low in the fetal circulation, less than 5% of these in mother (maximum 50 mIU/ml), suggesting that hCG secretion is directed into the maternal circulation and is prevented from entering the fetus. It may control fetal adrenal androgen synthesis, gonadal steroid production, brain growth and differentiation, protect the fetus from certain viral infection and may relax the umbilical vessels, keeping it from becoming too rigid.

- **Placenta:**

hCG was previously originated from the placenta. So, the production and presence of hCG and its subunits on placenta cell surface was detected by using various methods. These methods demonstrate hCG as a continuous layer on the surface of syncytiotrophoblast of the early and term placenta. hCG is responsible of giving the signals of invasion from syncytiotrophoblasts to the trophoblast cell layer of the placenta to invade the uterus to establish a vacuolar connection with the maternal circulation during pregnancy[(74)].

- **Central Nervous System (CNS):**

The Central Nervous System (CNS) is also one of the specific target tissues for hCG, by which it is able to elicit multiple effects in the (CNS) through binding to their receptors, which have been identified. hCG is involved in a multiple effect on the gonadotropin-releasing hormone (GnRH), constitute a negative short-loop feedback regulation of synthesis and secretion. Many observations administrated that gonadotropins can induce behavioral changes such as decreased feeding, exploratory activity and electrical activity of the brain.

- **Other organs:**

Studies on unicellular animals, some microorganisms and normal human tissue (ovary, testes, pituitary, lung, liver, kidney, colon and stomach) observed the presence of molecules immunologically similar to hCG.

b. Abnormal Production and Action of hCG

hCG belongs to the family of embryonically related marker proteins that include carcinoembryonic antigen and α-fetoprotein. Thus,a wid variety of trophoblast and nontrophoblast cancers and cancer cell contain intact hCG and/ or one of its subunits. The presence of them is probably due to synthesis rather than sequestration. Their expression increases in advanced cancer, suggesting that they might be involved in the progression of the disease.The regulatory mechanisms involved in the expression of hCG subuints genes in cancer cell are not known.Studies suggest that hCG may have dual role in cancers. It promotes some whereas it inhibits other, due to whether they produce intact hCG or just its subuint which may have a stimulatory effect, probably due to the formation of homodimers.

Trophoblastic Tumors

Detectable levels of hCG have been reported in conditions other than normal pregnancy, as originally described by Vaitukaitis and co-workers, included ectobic pregnancy, threatened abortion and trophoblastic disease. Its level is used for diagnosis of Down syndrome and genetically abnormal pregnancy. In ectobic pregnancy and threatened abortion hCG level are progressively decrease and it is not known whether it is a cause or consequence. Abnormally high levels of hCG are a risk factor for trisomy 21 and for the later onset of pre-eclampsia. Why hCG elevated is unclear, but such elevation could reflect either general trophoblast immaturity or increased syncytiotrophoblast turnover, both of which have been described in this disease. hCG and/or its subunits (in blood and urine) also play an important role in the diagnosis of gestational trophoblastic disease, as well as, for monitoring the success or failure of chemotherapy in these disease, making it a most effective tumor marker, including partial or complete moles and choriocarcinomas. Studies have demonstrated that choriocarcinomas produce excessive amounts of hCG, reach almost 100% sensitivity and specificity, due to the loss of self-regulation of biosynthesis and these high amounts promote tumor growth and metastasis in the host body. These findings suggest a potential treatment for this disease by selectivity inhibiting tumor hCG synthesis.

♦ Non trophoblastic Tumors

A great variety of non torphoblastic (gynecologic and nongynecologic) malignant tumors express hCG with a range of 19-30% of all tumors studied; moderately increased and showed variable correlation to tumor stage and histological grading. In sera, studies with highly specific sandwich procedures, indicated that βhCG is mainly elevated while intact hCG and αhCG are slightly elevated or within the normal range; in contrast with that observed in pregnant women who secrete intact hCG in larg excess.

The expression of hCG and/or one of its subunits increases in advanced cancer suggesting that they might be involved in the disease. Thus hCG measurement and other several circulating substances were used to determine whether patients have a high probability of metastasis, with up to 97% positive for hCG. The combined use of hCG and CEA together with cytology for better discrimination of benign from malignant effusions have been recommended. There is usually a parallel relation between these sera levels and the clinical evaluation of the disease under chemotherapy.

Employing two solid phase capture antibodies technique for urinary βhCG and hCGβcf ; demonstrated the presence of these forms in different malignancies . Using flowcytometry and immunohistochemical techniques ,showed a higher expression of βhCG than intact hCG and there was no relationship between the βhCG postivity and the histological type of tumor.

All above studies suggest that βhCG ,may be the major form, that serves as a valuable tool in diagnosis of malignant nontrophoblastic tumors in general.

• *Gynecologic Tumors:*

The early production of hCG by epithelial ovarian carcinoma in tissue culture has arise the possibility that this hormone may be a useful marker in gynecologic tumors. Most studies have focused on sera determination of hCG in this tumor. Low percentage, within the range of 18-25%, of total sera hCG level was found in early studies on patients with gynecologic cancer, using cut off level 5mlU/ml. While other recent studies have demonstrated that the percentage of intact and βhCG elevation was (48%) in comparable to that reported by Grossmann (37%) for gynecologic cancer; based on immunoradiometric design.

The elevation of βhCG was investigated with other tumor markers in women with primary epithelial ovarian cancer. This investigation has found a correlation between markers levels and cancer stage. It has also reported the importance of βhCG in following the ovarian cancer. While, the examination of selected epithelial ovarian cancer found no relation between the βhCG positivity and the histological type, or even the histological grade of the malignant tumors.

The usefulness of βhCG and hCGβcf has been demonstrated too in assessing the prognosis of primary cervical cancer. A detectable elevation of βhCG in 30% of cervical cancer has been observed while low elevation of intact hCG in the same patient (1.3%) has been seen. Non of the cervical adenocarcinoma has had elevated hCG level, but βhCG positive has been found in adenocarcinoma of the uterine cervix in which the cell of tumor have some histological resemblance to trophoblastic cell. Increasing of βhCG synthesis during progression of vulva cancer and its elevation has been found in 50% of patients with that cancer. This Increasing indicating that the patient with elevated sera βhCG had a worse progressive tumor compared with the group of normal βhCG.

For endometrial cancer, no base was found in the distribution of hCG level in patients with poorly and well differentiated endometrial cancer; suggesting that it might be involved in the progression, as promoter, of this disease.

- ***Non gynecologic Tumors:***

Like other hormones, hCG acts via binding to its receptors. Until about 18 years ago, hCG receptors had not been shown to be present in nongonadal tissues. The observation of their presence, including some cancers arising from nongonadal tissues, forced a change in the widely –held belief that hCG is only a gonadal regulating hormone.

In non gynecologic tumors, hCG has two different effects on tumor growth. It promotes some cancers, having a promoter property, such as cancer of the lung, whereas it inhibits prostate cancer. In fact, contraceptive hCG vaccine is now being tested, especially against cancer of colon and pancreas).

By employing radioimune assay (RIA), a great variety of malignant tumors have been found to be expresed hCG. Its levels were only moderately increased and showed variable correlation to tumor stage, histological grading and clinical course. The production of hCG or its subunits was found to be associated with tumors of high prognosis and greatest incidence such as tumors of (lung, liver, gastrointestinal tract tumor, meloma, testis, bladder, stomach, colon, pancreas, head and neck).

Different studies estimated a different percentage relation between βhCG level and variable tumor origin. Nevertheless, in all of them low elevation of intact hCG was demonstrate. It was found in 55% of patients with (pancreas, colon, stomach, bladder, and hepatocellular tumor). Other studies have observed the level of βhCG to be in the range of (30-47%) for bladder cancer, (32-72%) for pancreatic cancer and less than 15% in lung cancer. Studies on Urinary βhCG and hCGβcf revealed an over expression of them in malignant and benign tumors (bladder, pancreatic and Urothelial cancer).

The βhCG has been found to be a promising tumor marker in early diagnosis and following the cancer of (testis, bladder, liver, stomach, colorectal, pancreas and lung), indicating that patients with pancreatic cancer having highly elevated hCG sera levels than patients with normal hCG sera level are having wors progressive tumor.

Using immunohistochemical techniques various investigators have detected positive staining for hCG in many origin, such as, (colorectal cancer and Urothelial carcinoma). Other investigations have reported that extracts of cancer tissues from different origin tumors contained hCG-like material higher than normal tissue. More than 80 different cancer cell lines examined for the production of hCG and/or its subunits; demonstrated its expression in these different histological types and origins of cancer cell lines. The stimulation of growth of several of those cell lines by hCG has been reported. Thus cancer cells are able to regulate independently their own growth.

These findings with other results which confirmed that hCG has also chemical and physiological properties of growth factor; have provided scientific basis for studies of prevention and control of cancer by active or passive immunization against hCG and its subunits.

Potential Therapeutic Uses of hCG

Although still greater understanding is needed, gain made in the last decade has demonstrated that nongonadal actions of hCG are physiologically important and may have relevance to better understanding several diseases and their treatment.

- **Pregnancy:**

The statement on improving pregnancy rates in assisted reproductive technologies was based on the finding that hCG treatment of coculture of cow oviduct epithelial cells with two-cell cow embryos resulted in greater embryonic development. Coculture has been shown to improve pregnancy rate in women who failed to become pregnant in more than three previous cycles.

hCG treatment may save pregnancy if the threatened loss is not anatomic defects, infection or fetal anomalies. hCG treatment may work by increasing the placental endocrine activity, by preventing immunologic mechanisms that promote fetal rejection, by increasing uterine blood flow, by decreasing uterine activity, and so forth. Clinical trials are being conducted to assess objectively the therapeutic value of hCG treatment of women with threatened and habitual abortion.

The ability of hCG to maintain myometrial quiescence suggests it may be used in the treatment of preterm labor and delivery, unless it is caused by infection and premature rupture of membranes. In fact, administration of hCG has a tocolytic effect in a mouse preterm-labor model. This effect seems to be mediated by down-regulation myometrial gap junction.

- **Central Nervous System (CNS):**

HCG treatment has been shown to improve recovery of spinal cord-injured rats. Motor neurons, among other cells in spinal cord, contain hCG receptors. Although how hCG working through these receptors in healing spinal cord injuries is not known yet, it is noteworthy that hCG belongs to the same family as nerve growth factor, hCG neurotropic and neurotransmitter properties, and may suppress the immune responses perhaps through its action on Tcell, monocytes, and macrophages.

- **Non trophoblastic Tumors:**

Potential hCG isused in prostate cancer treatment is based on the finding that tumors contain hCG receptors and that hormone has antiproliferative and anti-invasive actions in prostate cancer cell. These finding suggest a potential use for hCG in the treatment of castrated prostate cancer patients.

hCG also has multiple anti-breast cancer action, which we will detail, that may explain the decreased breast cancer incidence in some women whom complete full-term pregnancy at a young age.

Human Chorionic Gonadotropin and Breast Tumor

The view on the relation between hCG and breast cancer partially turned from disease and tumor marker to disease and hormonal therapy relation in the last decade.

The main focus of many laboratories investigation is the prevention of breast cancer through the understanding of the endocrinological and molecular aspects of inhibition of cancer initiation and progression. So they focused on the inhibitory effect of pregnancy, which is mediated by the placental hormone (hCG). These studies encompass from the endocrinological influences that modulate the normal development of the mammary gland to the role that these influences play in determining the susceptibility or resistance of this organ to undergo malignant transformation when exposed to exogenous carcinogens. Breast cancer, like other cancers, is not one disease and has multiple etiologies and for these reasons, expectation that hCG treatment will help every women will be lowered.

Role of hCG in Breast Cancer

Several studies have established many evidence that there is very little probability for doubt that breast is another nongonadal target of hCG action and its action may be relevant to physiological changes in breast on preparation for lactation and at the same time confers protection against cancer.

- **Inhibition:**

Since pregnancy is a state of chorionic elevation of many hormones, the protective effect of pregnancy can be due to any of these hormones. Russo et.al., initiated a number of studies using a rodent breast cancer experimental model since women cannot be used to study. In this model, the administration of 7,12-dimethylbenz[α]anthracene (DMBA) resulted in a consistent induction of mammary tumors and was more effective in nulliparous than parous animals. It has been demonstrated that the inhibitory effect of pregnancy on mammary cancer initiation is mediated by hCG, depending on the observation of nulliparous animals which were greatly reduced prior to carcinogen administration when they ware pretreated or simultaneously treated with hCG.

This hormone may act on breast to promote nonreversible differentiation of proliferation-competent epithelial cell into secretory cells in terminal end buds. Coincidentally, this differentiation, which is a physiologic phenomenon to prepare the breast for lactation, also makes the cells less susceptible to carcinogenic transformation.

The effect of hCG on the development of mammary gland, not under the influence of pregnancy, has been evaluated to assess the effect of this hormone on two important endocrine organs: the ovary and pituitary gland. hCG showed no effect on ovarian-size, sera level of its hormones or sera level of pituitary hormones after cessation of the treatment; indicated that the hormonal milieu induced by hCG sufficed for differentiating the mammary epithelium. These findings predict that early hCG treatment of women plan to delay their first pregnancy may reduce their breast cancer risk.

- **Progression:**

The studies of the productive effect of hCG-induced differentiation on experimental mammary carcinogensis led to postulate the possibility that hCG may be useful in preventing the development of breast cancer in women. The Russo et.al., tested the effect of hCG on tumor progression, using rat experimental animal. They have found that treatment with hCG inhibited mammary carcinogenesis. This inhibition occurred by stopping the progression of early lesions (i.e, IDPs, CIS). Finding that indicated hCG has a significant potential as a chemoprventive agent not only before the cells were initiated by carcinogen, but after the carcinogenic process was vigorously progressing. Also, emphasizing the importance of hCG receptor in mediating anticancer actions of hCG in breast cancer cell. It has been shown that hCG can inhibit the tumor growth and invasion, increase the cyclic AMP and decrease ER levels in breast cancer cell lines which are positive ER, and can decrease the growth of ER negative breast cancer cell lines.

Mechanism of hCG Action in Breast Cancer

The differentiation effect, by hCG on the mammary gland, alone may not explain the protective role of this hormone. hCG may directly inhibit cell growth, invasion and proliferation. The protective role phenomenon of hCG was found to be in a great part mediated by several changes that have been observed in the mammary gland of rat treated with hCG, either alone or after DMBA treatment, but they were absent in the animals treated with DMBA alone, figure (49), such as:

- **Effect on Inhibin Expression:**

The hCG-induced differentiation of the mammary gland was found to be associated with synthesis of inhibin. The expression of both inhibin α and β was increased and there synthesis was accompanied by a significant activation of *c-myc* and *c-jun* gene, which remained activated even after the cessation of hCG treatment. The finding that *c-myc* and *c-jun* gene were also elevated indicates that early responsegenes could be involved in the pathway of hCG-inhibin-induced synthesis.

- **Effect on Programmed Cell Death (PCD) Genes Expression:**

It has been found that inhibition of progression of mammary carcinomas by hCG has been associated with the activation of genes known to be responsible of programmed cell death, cell growth arrest and apoptosis. There were remarkable inductions of several apoptotic genes such as P53, ICE . The effect of hCG on the activation of PCD gene was specific for mammary gland, since the hormonal treatment did not modify their expression in the ovary, even though it is the target organ of hCG action. In mammary and other epithelial cells in culture, both P53 dependent and P53 independent pathways have been identified. In hCG treatment the P53 dependent pathway was involved in the PCD process.

- **Effect on Insulin-like Growth Factor (IGF) Expression:**

hCG treatment decreases IGF and increases some of IGF binding proteins (IGFBPs). Since IGFs are potent mitogenes and antiapoptotic agents for breast cancer; thus decreasing of these factors, by hCG treatment, reduce the availability of IGF and its action.

- **Effect on New Genes Expression:**

The induction of the differentiated genes (β casein and whey acidic protein (WAP)), two of the major milk proteins in most specie, was observed in the treatment of mammary gland by hCG. Also a third gene called hormone-induced 1 (HI-1) was express in this process. hCG treatment was found to be enhancing the cellular DNA repair mechanisms of the mammary epithelium and increasing the expression of the above genes that might prevent and/or decrease the expression of other that might promot the carcinogenic transformation of breast epithelial cell.

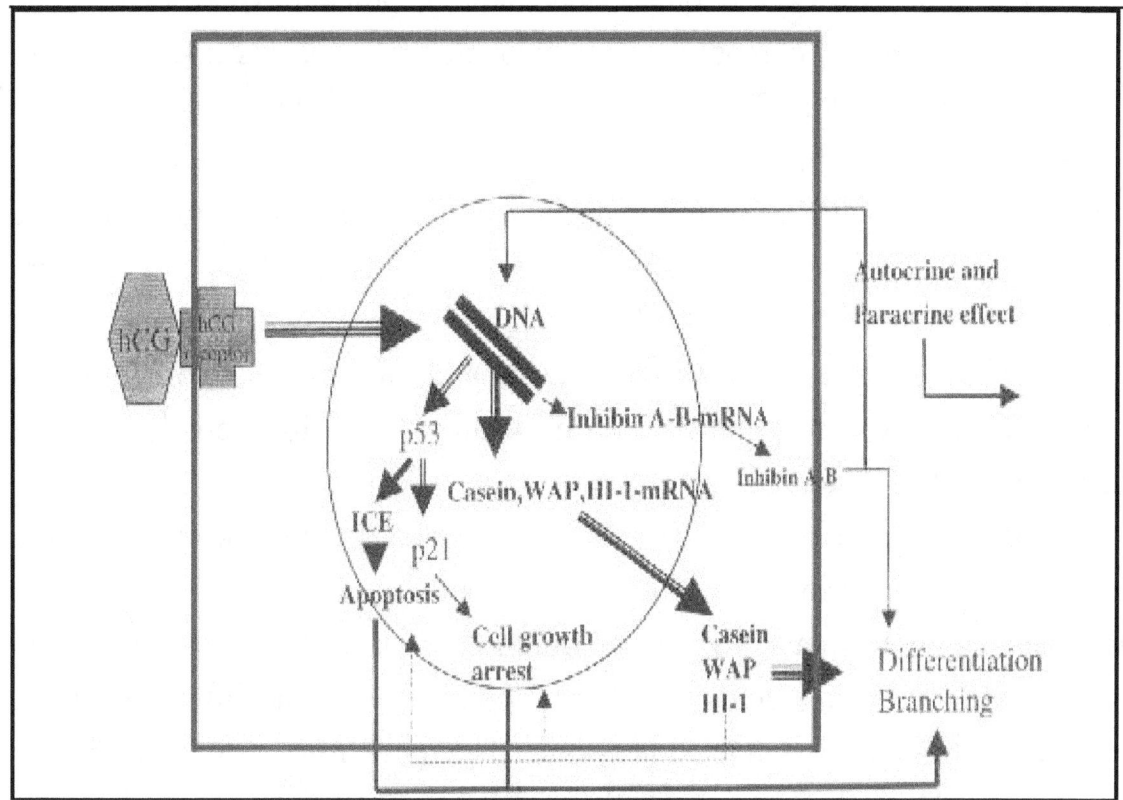

Figure (44): The postulated mechanism of action of hCG. The hormone binds to a specific membrane receptor, activating genes identified to be specific for pregnancy-or hCG-induced differentiation, and that have been found to be correlated with the lobular development of the mammary tissue. Thus, a pathway of activation of p53 and ICE may lead to apoptosis, or through p21 to cell growth arrest. Activation of inhibin α and β, and od the milk proteins, casein, whey acidic protein(WAP), and HI-1, may lead to the differentiation.

Human Chorionic Gonadotropin Receptor

There are two receptors for gonadotropins, one for FSH and another for hCG and LH. Although hCG and LH share a common receptor, not all the receptors binding characteristics of these two hormones are similar. This may be due to subtle structural difference between the two hormone molecules. Only intact hCG and LH but not any other hormones including alpha and beta subunits of hCG and LH, can bind to the receptors).

Structure of hCG Receptor

hCG receptor is a single chain transmembrane glycoprotein, which is a member of group II of the G protein-coupled receptor family. In human, the hCG receptor gene is composed of 11 exons and 10 introns and its coding region is over 60 kb long. The receptor protein contains 696 amino acids.

It composes of highly conserved seven transmamberane domains (I through VII) contain six potential sites for N-linked glycosylation and three sites for protein kinase C phosphorylation, as in figure (45). The mature receptor contains an extra cellular domain of 333 amino acids, a transmembrane domain of 266 amino acids and intra cellular domain of 70 amino acids. The extracellular domain is rather large in keeping with the large size of the ligand.

hCG receptors are present not only in cell surface membranes but also to a variable degree in lysosomes, rough and smooth endoplasmic reticulum, cis and trans Golgi elements, nuclear membranes and interface of dispersed and condensed chromatin. Some of the properties of intracellular organelle receptors are similar while others are different from those in plasma membranes. These differences may be partly due to differences in membrane environment of the organelles.

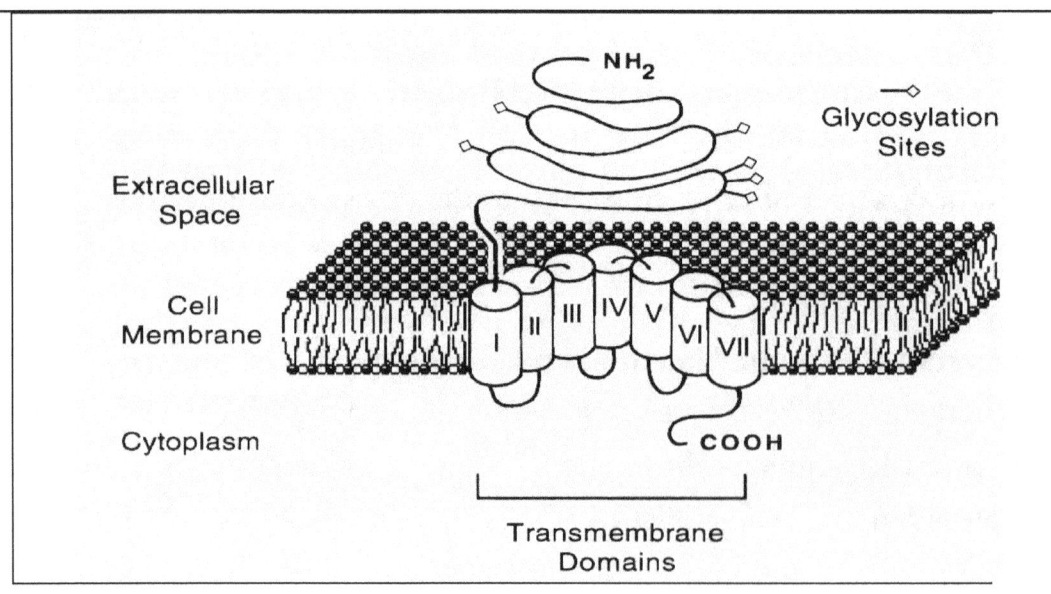

Figure (45):G-protein-coupled hormone receptor. HCG receptor is composed of seven highly conserved transmembrane domains (I to VII), a large extracellular domain with six potential glycosylation sites, and a relatively short cytoplasmic domain.,

Distribution and Function of hCG receptor

It was assumed for along time that only gonadal tissues could respond to hCG from human placenta and LH from the pituitary, because of the belief that the receptors which are required for their action were only present in the gonadal tissues.

This belief was challenged after the presnce of hCG receptor in the porcine uterus was first demonstrated.

Subsequently, several laboratories have employed many techniques for detecting hCG receptor by using Northern blotting, western blotting, ligand blotting, covalent receptor cross-linking, ligand binding, insitu hybridization, imunocytochemistry, gel mobility shift, gene transfecation, and laser scanning contocal microscopy. Not all techniques were used on all tissues; however more than one was used in most cases.

Tissues also contain lower levels of functional hCG receptors forcing a new look at old concepts that have been around for more than 50 years in reproductive biology and medicine.

Nongonadal hCG receptors have been described in a number of species; with identical affinity and specificity of binding to the hCG as gonadal receptors. These species include human, monkey, pig, cow, sheep, rat, rabbit, mice and turkey.

Even though more tissues than anyone ever imagined are now known to contain the receptors, other tissues like liver, kidney, lung, skeletal smooth muscle, heart and spleen are receptor negative.

Recent data have shown, while these in adult are receptor negative, some of these tissues in human fetus are receptor positive, suggesting that hCG may have been a developmental role.

Malignant transformation of nongonadal endocrine tissue can result in an inappropriate appearance of functional hCG receptor.

However not all kinds of tissues from all the species have yet been examined. The human nongonadal tissues are by far the most extensively studied among the species as lists in table (30), which included all tissues to date and some of their hCG receptor function.

Table (30): Nongonadal tissue distribution of hCG receptors and their functions.

Tissue	Function*
Oviduct	Release sperm bound to epithelial cell and enhance growth of early embryos in cocultures with epithilial cells
Uterus	Differentiation of stromal cells
Cervix	Increase COX expression
Placenta	Regulation of hCG biosynthesis
Fetal mambranes	Weaking of membranes through increase in COX-1 expression
Umbilical cord	Relax umbilical vessels
T-cells, monocytes, and macrophages	Increase monocyte chemoattractant protein-1 expression
Urinary bladder	Maintain normal itsfunctions
Skin	Regulate androgen metabolism and action
Bone	Turnover
Adrenal cortex-zona reticularis	Increase DHEAS secretion
Brain	Regulation of LH synthesis
Neural retina	Visual processing of information
Breast	Promote nonreversible differentiation of proliferative to secretory-type epithelial cells
Blood vessels in	Vasodilation through increasing prostacyclin

target tissues	
Prostate	Androgen metabolism and action
Male reproductive tract	Secretion and sperm maturation
*In many cases, functions are based on in vitro or in vivo data. In few cases, they are drevied from logical deduction	

hCG Receptor in Breast Tissue

Several laboratories have now demonstrated that rat and sow breast tissues, normal and malignant human breast tissues and human breast cancer cell lines contain hCG receptor transcript and receptor protein which can bind ^{125}I-hCG.

The receptor levels are the highest in breast epthelium, followed by blood vessels, stromal cell, smooth muscles and higher in the luteal phase than in proliferative phase, suggesting that ovarian steroid hormones may regulate breast hCG receptor levels. Sow breast receptor levels increase from the proliferative to the secretory phase, suggesting that progesterone may upregulate and/or estrogen may downregulate the receptors.

MCF-7 cells contain higher hCG receptor levels than other human breast cancer cell lines tested. This was due to an increased transcription of the gene of hCG receptor in the MCF-7cells.

In the breast cancer cell, hCG anticancer actions were found to be mediated by hCG receptor. So the present study was undertaken as a complementary study to characterize the binding condition between hCG and its receptor.

Materials and Methods

- **Patients**

Three groups of breast tumor patients were included in this study.

Group I : Consisted of 25 patients with benign breast tumors

Group II : Consisted of 21 premenopausal patients with breast cancer.

Group III : Consisted of 14 postmenopausal patients with breast cancer.

Group IV : Consisted of 12 controls.

The patients were newly diagnosed and were not undergone any type of therapy. Patients suffered from any disease that may interfere with our study were excluded.

Table (31): The host information of breast tumor patients and healthy subjects studied.

Group	Patients	No.	Age	Type of tumor
I	Benign breast tumor	25	14-41	Fibrocystic chandes (adenosis)
II	Premenopausal malignant breast tumor	21	30-47	Infiltrattive ductal carcinoma
III	Postmenopausal malignant breast tumor	14	54-72	Infiltrattive ductal carcinoma
IV	Control	12	14-37	-

- **Blood Sampling**

Blood samples (7ml) were obtained from the patients just before surgery by veinpuncture. Age matched sera were obtained from (12) healthy premenopausal women.

Blood samples were centrifuged at 1500xg for 10 minutes after allowing the blood to clot at room temperature. The sera were aliquoted and frozen at –20 °C until assaying.

- **Collection of Breast Tissue Specimens**

The tumor tissues were surgically removed from breast tumor patients by either mastectomy (cancer patients) or lumpectomy (benign tumor patients). The specimens were cut off and immediately rinsed with ice-cold isotonic saline solution. They were collected individually in plastic receptacles and stored at −20 °C until homogenization.

- **Preparation of Breast Tumor Tissue Homogenate**

The frozen tissue were thawed, weighed, pulverized finely with a scalpel in petri dish standing on ice bath, and then homogenized at 4°C in a buffer solution with a ratio of 1:3 (weight-volume) by using a manual homogenizer.

The buffer used was Tris/HCl (0.05M, pH 7.2) containing (0.25M sucrose, 5mM EDTA and glycerol 20%). The homogenate was filtered through several layers of nylon gauze to eliminate fibers of connective tissue, then centrifuged at 1000xg for 15 minutes at 4°C. The supernatants were used throughout our study.

Methods

- **Determination of hCG Levels in Sera of Benign and Malignant Breast Tumor Patients and Controls**

Sera levels of hCG were measured in samples collected from pre- and post-menopausal patients and healthy premenopausal women by immunoradiometric assay (IRMA).

The assay details were carried out according to the following procedure:

1. A set of coated tubes containing 50μl of hCG standards or sera of patients and controls, 200 μl of ^{125}I-anti hCG antibody was added to each tube, then mixed well by hand.

2. Two additional non-coated tubes containing only 200 μl of ^{125}I-anti hCG antibody, for total activity computation, were set aside until counting.

3. All tubes were incubated for 60 minutes at 25°C on a horizontal shaker.

4. Aspirate thoroughly the contents of all the tubes except of those for total activity (T) measurement.

5. The tubes were wash twice with 2ml of washing solution and aspirate the solution immediately.

6. The radioactivities of all tubes were measured by gamma counter.

7. *The assay protocol of sera hCG was described in Table (32).*

Table (32) IRMA assay protocol of sera hCG (IU/L)

Step 1 Pipetting	Step 2 Incubation	Step 3 Counting
Add to coated tubes Successively: 50 µl of calibrator or sample and 200 µl of tracer then Mix.	Incubate the mixture 1 hour at 18-25°C with shaking.	Aspirate the content of the tubes carefully, except 2 tubes for total activity (T). Rinse the tubes with 2ml of wash solution and aspirate twice, then count activity (cpm).

Calculations

1. The mean of c.p.m was determined for each pair of duplicate tubes of standard and unknown sample.

2. A standard curve was drawn by plotting the c.p.m for each standard on the Y-axis against the corresponding concentration on the X-axis, as shown in figure (51).

3. The unknown concentration of the sample was calculated from the standard curve.

- Determination of Total Protein Content in Breast Tumor Homogenate

The total protein content of breast tissue homogenate was determined by the method of Lowry et.al, using bovine serum albumin (BSA) as the standard protein and the absorbence of the developing color was read at 600 nm against the appropriate blank.

Binding Studies of hCG in Breast Tumor Homogenate with ^{125}I- anti hCG Antibody

. Preliminary Tests of hCG Binding in Breast Tumor Homogenate with ^{125}I-Anti HCG Antibody

1. One hundred microliters of breast tumor homogenate containing 200μg protein for benign, pre- and post-menopausal malignant breast tumor respectively was added to 5 μl (for benign breast tumor) and 10 μl (for pre- and post-menopausal breast tumor) of ^{125}I-anti hCG antibody (1470μg.ml^{-1}), the volume of the mixtures were made up to 250 μl with Tris/HCl buffer (0.05M, pH7.4) containing 0.1% BSA.

2. Two additional tubes containing only 5,10 μl of ^{125}I-anti hCG antibody, were set aside until counting for total activity.

3. The tubes were incubated at 25°C for 60 min.

4. After incubation, 500 μl of PEG 6000(10%) were added to the tubes and incubated again for 150min at 4°C.

5. After incubation, the tubes were centrifuged at 1500xg for 30min at 4°C.

6. The supernatant was discarded by decanting the assay tubes, then the tubes were inverted on a filter paper for 10 min.

7. The rims of the tubes were swabbed with a cotton piece and the amount of bound radioactivity (c.p.m) was counted using gamma counter.

Calculations

1. The bound fraction (B) represents the counted radioactivity in each tube, expressed in c.p.m i.e (^{125}I-anti hCG antibody/hCG) complex.

2. Total activity (T) represents the counted radioactivity in the tubes containing only ^{125}I-anti hCG antibody.

3. The (B/T)% ratio for each tubes were calculated as follows:

$$(B/T)\% = \frac{\text{Sample mean counts (B)}}{\text{Total activity mean counts (T)}} * 100$$

Most Appropriate Conditions of the Binding of hCG in breast Tumor Homogenate with ^{125}I-anti hCG Antibody.

- *The Effect of Different Protein Amount of Breast Tumor Homogenate on the Binding of hCG with ^{125}I-anti hCG Antibody*

1. A volume of 5 µl (for benign breast tumor) and 10 µl (for malignant pre- and post- menopausal breast tumor) of ^{125}I-anti hCG antibody were added to 100µl containing increasing amounts (100, 200, 300, 400, 500, 600, 700 and 800 µg protein) for benign, malignant pre- and post- menopausal homogenate of breast tumor respectively in a final volume of 250 µl which made up with Tris/HCl buffer (0.05M, pH 7.4) containing 0.1% BSA.

2. Two additional tubes containing only 5 and 10 µl of ^{125}I-anti HCG antibody were set aside until counting for total activity.

3. The tubes were incubated at 25°C for 60 minutes.

4. After incubation, the (^{125}I-anti hCG antibody/hCG) complex was estimated.

calculations

1. The (B/T) percent values were determined

2. The percent of binding values (B/T)% were plotted against the concentration of ^{125}I-anti hCG antibody.

- ***The Effect of Different Concentrations of ^{125}I-anti hCG Antibody on the Binding with hCG in Breast Tumor Homogenate***

1. One hundred microliters of the optimum amounts of breast tumor homogenate (700, 600 and 600µg proteins of benign, malignant pre- and post-menopausal, breast tumors homogenate respectively), were added to increasing volumes 2, 3, 4, 5, 6, 8 and 10µl (0.012-0.059 µg.ml^{-1}) for benign and malignant postmenopausal breast tumors and 10, 15, 20, 25, 30, 35, 40 and 45µl (0.059-0.2646 µg.ml^{-1}) for malignant premenopausal breast tumors of ^{125}I-anti hCG antibody (1470µg.ml$^{-1)}$ then the volume were made up to 250 µl with Tris/HCl buffer (0.05M, pH 7.4) containing 0.1% BSA.

2. A set of tubes containing only the same increasing volumes of ^{125}I-anti HCG antibody (2, 3, 4, 6, 8, 10, 15, 20, 25, 30, 35, 40 and 45 µl) were set a side until counting for total activity computation.

3. The tubes were incubated at 25°C for 60 minute.

4. After incubation, the (^{125}I-anti hCG antibody/hCG) complex was estimated by following the steps 4, 5,6 and 7 in section (2.3.1).

Calculations

3. The (B/T) percent values were determined

4. The percent of binding values (B/T)% were plotted against the concentration of ^{125}I-anti hCG antibody.

- ***The Effect of pH on the Binding of hCG in Breast Tumor Homogenate with ^{125}I-anti hCG Antibody***

1. One hundred microliters of the optimum amounts of breast tumor homogenate (700, 600,600 µg Proteins of benign, malignant pre- and post-menopausal, breast tumors homogenate respectively),

were added to (3 μl (0.01764μg.ml^{-1}); 25 μl (0.147μg.ml^{-1}) and 4 μl (0.02353μg.ml^{-1}) for benign and malignant pre-, post-menopausal breast tumors homogenate respectively) of ^{125}I-anti hCG antibody (1470μg.ml^{-1}) the mixtures volumes were made up to 250 μl with Tris/HCl buffer (0.05M) containing 0.1% BSA of different pH (6.8, 7.0, 7.2, 7.4, 7.6, 7.8 and 8.0).

2. Three additional tubes containing only (3, 4 and 25) μl of the ^{125}I-anti hCG antibody were set aside until counting for total activity computation.

3. The tubes were incubated at 25°C for 60 minute.

4. After incubation, the (^{125}I-anti hCG antibody/hCG) complex was estimated.

Calculations

1. The values of (B/T) % were determined

2. The percent of binding values (B/T)% were plotted against their corresponding pH values.

• *Time Course of the Binding of hCG in Breast Tumor Homogenate with ^{125}I-anti hCG Antibody*

1. One hundred microliters of the optimum amounts (700, 600,600 μg for benign, malignant pre- and post-menopausal, breast tumors homogenate respectively), were added to 3μl (0.01764μg.ml^{-1}), 25μl (0.147μg.ml^{-1}) and 4μl (0.02353μg.ml^{-1}) for benign and malignant pre-, post-menopausal breast tumors homogenate respectively of ^{125}I-anti hCG antibody (1470μg.ml^{-1}). The volumes were completed to 250 μl with Tris/HCl buffer (0.05M, pH 7.2) containing 0.1% BSA.

2. All tubes were incubated at 25°C for different time interval (30, 60, 90, 120, 150 and 180) min.

3. Three additional tubes containing only (3, 4 and 25) μl of the ^{125}I-anti HCG antibody were set aside until counting for total activity computation.

4. After incubation, the (^{125}I-anti hCG antibody/hCG) complex was estimated by following the steps 4, 5,6 and 7 in section (2.3.1).

5. To determine the time course of hCG binding to ^{125}I–anti hCG antibody at different temperatures. Steps 1, 2 and 3 in the same experiment were repeated at different temperatures (5, 37, 45 °C).

Calculations

1. The (B/T) percent values were determined

2. The values of (B/T)% were plotted against the corresponding incubation time.

- ***The Effect of Halides on the Binding of hCG in Breast Tumor Homogenate with ^{125}I-anti hCG Antibody***

1. One hundred microliters of the optimum amount of (700, 600, 600 µg proteins of benign, malignant pre- and post-menopausal, breast tumors homogenate respectively), were added to 3 µl (0.01764 µg.ml^{-1}), 25µl (0.147µg.ml^{-1}) and 4µl (0.02353µg.ml^{-1}) for benign and malignant pre-, post-menopausal breast tumors homogenate respectively) of ^{125}I-anti hCG antibody in a final volume of 250 µl (completed with Tris/HCl buffer (0.05M, pH 7.2, 0.1% BSA) containing 0.1M of each of the following halides: NaF, NaCl, NaBr and NaI. A sample without the addition of any halides was used as a control.

2. Three additional tubes containing (3, 4 and 25 µl) of ^{125}I-anti hCG antibody only, for total activity computation, were set aside until counting.

3. The tubes were incubated for 150 min at 4°C (benign breast tumor homogenate), 120 and 150 min at 45°C (pre-and post-menopausal malignant breast tumor homogenate).

4. After incubation, the (^{125}I-anti hCG antibody/hCG) complex was estimated.

Calculations

1. **The values of (B/T)% were determined**

2. The values of (B/T)% were plotted against halides concentrations.

Determination of Affinity Constant (Ka) and the Maximal Binding Capacity (B_{max}) of hCG in malignant Premenopausal Breast Tumor Patients Associated with ^{125}I-anti hCG Antibody

1. One hundred microliters of the optimum amount of (600µg protein) of malignant premenopausal homogenate was incubated with increasing volumes (5, 10, 15, 20 and 25 µl) of ^{125}I-anti hCG antibody (0.0294-0.147µg.ml^{-1} protein). The final volumes were made up to 250 µl with Tris/HCl buffer (0.05M, pH 7.2, 0.1% BSA).

2. A set of additional tubes containing only increasing volumes (5,10,15,20 and 25µl) of ^{125}I-anti hCG antibody were set aside until counting for total activity computation.

3. The tubes were incubated for 120 min at 45°C.

4. The previous steps were performed at different temperature (4, 25 and 37°C), the time of incubation needed to get the equilibrium state were (120min at 5 °C, 90 min at 25°C and 150min at 37°C).

Calculations

1. The values B/F ratio were determined:-

B: The bound radioactivity (mean of counts c.p.m), which represents the (^{125}I-anti hCG antibody/hCG) complex.

F: The free radioactivity (mean of the counts c.p.m), which represented the non-bound ^{125}I-anti hCG antibody.

T: The total radioactivity mean of the counts.

F = Total counts (T) – Bound radioactivity (B)

2. The concentration of the (^{125}I-anti hCG antibody/hCG) complex in (mg/ml) that formed after time (t) was calculated from the following

$$B(ug/ml) = \frac{B(c.p.m)}{T(c.p.m)}$$

equation:-

3. The affinity constant and the maximal binding capacity were determined according to Scatchard equation: -

$$\frac{B}{F} = \frac{1}{Kd}(B_{max} - B)$$

$$Ka = \frac{1}{Kd}$$

Where:

Ka = Affinity constant

Kd = Dissociation constant

B_{max} = Maximal binding capacity

4. The values of the ratio B/F were plotted against the values of B in (μg/ml), gives a linear relationship. The values of the affinity constant of the binding (Ka) at each temperature can be calculated from the slop of the straight line, while the value of the total concentration of hCG (B_{max}) in breast tumor tissue was calculated from the intercept with the x-axis.

Results and Discussion

Three groups of breast tumor patients were included in this study. These groups were classified according to the type of the tumor, as confirmed by histopathological examination. Tissue homogenization was carried out in 0.25M sucrose, 5mM EDTA, 20% glycerol and in cold medium in order to avoid protein denaturation and to decrease the proteolytic enzymes activity.

Sucrose is a hypotonic solution that enhances the rupture of plasma cell membranes and preserves other cell organelles. The proteolysis of the proteins, by endogenous protease, has been inhibited by the inclusion of EDTA as an inhibitor in extraction buffer, while supplementation of the buffer with glycerol will improve the stability of the crude membrane receptor.

Determination of hCG Levels in Sera of Breast Tumors Patients

With a variety of techniques, hCG has been stated to be an appropriate product of human breast carcinoma. It has been proposed as a marker for breast cancer on the basis that raised levels were found in the peripheral blood of some patients. The first known reference of an association of hCG with breast cancer was made by McArthur. But, at that time no biological test specific for hCG was available to substantiate the claim. With the generation of antisera to the β-subunit of hCG, a radioimmunoassay (RIA) was set up which has been stated not to cross-reaction with hLH and enable small amounts of hCG to be assayed in blood. A number of studies using peripheral blood of breast cancer patients with different clinical tumor status have since been published in which a wide variety in the prevalence of raised levels was noted. In this study, the hCG levels in sera of patients with benign and pre and post-menopausal malignant breast tumor were measured by IRMA method with matching with a control group subjects. Table (34) summarized the groups and the mean concentration of hCG in the four groups. For the mean serum hCG level showed a slight elevation (5.8±0.5) as compared to the control mean serum. For malignant pre-menopausal, hCG level showed detectable elevation than control group but lower than postmenopausal malignant group, which showed highest elevation than the other groups. These results are similar, in general, to previous studies in which the presence, distribution and levels of hCG and/or its subunits and its genes have been previously studied in sera, tissues, cyst fluid, cytosol and cell lines of different types of breast tumors. In these studies, high levels of hCG or its subunits have been found in post-menopausal in comparison with pre-menopausal women. Other study demonstrated that all βhCG producing tumors were ductal carcinomas of the breast. But in general these studies have found no positive correlation neither between the levels of hCG and/or its subunits and clinical stage nor the mass of the tumor. For sera investigation, different percentage of hCG levels from these studies have been estimated. Some studies gave a percentage with the range of (13-17%), while other give it with the range (30-50%). The explanation for the synthesis of the hCG hormone by non endocrine tumors is uncertain. Many hypothesis appeared to explain this phenomena. One, that ectopic production of the hormone represent gene depression associated with malignancy). Other hypothesis that in a few cases the tumor might produce hormone of random peptide synthesis. Immunohistochemical localization of hCG and its γ and β subunits in breast cancer cell suggests that most of the βhCG measured in the protein extracts was tumorigenic in origin. Because of continuing need to derive markers demonstrable in breast tumor or in the sera of breast cancer patients, recent RT-PCR assay for multiple markers has been used to detect circulating breast cancer cells using hCG. This method has shown that positive RT-PCR signal in blood samples of affected patients correlated with stage, in particular for βhCG. Also recent multimarker revers trancription-PCR assay has shown that a combination of βhCG and other marker correlated significantly with tumor size

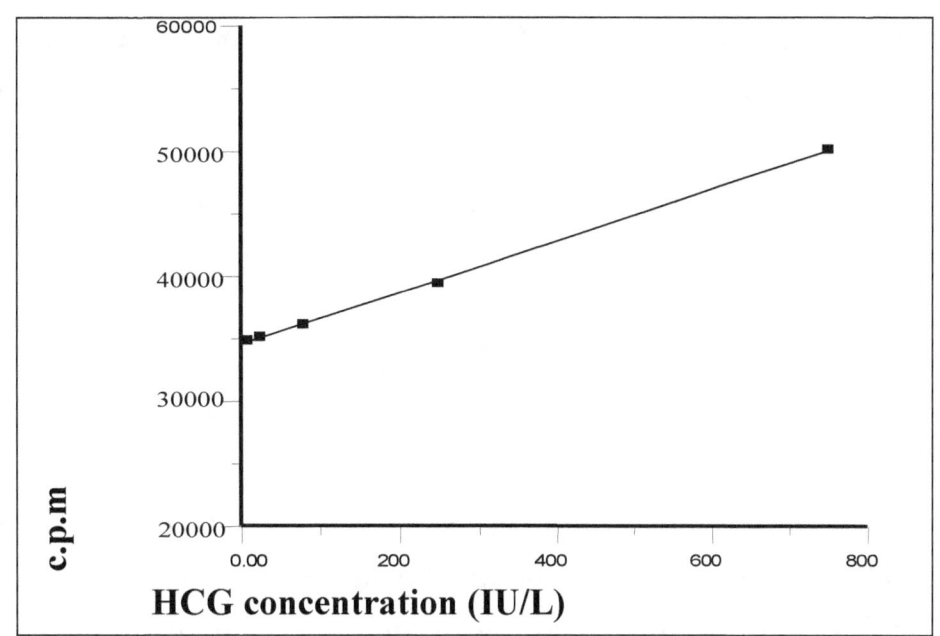

Figure (46): Standard curve of hCG determination in human sera by IRMA method.(all other details are explained in the text).

Table (33): Sera hCG levels (IU/L) in patients with benign and malignant breast tumors. (All details are explained in the text).

Group	Patients	No. of cases	Age range (Year)	Sera HCG IU/L (mean ± SD)	P values
I	Benign breast tumors	25	14-41	5.8 ± 0.5	P>0.05
II	Premenopausal malignant breast tumors	21	30-47	9.3 ± 1.9	P<0.05
III	Postmenopausal malignant breast tumors	14	54-72	18.7 ± 2.1	P<0.05
IV	Control	12	14-37	5.2±0.8	--

Binding Studies of hCG in Breast Cancer Tumor Homogenate with ^{125}I-anti hCG Antibody

Preliminary Tests of hCG Binding in Breast Tumor Homogenate with ^{125}I-anti hCG Antibody

Binding study of radioliabled anti hCG antibody and hCG in breast tumor tissues was tested by preliminary binding. This test was made by incubation of ^{125}I-anti hCG Antibody with homogenate for 60 min at 25°C.

These conditions, as a beginning, were used for incubation according to the conditions that mention in the kit. For the separation (precipitation) of the complex from binding mixture, at first we used the centrifugation at 1500xg for 30 min only without using any precipitating reagent.

The percent of binding values (B/T%) were very low, about (1.5- 2.0 %) for the three groups. So, the try to use precipitating reagent was made. PEG 6000 was used to precipitate the complex with testing its precipitation effect on ^{125}I-anti hCG Antibody alone.

It has been found that percent of binding values (B/T%) were rise to (3.8%) for benign, (22.8%) for malignant premenopausal and (22.5%) for malignant postmenopausal breast tumors, without precipitation of the free tracer.

In these experiments, the obtained data revealed that precipitation of complex by centrifugation alone is not efficient, which lead to use another way to obtained better yield. It also reveals that malignant breast tumors included higher incidence of hCG than those of benign breast tumors.

Most Appropriate Conditions of the Binding of hCG in breast Tumor Homogenate with ^{125}I-anti hCG Antibody.

The Effect of Different Protein Concentration of Breast Tumor Homogenate on the Binding of hCG with ^{125}I-anti hCG Antibody

The binding of antigen(hCG) to antibody (^{125}I-anti hCG Antibody) is not static, it is instead an equilibrium reaction that proceeds in three phases as in the following equation:

$$Ag_n + Ab \underset{k-1}{\overset{k+1}{\rightleftharpoons}} Ag_n Ab \underset{k-2}{\overset{k2}{\rightleftharpoons}} Ag_a Ab_b \quad \cdots\cdots\cdots\cdots (1)$$

Depending on the relative concentration of Ag and Ab, the complex may cross-link in the third phase to form larger complexes, which then precipitate out of the solution, which namely more precipitation complex, higher binding. So, to determine the effect of different Ag(hCG) concentration from breast tumor homogenate on the binding, increasing amount of homogenate was incubated with ^{125}I-anti hCG Antibody. The binding of hCG in breast tumor with ^{125}I-Anti hCG Antibody, as shown in figure (53), was detected only in highly concentration of the homogenate (300 µg) for benign tumor and increases with homogenate increasing until it reaches the saturability at (700 µg) with little increase of the next concentrations. In malignant tumors, bindings were higher from the primary concentration, with the same curve behavior. Previous studies have shown that besides circulatory hormone, mammary glands may synthesize hCG or similar peptides which may act in an autocrine and paracrine manner. They found that human breast cancer tissues immunostain for hCG and breast cyst fluid contains very high levels of biological active hCG as compared to blood.

Six hundred microgram of malignant pre- and postmenopausal breast tumors and (700 µg) for benign breast tumors were used in the next experiments since they gave maximum value of binding (B/T%).

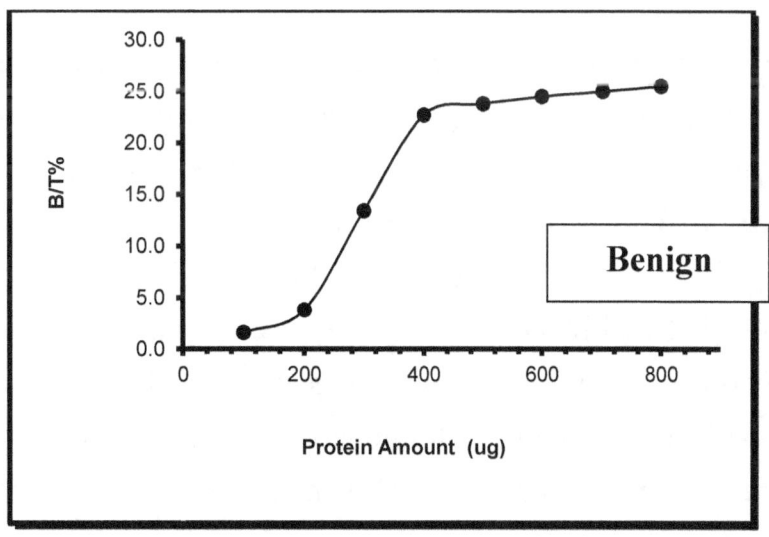

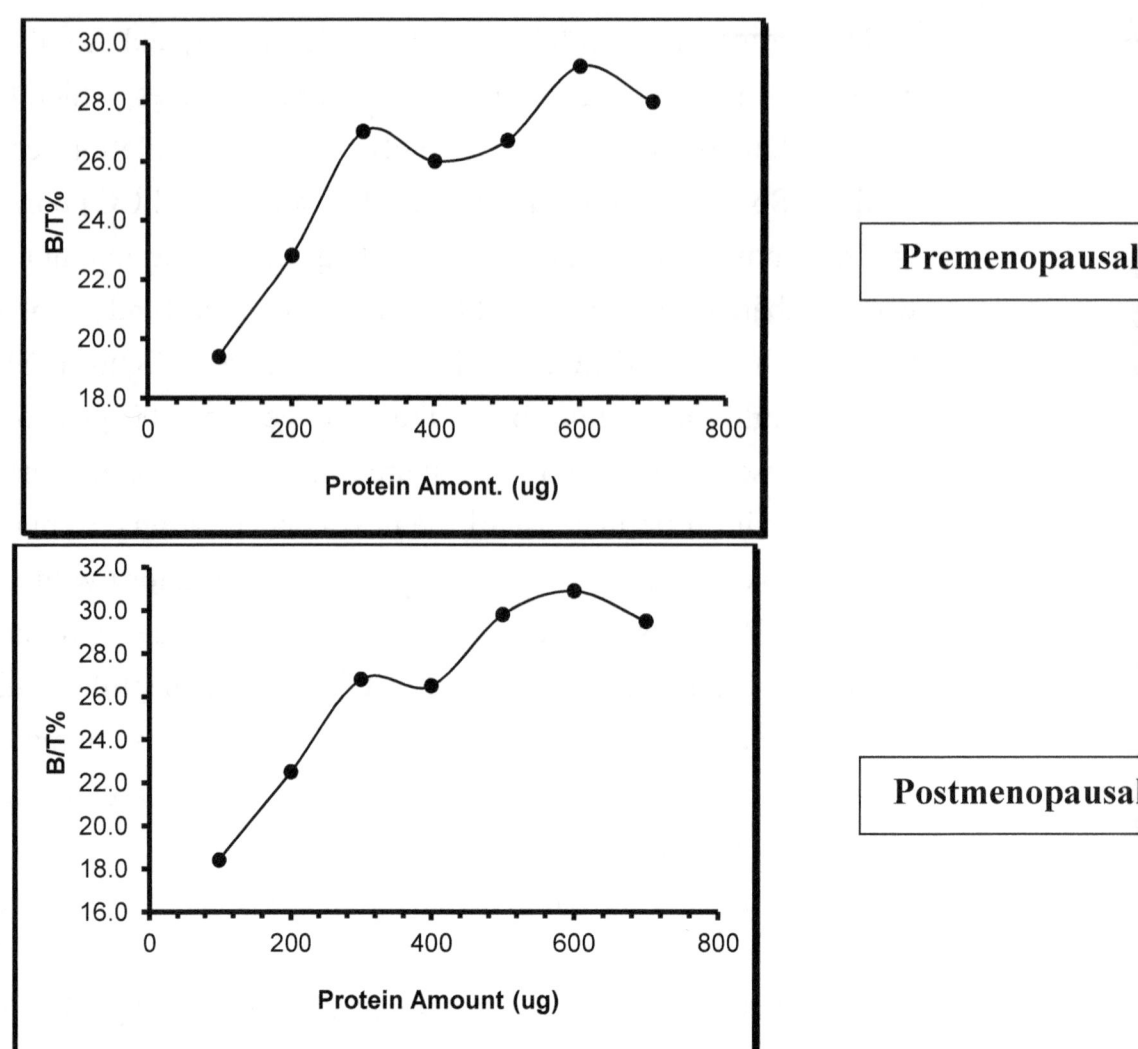

Figure (47): Influence of increasing protein amount on the binding with ^{125}I-anti hCG Antibody. (All other details are explained in the text).

The Effect of Different Concentrations of ^{125}I-anti hCG Antibody on the Binding with hCG in Breast Tumor Homogenate

As in the previous equation (1), the binding reaction may also be antibody (^{125}I-anti hCG Antibody) dependent process. To investigate that, increasing concentration of antibody, presents by ^{125}I-anti hCG antibody, was incubated with fixed amount of homogenate protein. Binding curve behavior for the three groups is similar, as shown in figure (47). For benign tumor, binding reached its maximum level from the primary ^{125}I-anti hCG antibody concentration (3μl, 0.0176μg.ml^{-1}); clarifying that the presence of hCG in benign breast tumor is very little and its detection need high concentration level of homogenate with low concentration of ^{125}I-anti hCG antibody to reaches equilibrium, which presents by maximum binding. This finding may explain some previous works, which have shown the presence of hCG in urine and normal tissue only be concentration of extracts to achieve hCG levels sufficient for measurement. Also unable to detect measurable level of hCG in many breast tumors by using the standard condition of the hCG kits which use low amount of sera and much higher excess amount of the ^{125}I-anti hCG antibody (as in IRMA kit). Premenopausal group has shown higher ^{125}I-anti hCG antibody concentration need (25μl, 0.147μg.ml^{-1}) with less maximum binding (32.4%) than postmenopausal group (33.2%, 4μl(0.0235μg.ml^{-1})). In malignant breast tumor, it was not arrived in the previous studies to a conclusion if breast cancer females at postmenopausal age have a larger level or number of hCG secreting tumor than of younger age group. Beside that, although the relation between the size and stage of tumor with the secretion of hCG is not significant, some studies declared that some breast tumors with large volumes secreted large amount of hCG. These finding may explain the need of premenopausal tumor group, which having a large and bigger tumor volume than postmenopausal group, to more amount of ^{125}I-anti hCG antibody concentration.

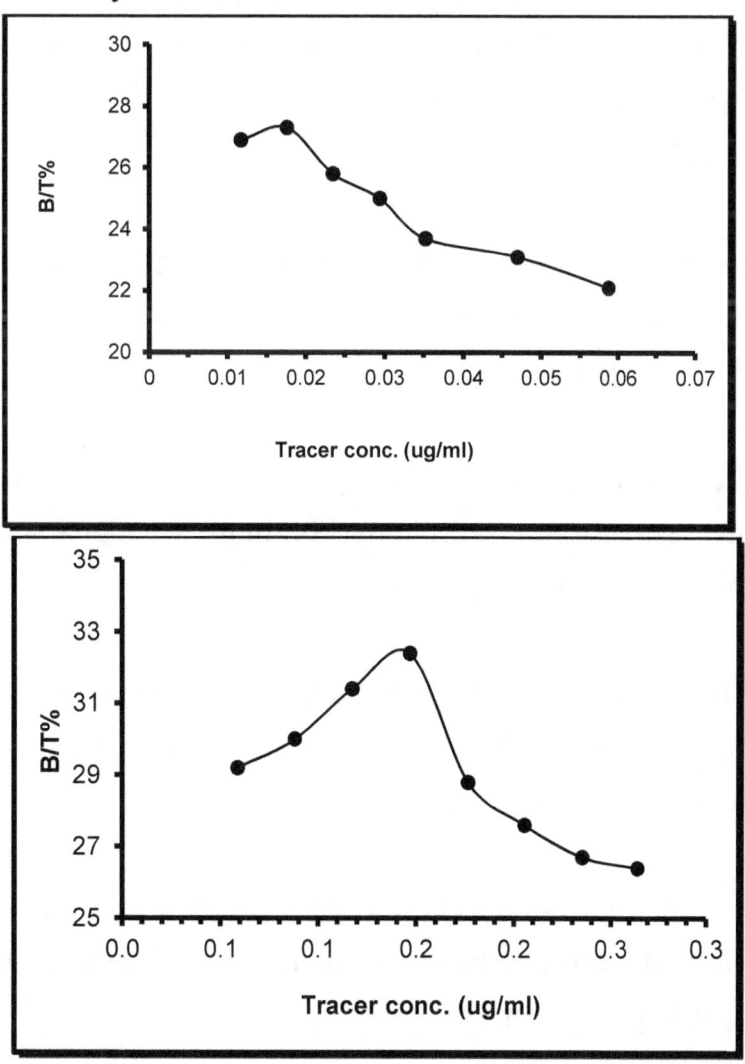

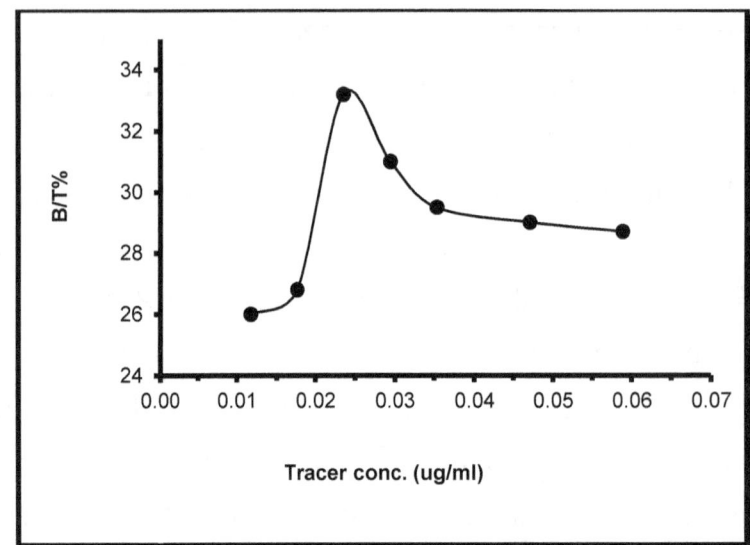

Figure (48): Influence of increasing ^{125}I-Anti hCG Antibody concentration on the binding with hCG in the homogenate. (All other details are explained in the text.

The Effect of Different pH on the Binding of hCG in Breast Tumor Homogenate with ^{125}I-anti hCG Antibody

One of the factors that affect complex precipitation is pH. To determine the optimal pH for hCG activity, the binding of fixed concentration of ^{125}I-anti hCG antibody was performed with fixed amount of tumors homogenate at different pH.

The binding value is stated in figure (49), showing that the three groups reached maximum binding at pH 7.2 with decrease in binding percent at pH higher or lower than optimum pH.

The attraction between antigen and antibody molecules involves electrostatic attraction, hydrogen bonding, vander waals forces and hydrophobic interactions.

In previous studies, a short exposure of hCG solution to pH less than pH 5.0 caused a slight reduction in the biologic activity, indicating that neuraminic acid was not released.

In solution, optimum pH is the most important contributors to nonequivalent attraction between antigen and antibody making it one of the factors that influence reaction speeds.

The optimum pH in this experiment may charged polar groups on the amino acids residues of hCG and/or ^{125}I-anti hCG Antibody, making them strongly attracted to each other.

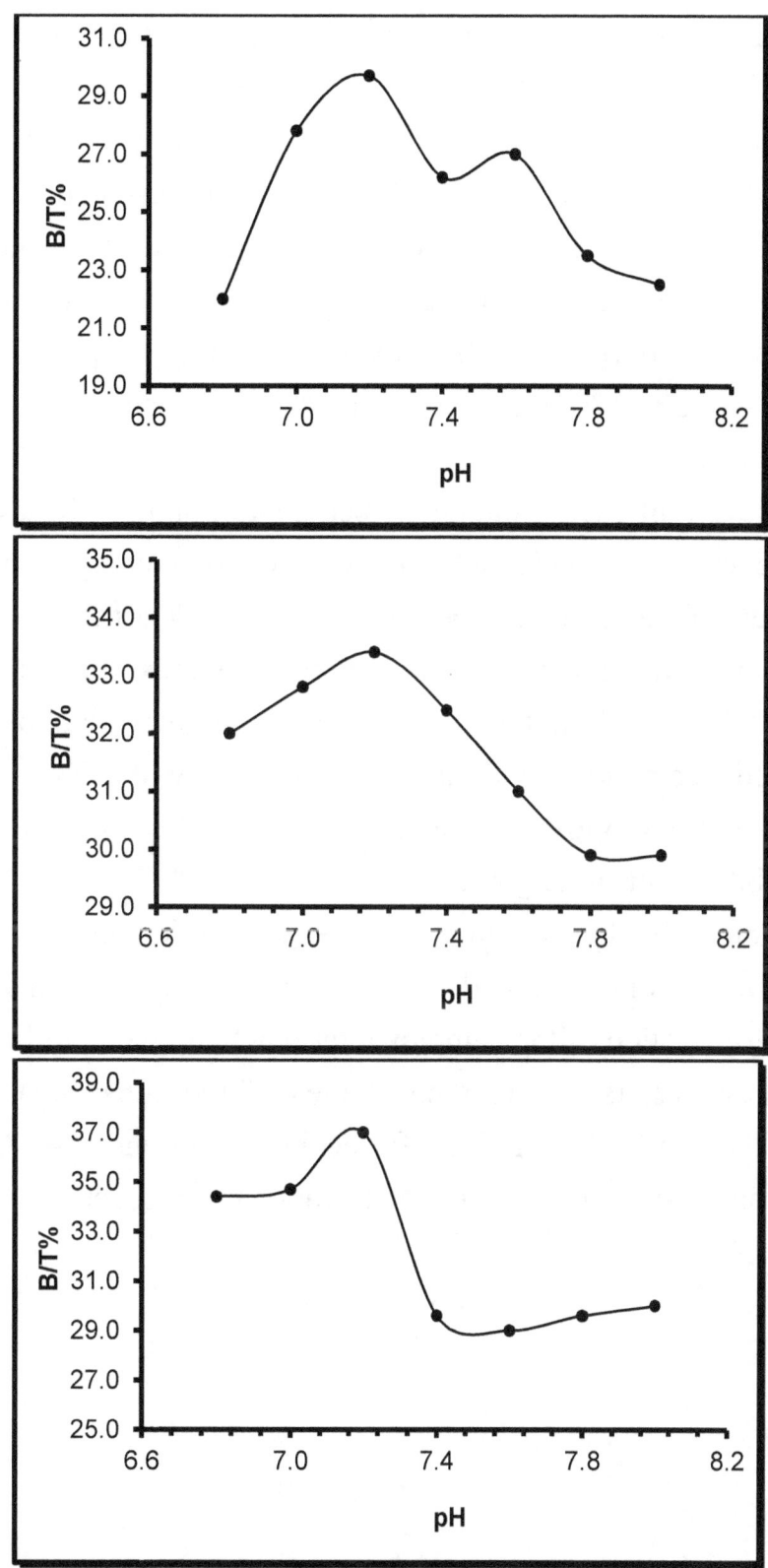

Figure (49): Influence of pH on the binding of ^{125}I-anti hCG Antibody with hCG in the homogenate. (All other details are explained in the text).

Time Course of the Binding of hCG in Breast Tumor Homogenate with ^{125}I-anti hCG Antibody

To determine whether the rate of hCG binding to ^{125}I-anti hCG antibody is temperature and time dependence, the bindings were cared out at different incubation time in four temperatures (5,25,37,45) °C; as shown in figure (50).

For benign tumors group, the binding shows inverse relation between elevation of temperature and percent of maximum binding, which was greatest (35.9%) at 5°C in 120 min. Malignant tumor groups exhibited variant behavior, the relation between elevation of temperature and present of binding was in conformance. The maximum binding for premenopausal was (46.2 %) at 45 ° C in 120min and was (41.7%) for postmenopausal at 45°C in 160min of incubation.

The behavior of hCG binding in benign tumor may be explain that it needs less energy than malignant tumor to over come energy barrier and give the maximum binding or may be because of the degradation of hCG form benign tumor with increasing temperature. Previous study has shown an agreement with this hypothesis by declare that elevation of temperature markedly reduced biologic and immunologic activity of hCG.

For the three groups the binding at 25°C do not follow the order mentioned before for these groups.

Incubation of binding mixture for time periods longer than that required for maximum binding resulted in decrease binding; this may be due to reversible dissociation of the complex after reaching the equilibrium state.

From these results, the binding studies of the subsequent experiments were carried out at 5 °C for 150min for benign tumor and at 45°C for 120 and 150 min of incubation for pre- and postmenopausal respectively.

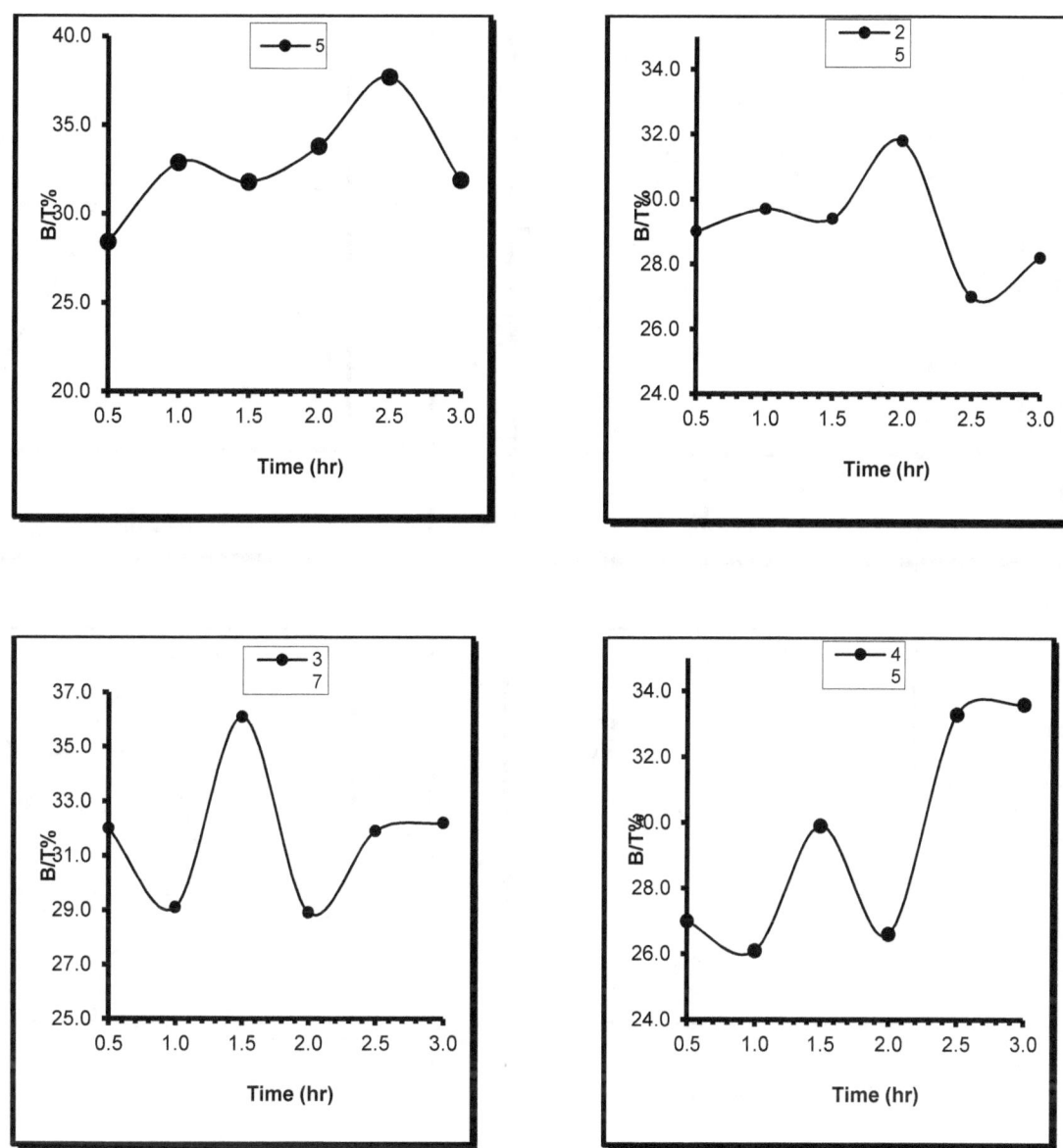

Figure (50): Time course of the binding of ^{125}I-anti hCG Antibody with hCG in benign breast tumors. (All other details are explained in the text).

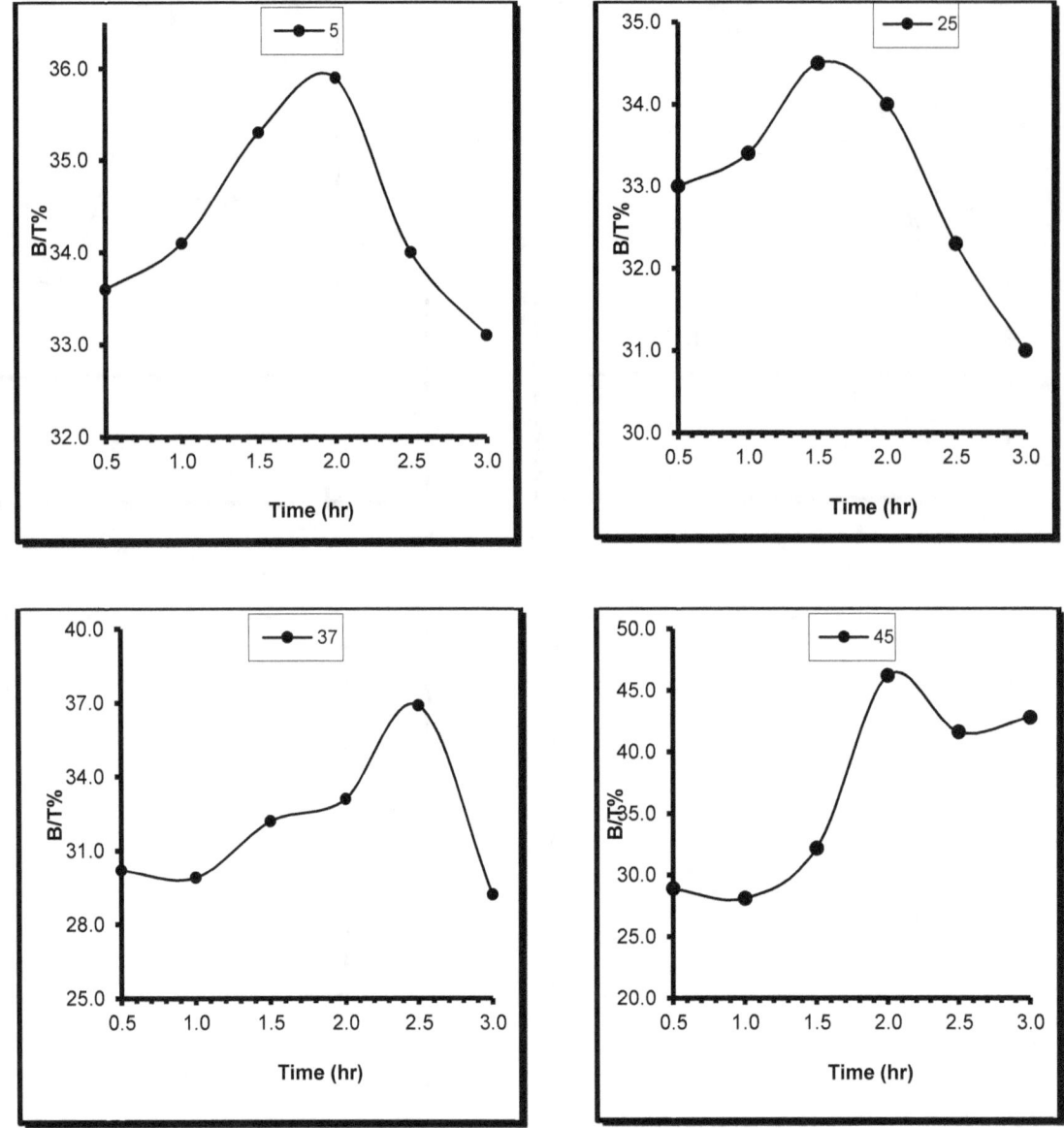

Figure (51): Time course of the binding of ^{125}I-anti hCG Antibody with hCG in malignant premenopausal breast tumors. (All other details are explained in the text).

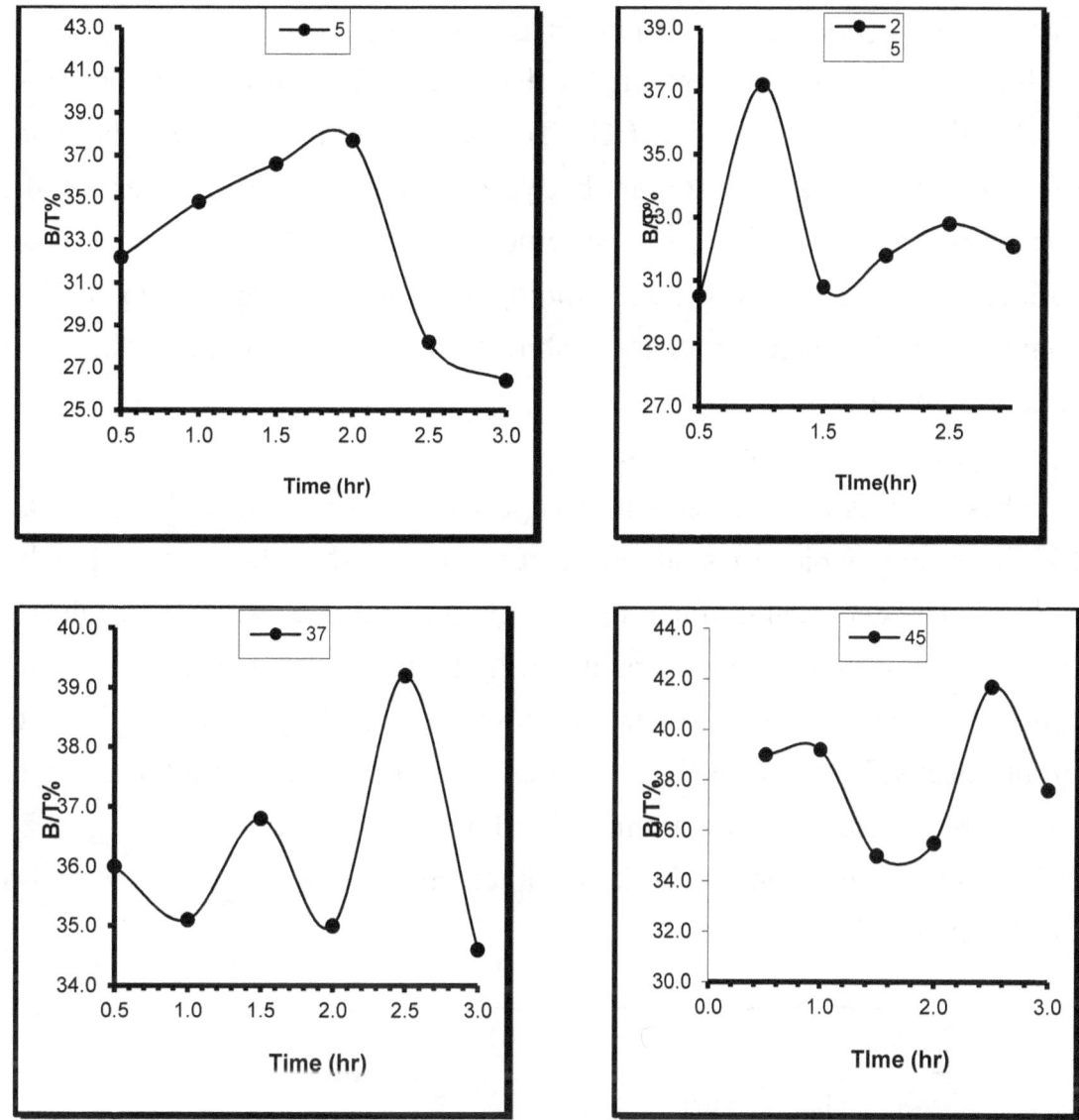

Figure (52): Time course of the binding of ^{125}I-anti hCG Antibody with hCG in malignant postmenopausal breast tumors. (All other details are explained in the text).

The Effect of Different Halides on the Binding of hCG in Breast Tumor Homogenate with ^{125}I-anti hCG Antibody

Ionic species and ionic strength affect the binding of the antigen(^{125}I-anti hCG antibody) and antibody(hCG). To investigate this effect on the binding of hCG and ^{125}I-anti hCG antibody, different sodium halides at 0.1M concentration were added to the binding mixture. It seems that sodium halides decrease the maximum binding of (hCG/^{125}I-anti hCG antibody) complex in the three groups, as shown in figure (53), according to the following order:

NaI > NaBr > NaCl > NaF

These results are in greement with pervious study made by Lanja.E.O. on LH hormone, which has identical α subunit and 80% identical β subunit of hCG. The decreasing of maximum binding could be as a result of decrease of ionic radius and increasing radius of hydration for anionic salts, leading to greater interaction of the salt, having lower degree of hydration, with an ionic group located in the antibody or antigen combining site. From these results it seems that decreasing of maximum binding could be due to the large size of iodine ion, which could inhibit the interaction between hCG and ^{125}I-anti hCG antibody.

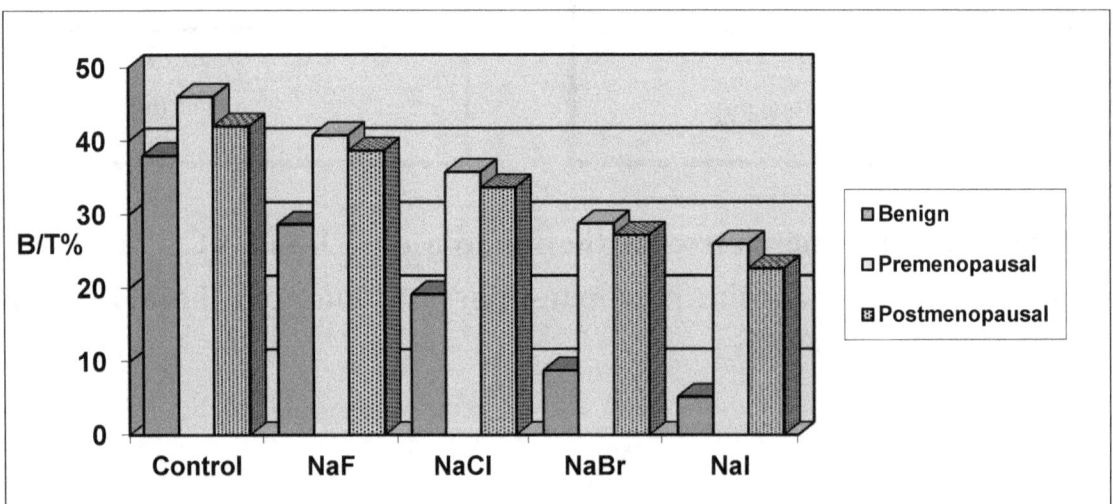

Figure (53): The Effect of Different Halides on the Binding of hCG in Breast Tumor Homogenate with ^{125}I-anti hCG Antibody.

Determination of Affinity Constant (K_a) and the Maximum Binding Capacity (B_{max}) of hCG in Pre-menopausal Malignant Breast Tumor Homogenate Associated with ^{125}I-anti hCG Antibody

The simplest proposed model representing this interaction is:

$$^{125}\text{I-anti hCG Antibody} + \text{hCG} \underset{K_{-1}}{\overset{K_{+1}}{\rightleftharpoons}} [^{125}\text{I-anti hCG Antibody/hCG}] \quad \ldots\ldots\ldots(1)$$

Where:

K_{+1}: is the association rate of ^{125}I-anti hCG to hCG.

K_{-1}: is the dissociation rate of (^{125}I-anti hCG/hCG) complex formed.

At equilibrium:

$$K_a = \frac{[^{125}\text{I-anti hCG Antibody/hCG}]}{[^{125}\text{I-anti hCG Antibody}][\text{hCG}]} \quad \ldots\ldots\ldots(2)$$

$$K_d = \frac{[^{125}\text{I-anti hCG Antibody}][\text{hCG}]}{[^{125}\text{I-anti hCG Antibody/hCG}]} \quad \ldots\ldots\ldots(3)$$

Thus

$$K_a = \frac{1}{K_d} = \frac{K_{+1}}{K_{-1}} \quad \ldots\ldots\ldots(4)$$

Where:

K_a: is the equilibrium constant of the association (affinity constant).

K_d: is the equilibrium constant of the dissociation ^{125}I-anti hCG antibody/hCG) complex.

In this experiment, scatchard plot analysis was used to measured the concentration of hCG in pre-menopausal malignant breast tumor homogenate (B_{max}) and the affinity constant (K_a) of the binding with ^{125}I-anti hCG antibody, as in figure (54), by using the optimal conditions, which were obtained in previous experiments.

Results in table (34) show that affinity constant (Ka) is temperature depended. It increased with increased temperature from (8.8809 µg^{-1}.ml) at 5°C to (13.417 µg^{-1}.ml) at 45 °C.

Whereas the values of dissociation constant (K_a), which calculated by using equation (4), shows the lowest K_d value at 45 °C with the following order 25<5<37<45 °C.

The straight line which obtained from scatchard plot analysis, as shown in figure (54), indicate the presence of only one species of hCG site, or more but with the same affinity and number of binding site.

Table (34): the kinetic parameter of ^{125}I-anti hCG antibody binding to hCG in pre-menopausal malignant breast tumor homogenate. (All other details are explained in the text).

Temp °C	Binding Capacity $B_{max}(\mu g.ml^{-1})$	K_a ($\mu g^{-1}.ml$)	K_d ($\mu g.ml^{-1}$)
5	0.1405	8.8809	0.1125
25	0.1474	7.6865	0.1300
37	0.1481	10.613	0.0942
45	0.1905	13.417	0.0745

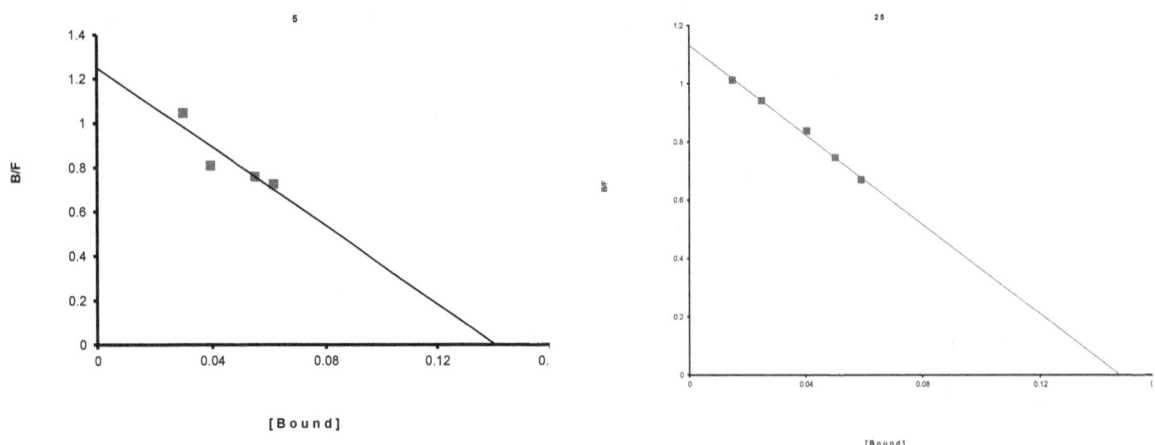

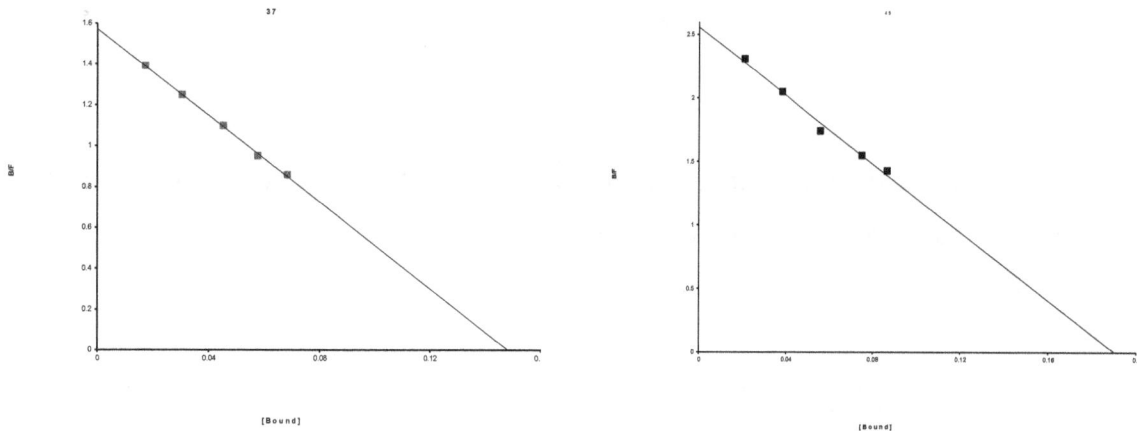

Figure (54): Scatchard plot of ^{125}I-anti hCG antibody binding to the partially isolated hCG in pre-menopausal malignant Breast Tumor Homogenate. (All other details are explained in the text).

A quantitative radioreceptor assay for detection and analysis of membrane-associated hCG receptor

Introduction

A quantitative radioreceptor assay for detection and analysis of membrane-associated hCG receptor of human breast tissues was developed by binding highly specific and biological active radiolabed hormone (^{125}I-hCG) to particulate membrane preparations. The optimum conditions for ^{125}I-hCG-receptor binding were obtained as follows: the optimum protein amount was (2.5, 10 and 10 µg) for benign, pre- and post-menopausal malignant breast tumor homogenate respectively. ^{125}I-hCG concentration was (18×10^{-4} ng.ml^{-1}) for benign, while it was (16×10^{-4} ng.ml^{-1}) for both pre- and post-menopausal malignant breast tumor homogenate. The optimum pH was 7.0 for benign group while it was 7.6 for the two malignant groups. It was found that the binding was time and temperature dependence that the optimum binding temperature and time for benign group were at 45°C in 60 minutes. For pre-menopausal malignant group it was at 37 °C for 30 min while it was at 5 °C and 60 min for post-menopausal malignant group. The use of different halides result in the decrease of maximum binding of benign and malignant groups. The concentration of hCG-receptor (B_{max}) and affinity constant (K_a) in benign and pre-menopausal malignant groups were measured by Scatchard analysis indicating that they are higher in benign group than malignant group.

hCG and LH are structurally and functionally similar hormones. They bind to a common receptor. The receptor is a single chain transmembrane glycoprotein that belongs to the family of G-protein-coupled receptors. It is consists of a large extracellular ligand-binding domain, seven transmembrane-spanning regions and a short cytoplasmic tail.

For many years, the knowledge about the localization and function of hCG receptors were constrained to only the gonadal tissues. Neither their localization nor function in nongonadal tissues of human body was suspected until about 18 years ago. The first reports on nongonadal hCG receptors initiated further studies to characterize them in detail.

The nongonadal receptors were investigated by using a wide range of traditional ligand binding as well as ligand blotting and covalent receptor cross-linking, that detect the receptor from gene transcription to signaling pathways and ultimately the biological response.

The hCG receptor mRNA and protein were found and characterized in the human myometrium, placenta, fallopian tube, breast, brain, as well as other tissues. Nongonadal distribution of hCG receptors are not species specific, as they have been found in human, monkey, rat, rabbit, mice, pig, cow, sheep, and even turkey.

The presence of these nongonadal receptors raised the possibility of their functions and many subsequent studies were directed to address this possibility. These studies revealed the presence of these receptors not only in tumor tissue but also found association between receptor density and oncologic process. The association between structural change in the receptor and the defective regulation was also found.

In the breast cancer cell, hCG anticancer actions were found to be mediated by hCG receptor. So the present study was undertaken as a complementary study to develop a radioreceptor assay for hCG receptor determination in benign and malignant breast tumors. These determinations were undertaken to characterize the optimum binding condition between hCG and its receptor.

Materials and Methods

Binding Studies of hCG Receptor in Breast Cancer Tumor Homogenate with ^{125}I- hCG

Preliminary Tests of hCG Receptor Binding in Breast Tumor Homogenate with ^{125}I-hCG

1. One hundred microliters containing (30 µg of protein for benign, malignant pre-and post-menopausal breast tumors respectively) was added to 40 µl of ^{125}I- hCG (10ng.ml^{-1}), the volume of the mixture was made up to 250 µl with Tris/HCl buffer (0.05M, pH 7.2) containing 0.1% BSA.

2. One additional tube containing only 40 µl of ^{125}I-hCG were set aside until counting for total activity.

3. The tubes were incubated at 25°C for 60 min.

4. After incubation, the tubes were centrifuged at 1500xg for 30 min at 4°C.

5. The supernatant was discarded by decanting the assay tubes, then the tubes were inverted on a filter paper for 10 min.

6. The rims of the tubes were swabbed with a cotton piece and the amounts of bound radioactivity (c.p.m) were counted in a gamma counter; this (c.p.m) is refer to the total binding.

7. Non specific binding was accounted for by preparing the same incubation with addition of 250 fold excess of unlabeled hCG .

Calculations

1. The counted of radioactivity in each tube (expressed in c.p.m) represents the total binding fraction (TB), i.e (^{125}I-hCG /hCG receptor) complex.

2. The counted of radioactivity in each tube containing ^{125}I- hCG and excess of unlabeled hCG represents the nonspecific binding (NSB).

3. The specific binding (SB), expressed in c.p.m, was calculated by subtracting the radioactivity, expressed in c.p.m, obtained in the

presence of unlabeled hCG from that produced in the absence of unlabeled hCG.

$$SB(c.p.m) = TB(c.p.m) - NSB(c.p.m)$$

4. The precent of specific binding (SB%) can be calculated from the following formula:

$$SB\% = \frac{SB(c.p.m)}{T(c.p.m)} \times 100$$

Where: T is the total count of the ^{125}I- hCG, expressed in c.p.m,

Most Appropriate Conditions of the Binding of hCG Receptors in Breast Tumor Homogenate with ^{125}I- hCG.

- *The Effect of Different Protein Amount of Breast Tumor Homogenate on the Binding of hCG Receptors with ^{125}I-hCG*

1. Forty microliters of ^{125}I-hCG (10ng.ml^{-1}) to 100 µl containing increasing protein amounts (1.5, 2, 2.5, 5, 10, 15, 20, 25, 30 µg of benign and 2.5, 5, 10, 15, 20, 25, 30, 35 µg.ml^{-1} of pre- and post- menopausal malignant breast tumors respectively). The volume of the mixture made up to 250 µl with Tris/HCl buffer (0.05M, pH 7.2) containing 0.1% BSA.

2. One additional tube containing only 40 µl of ^{125}I-hCG was set aside until counting for total activity computation.

3. The tubes were incubated at 25°C for 60 min.

4. After incubation, the (^{125}I-hCG /hCG receptor) complex was estimated.

Calculations

1. The percent of specific binding (SB%) was calculated.

2. The percent of specific binding value (SB%) were plotted against the increasing protein amounts.

- *The Effect of Different Concentrations of ^{125}I-hCG on the Binding with hCG Receptors in Breast Tumor Homogenate*

1. One hundred microliters containing (2.5, 10, 10 µg proteins of benign, malignant pre- and post-menopausal breast tumors homogenate respectively) were added to increasing volumes (10, 20, 25, 30, 35, 40,

45, 50 and 55 µl) containing (4×10^{-4} - 22×10^{-4} ng.ml^{-1}) of ^{125}I-hCG (10ng.ml^{-1}) then the volume were made up to 250 µl with Tris/HCl buffer (0.05M, pH 7.2) containing 0.1% BSA.

2. A set of tubes containing only the same increasing volumes of ^{125}I-hCG (10, 20, 25, 30, 35, 40, 45, 50 and 55 µl), were set a side until counting for total activity computation.

3. The tubes were incubated at 25°C for 60 min.

4. After incubation, the (^{125}I-hCG /hCG receptor) complex was estimated.

Calculations

1. The percent of specific binding (SB%) was calculated according to the formula. The percent of specific binding values (SB%) were plotted against the concentration of ^{125}I-hCG.

- ***The Effect of Different pH on the Binding of hCG Receptors in Breast Tumor Homogenate with ^{125}I-hCG***

1. One hundred microliters containing (2.5, 10, 10 µg proteins of benign, malignant pre- and post-menopausal breast tumors homogenate respectively) were added to 45µl(18×10^{-1} ng.ml^{-1}), 40µl(16×10^{-1} ng.ml^{-1}) and 40µl(16×10^{-1} ng.ml^{-1}) for benign and malignant pre-, post-menopausal breast tumors homogenate respectively) of ^{125}I-hCG (10ng.ml^{-1}), the mixtures volumes were made up to 250 µl with Tris/HCl buffer (0.05M) containing 0.1% BSA of different pH (6.8, 7.0, 7.2, 7.4, 7.6, 7.8, 8.0).

2. Two additional tubes containing only (40 and 45) µl of the ^{125}I-hCG were set aside until counting for total activity computation.

3. The tubes were incubated at 25°C for 60 min.

4. After incubation, the (^{125}I-hCG /hCG receptor) complex was estimated.

Calculations

1. The percent of specific binding (SB%) was calculated according to the formula. The percent of specific binding values (SB%) were plotted against their corresponding pH values.

- ***Time Course of the Binding of hCG Receptors in Breast Tumor Homogenate with ^{125}I-hCG***

1. One hundred microliters containing (2.5, 10, 10 μg proteins of benign, malignant pre- and post-menopausal breast tumors homogenate respectively) were added to 45μl(18×10^{-1} ng.ml^{-1}), 40μl(16×10^{-1} ng.ml^{-1}) and 40μl(16×10^{-1} ng.ml^{-1}) for benign and malignant pre-, post-menopausal breast tumors homogenate respectively) of ^{125}I-hCG (10ng.ml^{-1}), the mixtures volumes were made up to 250 μl with Tris/HCl buffer (0.05M, pH 7.0 and 7.6) containing 0.1% BSA for benign and malignant tumor respectively.

2. All tubes were incubated at 25°C at different time interval (30, 60, 90, 120, 150 and 180) min.

3. Two additional tubes containing only (40 and 45) μl of the ^{125}I-hCG were set aside until counting for total activity computation.

4. After incubation, the (^{125}I-hCG /hCG receptor) complex was estimated by following the steps 4, 5, 6 and 7 in section (3.3.1).

5. To determine the time course of HCG receptors binding to ^{125}I–hCG at different temperatures. Steps 1, 2 and 3 in the same experiment were repeated at different temperatures (5, 37, 45 °C).

Calculations

1. The percent of specific binding (SB%) was calculated according to the formula. The percent of specific binding values (SB%) were plotted against the incubation time.

- ***The Effect of Different Halides on the Binding of hCG Receptors in Breast Tumor Homogenate with ^{125}I-hCG***

1. One hundred microliters containing (2.5, 10, 10 μg proteins of benign, malignant pre- and post-menopausal breast tumors homogenate respectively) were added to 45μl(18×10^{-1} ng.ml^{-1}), 40μl(16×10^{-1} ng.ml^{-1}) and 40μl(16×10^{-1} ng.ml^{-1}) for benign and malignant pre-, post-menopausal breast tumors homogenate respectively) of ^{125}I-hCG (10ng.ml^{-1}), the mixtures volumes were made up to 250 μl with Tris/HCl buffer (0.05M, pH 7.2) containing 0.1% BSA containing 0.1M of each of the following halides: NaF, NaCl, NaBr and NaI). A sample without the addition of any halides was used as a control.

2. Two additional tubes containing only (40 and 45) μl of ^{125}I-hCG were set aside until counting for total activity computation.

3. The tubes were incubated for 60min at 45°C (benign breast tumor homogenate), 30min at 37°C and 60min at 5°C (pre-and post-menopausal malignant breast tumor homogenate).

4. After incubation, the (^{125}I-hCG /hCG receptor) complex was estimated

Calculations

1. The percent of specific binding (SB%) was calculated according to the formulaThe percent of specific binding values (SB%) were plotted against halides concentrations.

Determination of the Concentration of hCG-Receptor and the Affinity concentration of ^{125}I-hCG Association with its Receptors in Benign and Malignant Breast Tumor.

1. One hundred microliters containing (2.5, 10, 10 μg proteins of benign, malignant pre- and post-menopausal breast tumors homogenate respectively) were incubated with increasing volume 25, 30, 35, 40, 45μl (1-1.8ng) and 20, 25, 30, 35, 40μl (0.8-1.6ng) of ^{125}I-hCG (10ng.ml^{-1}) for benign and pre-malignant breast tumors homogenate respectively, the mixtures volumes were made up to 250 μl with Tris/HCl buffer (0.05M, pH 7.0 and 7.6) for benign and malignant tumor respectively.

2. A set of tubes containing only (20, 25, 30, 35, 40, 45μl) of ^{125}I-hCG were set aside until counting for total activity computation.

3. The tubes were incubated for 60min at 45°C for benign breast tumor homogenate and 30min at 37°C for pre- menopausal malignant breast tumor homogenate.

Calculations

The values of ^{125}I-hCG which is bound specifically were calculated by using the following formula:

$$B = \frac{\text{Total binding - Non specific binding}}{\text{Total count}} \times \text{Concentration of } ^{125}\text{I-HCG in each assay tube}$$

4. The concentration of receptors and affinity constant were determined according to Scatchard equation: -

$$\frac{B}{F} = \frac{1}{K_d}(B_{max} - B)$$

$$K_a = \frac{1}{K_d}$$

Where:

K_a = Affinity constant

K_d = Dissociation constant

B_{max} = Maximal binding capacity

4. The values of the ratio B/F were plotted against the values of B in (ng /ml), gives a linear relationship. The values of the affinity constant of the binding (K_a) can be calculated from the slop of the straight line, while the value of the total concentration of hCG-receptors (B_{max}) was calculated from the intercept with the x-axis.

Results and Discussion

Membrane preparation

As described the experiment of Preparation of breast tumor tissue homogenate in chapter two, the method used for preparation the crude homogenate for hCG binding studies was the same for membrane preparation for obtaining hCG receptor. In previous studies the extracted hCG receptor that used for binding studies could be obtained as either particulate membrane by homogenate centrifugation at different speeds or soluble receptors extracted from the membrane by using a nonionic detergent.

Binding Studies of hCG Receptor in Breast Cancer Tumor Homogenate with ^{125}I-hCG

Preliminary Tests of hCG Receptor Binding in Breast Tumor Homogenate with ^{125}I-hCG

Benign and malignant breast tumor were investigated by using breast tumor homogenate as a source of hCG receptors. These receptors were detected through the incubation of crude homogenate with ^{125}I-hCG with and without added of non-radioactive hCG in order to demonstrate whether the specific binding was proportional to the concentration of hCG receptor. The incubation was carried out at 25°C for 60 minute, as a beginning, according to the condition that mention in the RIA kit; then separate the formed complex at 1500xg for 30 minute. Although many previous studies used PEG 6000 for precipitate (^{125}I-hCG /hCG receptor) complex, but in this experiment the use of this precipitating reagent was neglected due to its precipitation of the free ^{125}I-hCG. The specific binding was found to be (3%) in benign tumor, (2.2%) and (3.3%) for malignant pre- and post-menopausal breast tumor respectively. These data revealed the presence of hCG receptor in human breast tumors. The low binding value probably due to the relatively large size of the complex that restricted the separation procedure.

Most Appropriate Conditions of the Binding of hCG Receptors in Breast Tumor Homogenate with ^{125}I-hCG.

The Effect of Different Protein Amount of Breast Tumor Homogenate on the Binding of hCG Receptors with ^{125}I-hCG

Increasing amount of breast tumor homogenate were incubated with fixed amount of ^{125}I-hCG with and without added of non-radioactive hCG. The specific binding percent was increased with the increasing concentration of the homogenate in the incubation mixture, as shown in figure (55). This nonlinearly may be due to the heterogeneously between the ligand and receptor preparation. In this experiment benign tumor required lower amount (2.5μg) than other groups for reaching the maximum binding (4.6%), while malignant pre- and post- menopausal groups required (10 μg) to reach there maximum binding (4.1%, 3.8%) respectively. Although many previous studies preformed that malignant breast cell lines contacting significantly higher level and higher relative transcription rate of hCG receptors genes than normal breast cell line; but from the initial researches until now, there was no sign to the presenting values of hCG receptor binding with ^{25}I-hCG in each kind. In other hand, other previous studies estimated that hCG level are higher in malignant breast tumor (infiltrated ductal carcinoma) than benign breast tumor. So, the obtunded results could be explained that even the malignant breast tumors have higher hCG receptors, it also have higher hCG level. This may lead to a conclusion that the number of hCG receptors that occupied by hCG in malignant tumor is higher than benign tumor in which higher free receptors are available for binding with ^{25}I-hCG. According to these results the optimum proteins amount in this experiment were used in all subsequent experiments.

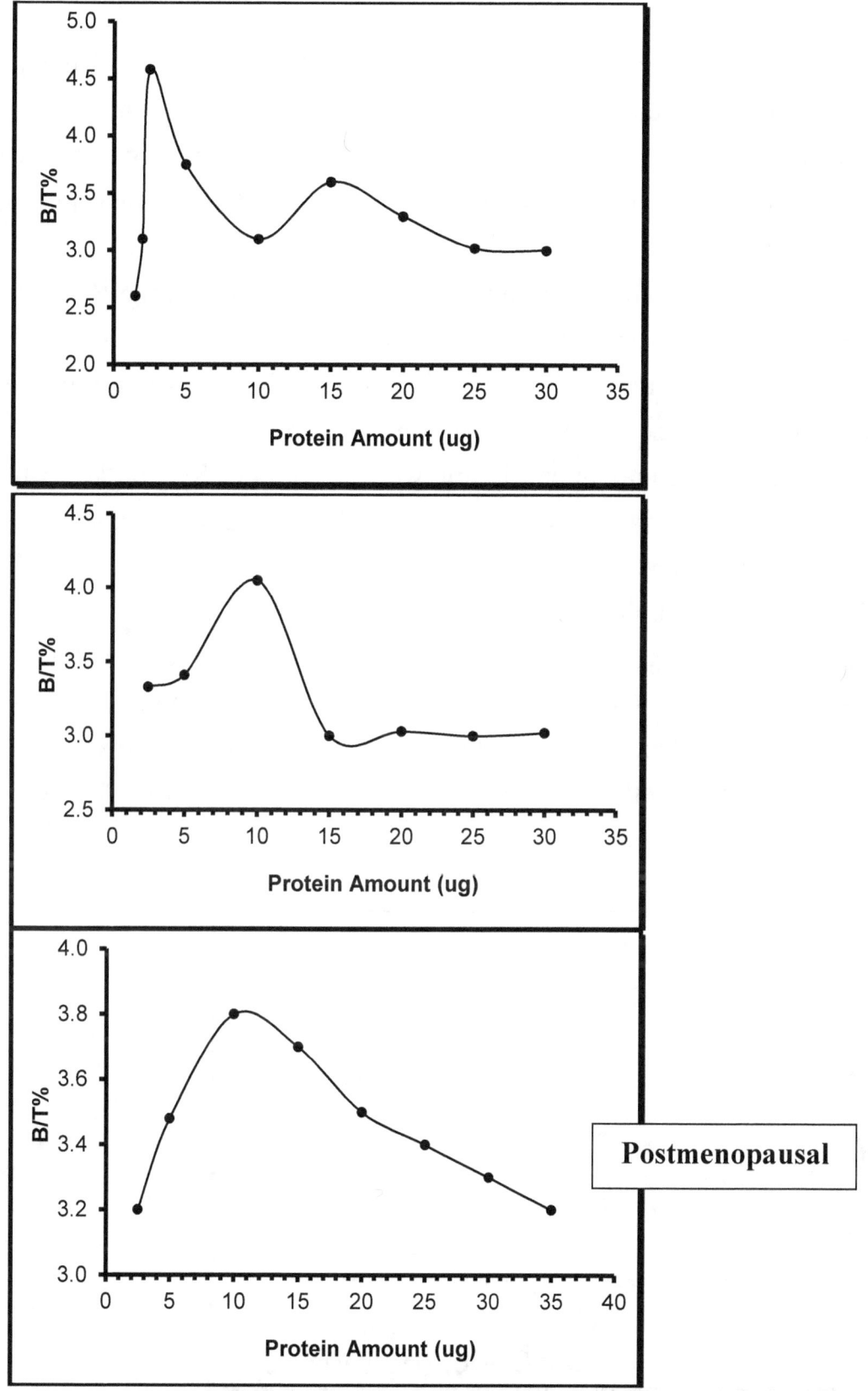

Figure (55): Influence of protein amount on the binding with ^{125}I-hCG. (All other details are explained in the text).

The Effect of Different Concentrations of ^{125}I-hCG on the Binding with hCG Receptors in Breast Tumor Homogenate

To estimate another factor affects the (^{125}I-hCG /hCG receptor) complex formation, fixed concentration of benign (2.5 µg), malignant pre- (10 µg) and post-menopausal (10 µg) breast tumor homogenate were incubated with increasing concentration of ^{125}I-hCG for 60min at 25°C, as shown in figure (56).

In this figure, binding increased with the added amount of the ^{125}I-hCG in the three groups. This is probably due to the cross-linking of hCG and ^{125}I-hCG is more likely with the increasing ^{125}I-hCG in the incubation mixture to perform large complex until reached its maximum binding.

For benign tumor maximum binding was (5.5%) at (1.8 ng.ml^{-1}) ^{125}I-hCG antibody concentration, (4%) and (3.7%) for malignant pre-and post-menopausal breast tumors receptively at (1.6 ng.ml^{-1}).

After the maximum binding the increasing of ^{125}I-hCG concentration lead to decrease in binding percent, probably because all the hCG receptors sites are covered with ^{125}I-hCG and inhibiting the formation of the complex. This indicates the dependence of the binding on ^{125}I-hCG in binding mixture.

According to these results the above concentration of ^{125}I-hCG was used in the subsequent experiments.

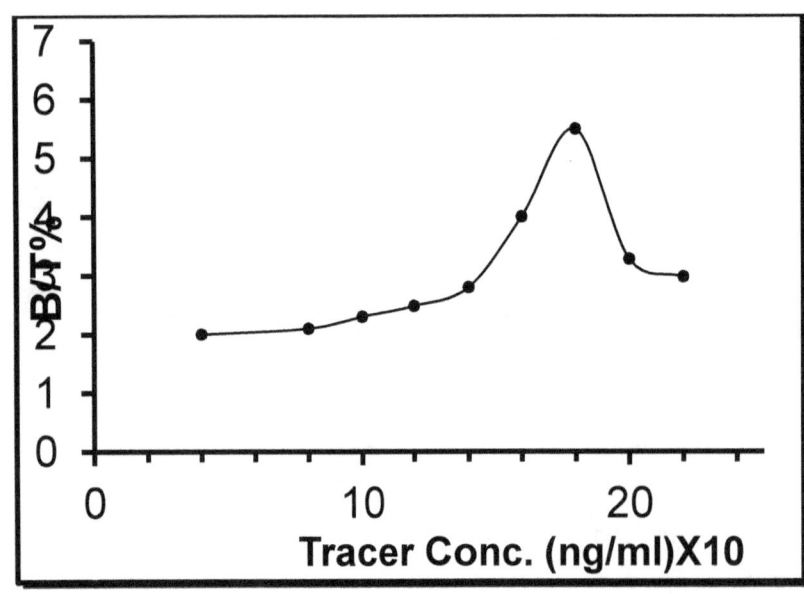

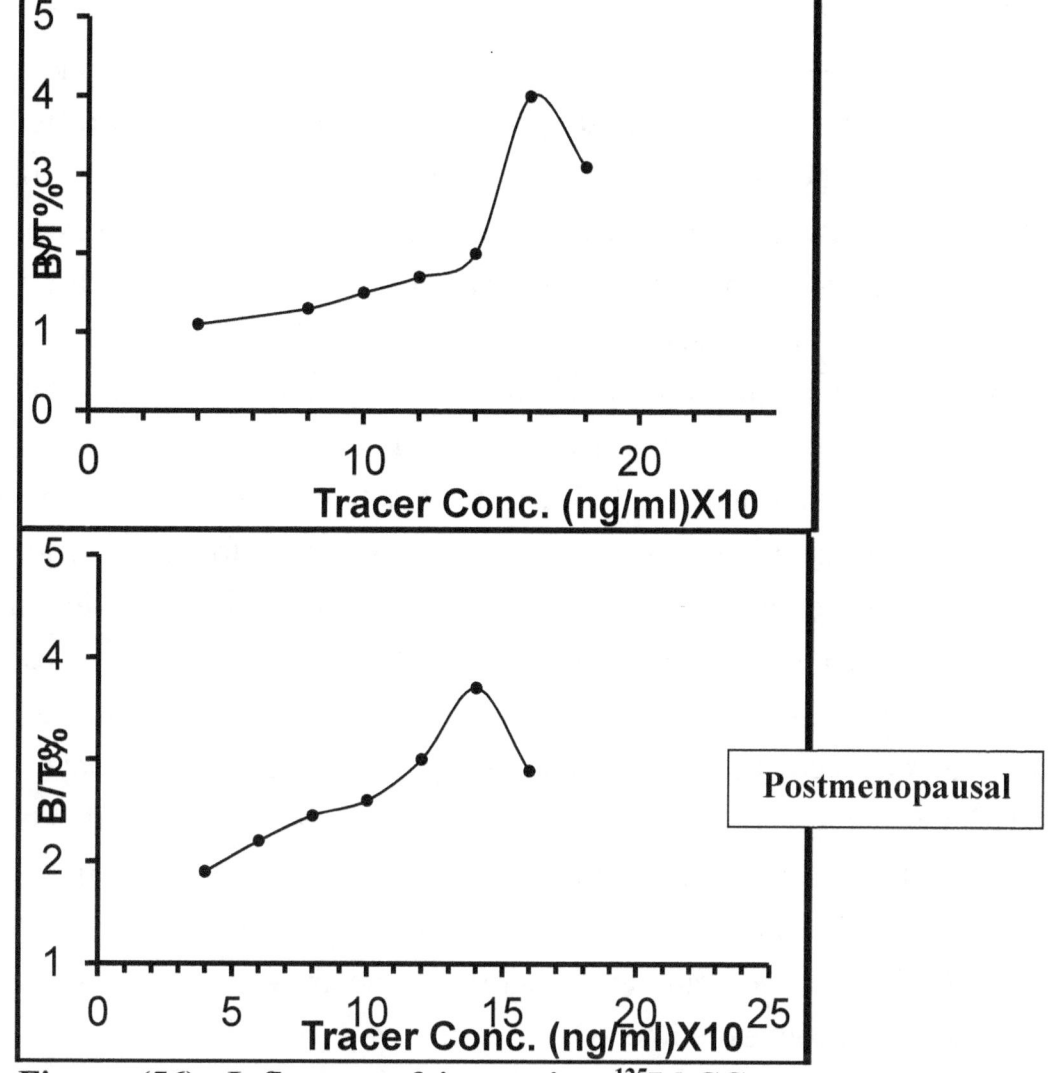

Figure (56): Influence of increasing ^{125}I-hCG concentration on the binding with hCG receptors.

The Effect of Different pH on the Binding of hCG Receptors with ^{125}I-hCG in Breast Tumor Homogenate.

To determine the optimal pH for receptor activity, the binding was preformed at different pH values, as shown in figure (57). In this figure, benign tumor showed two sharp peaks obtained at pH 7.0 (6.7%) and pH 7.4 (6.0%); while the malignant breast tumors showed two sharp peaks at pH 7.2 (pre- 4.1%, post- 3.7%) and pH 7.6 (pre- 5.0%, post- 4.6%). From these results, the bindings were found to be pH dependent and the shift in the pH of the environment may affect the properties of the macromolecules involved in the binding. It was not determined whether the decreased binding at suboptimum pH values resulted from a decrease in binding affinity, binding rate or inactivation of hCG receptor. The presence of two peaks for each group may be refers to the presence of more than one kind of hCG receptors which have different property. Previous studies support this hypothesis by revealing that human breast cell lines contain multiple hCG receptors transcripts and three proteins of different molecular sizes which can bind ^{125}I-hCG. Whether the multiple transcripts were the products of the different transcription initiation sites or from alternate splicing of a single transcript or due to the differences in polyadenylation is not known. Also unknowns is whether the receptor proteins were translational products of different transcripts or are the result of proteolytic degradation despite careful handling and the presence of protease inhibitors in the buffers. The presence of multiple receptor transcripts and more than one receptor protein have previously been found in other hCG receptor-positive tissues. Other previous study revealed disparate results of pH effect on hCG receptors from different sources. In bovine corpus luteal plasma membrane preparation, the maximum binding was at pH 7.2-7.4 and declined both at high and low pH values. In rat testis the binding exhibited a broad pH optimum between pH 6.8 and 7.3, while in another study the equilibrium binding of rate testis homogenate with ^{125}I-hCG showed a relatively sharp pH optimum at 7.4[238]. Rat ovarian hCG receptor binding have been carried out almost exclusively at 7.4[240]. In uterus hCG receptor, the effect of pH on the binding exhibited optimum pH between 7.0-7.4[42]. According to these results, the above optimum pH were used in the subsequent experiments.

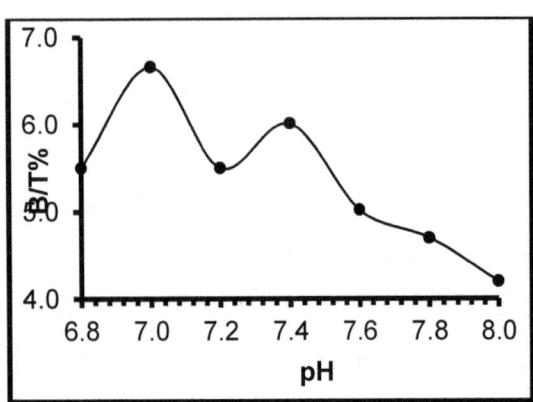

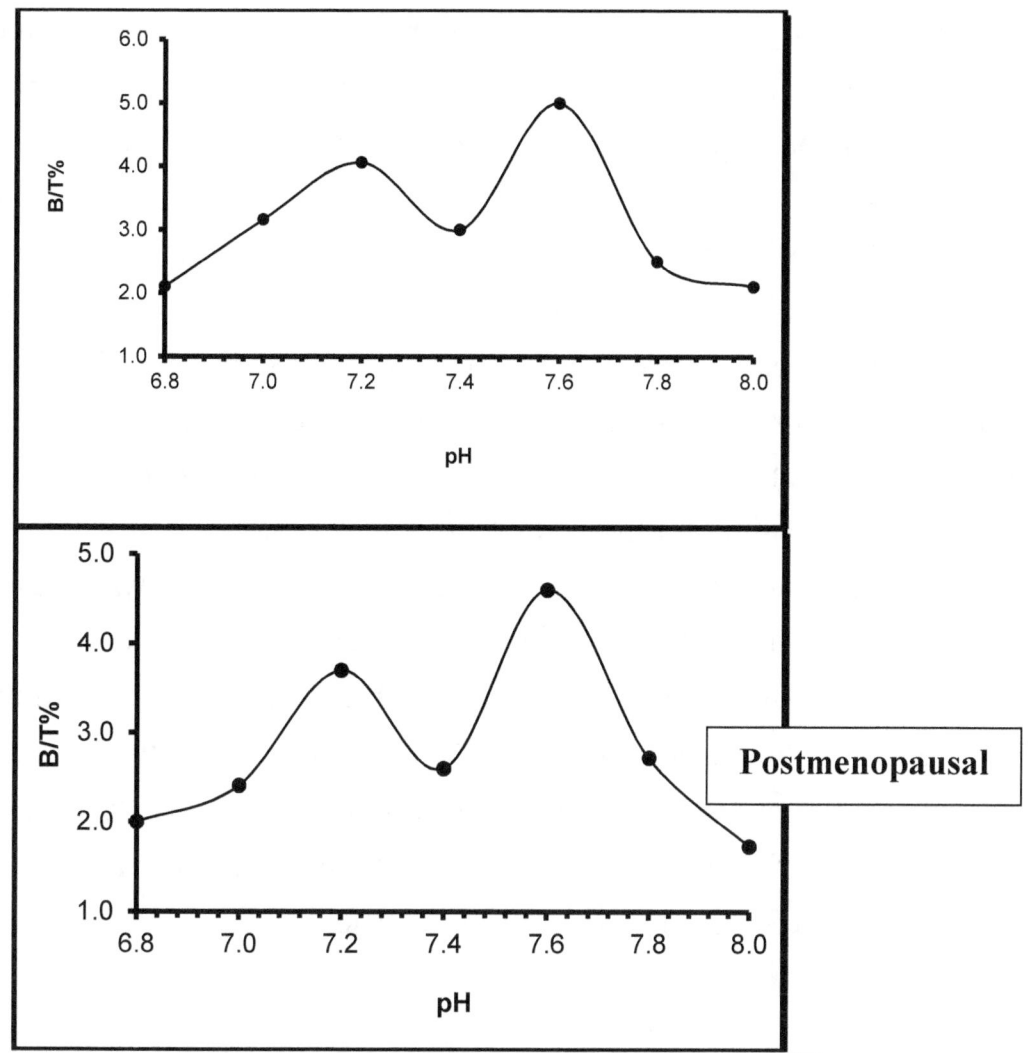

Figure (57): Influence of pH on the binding of ^{125}I-hCG with hCG receptors. (All other details are explained in the text).

Time course of The Binding

Time course pattern for ^{125}I-hCG binding with hCG receptor in breast tumors homogenate at different temperatures (5, 25, 37, 45) was found to be markedly time and temperature dependent, as stated at figures (58), (59), (60). The maximum binding was occurred in less temperature when binding move from benign to postmenopausal group. The maximum binding (8.1%) in benign tumor was obtained at 45°C after incubation for 60 minute. For malignant premenopausal tumor it was (6.2%) at 30 min., whereas it was (5.8%) at 60 minute for postmenopausal breast tumor. This could be due to the presence of different active forms of the receptor that each form is active in different kind of tumors. In the three groups, incubation of ^{125}I- hCG with hCG receptor in breast tumor homogenate for time periods longer than required for maximum binding result in decreasing the bonding. This could be due to partial inactivation of hCG receptors or partial degradation of ^{125}I-hCG occurring during the binding assay. For benign group the maximum binding was obtained at 45°C and this could be that hCG receptor need more energy than malignant tumor to over come energy barrier and give the maximum binding. For malignant premenopausal group, the loss of binding was more rapid at 45 °C than 37°C and this could partially explain the lower maximum values obtained for binding at 45°C compared with 37°C. While in malignant postmenopausal group, greater binding occurred at the lower temperature 5°C, this may be due to the more rapid degradation of hCG receptors during incubation at higher temperatures previous studies showed variable results depending on the source of hCG receptors. In studying the presence of hCG receptors in Porcine uterus, maximum binding was at 24°C in 16 hour, while the maximum binding of hCG receptors for cervical porcine was at 36°C in 4 hours[244]. For rat testis hCG receptor, the maximum binding in the crude homogenate was at 25°C in 4 hours, while the maximum binding for soluble rat testis hCG receptor was found to be at 25°C in 10 hours. In human ovarian and uterine homogenates, hCG receptor reached its maximum binding at 37°C for 1 hour for each tissue.

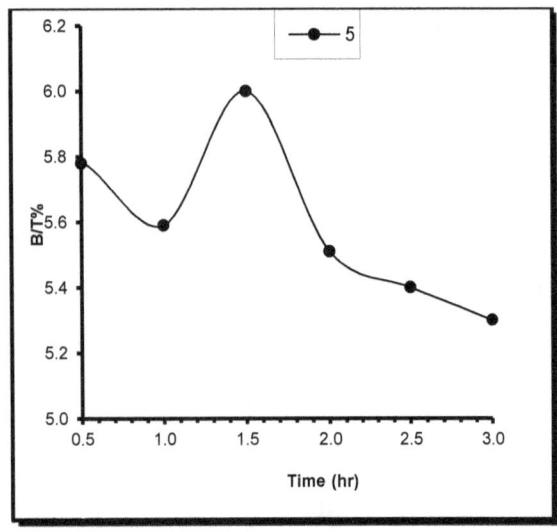

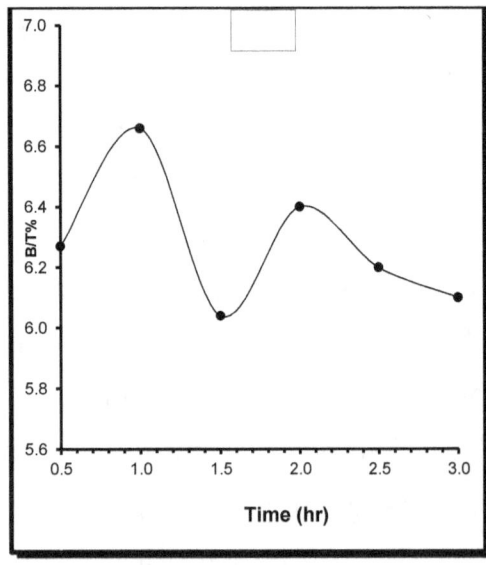

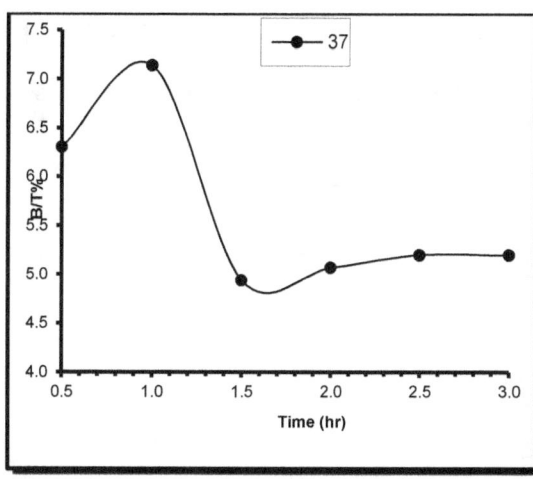
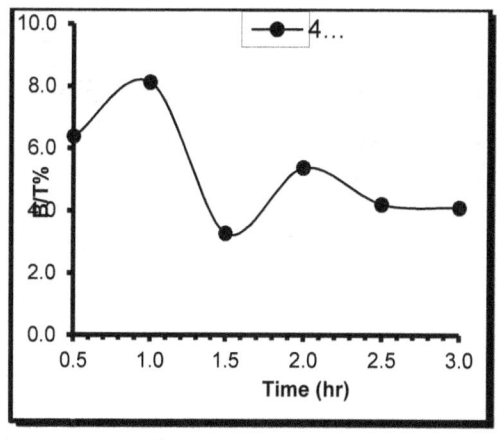

Figure (58): Time course of the binding of ^{125}I- hCG with hCG receptor in benign breast tumors. (All other details are explained in the text).

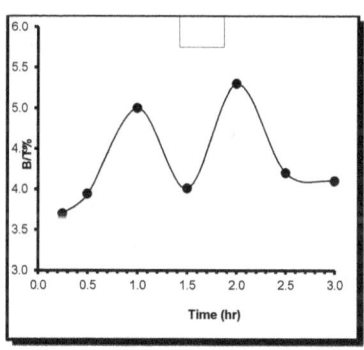
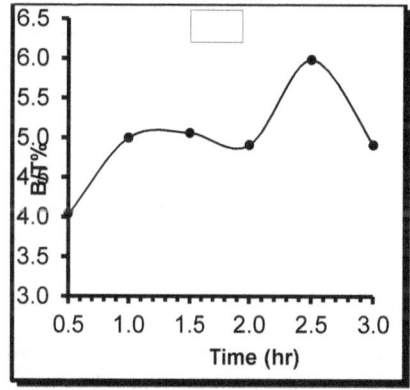
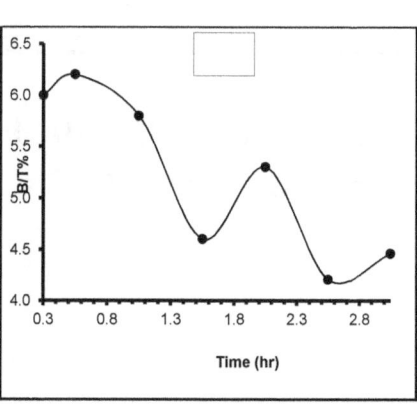
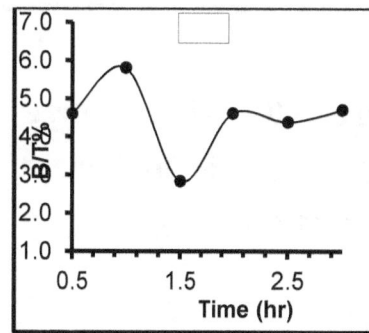

Figure (59): Time course of the binding of ^{125}I- hCG with hCG receptor in pre- malignant breast tumors. (All other details are explained in the text).

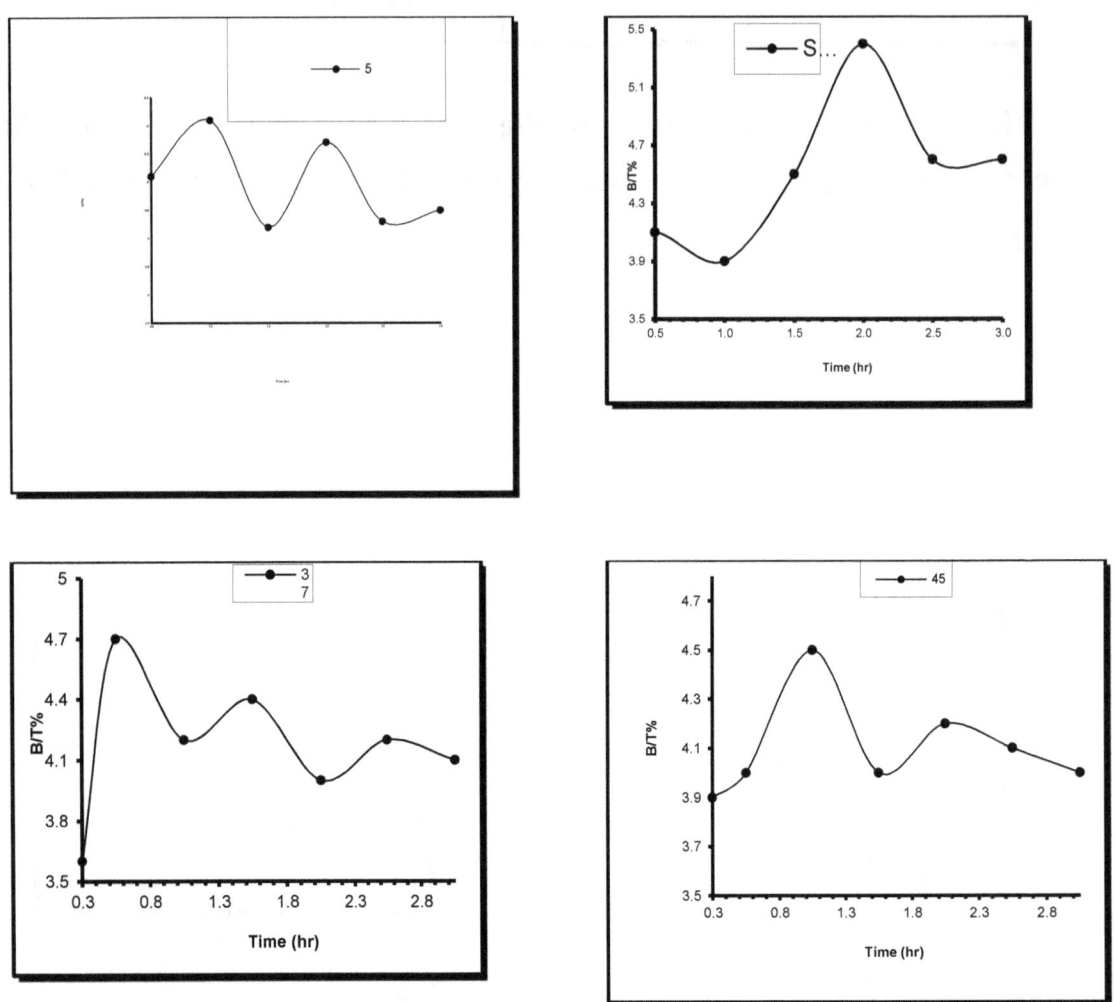

Figure (60): Time course of the binding of ^{125}I- hCG with hCG receptor in post- malignant breast tumors. (All other details are explained in the text).

The Effect of Different Halides on the Binding

Different sodium halides at 0.1M concentration were investigated to study their action on the binding of ^{125}I- hCG with hCG receptor in breast tumor homogenate, as shown in figure (61). Even It seems that sodium Florid made a slight increase, but in general the sodium halide decrease the binding of (^{125}I- hCG/hCG receptor) complex in the two groups according to the following order:

Benign breast tumor   NaF < NaBr < NaI < NaCl

Malignant premenopausal breast tumor   NaF < NaBr < NaCl < NaI

Sodium iodide causes lower specific binding in the two groups; this could be due to that NaI has less degree of hydration permits. So, greater interaction of Iodine salt with an ionic group located in the hCG receptor or ^{125}I- hCG combining site happen.

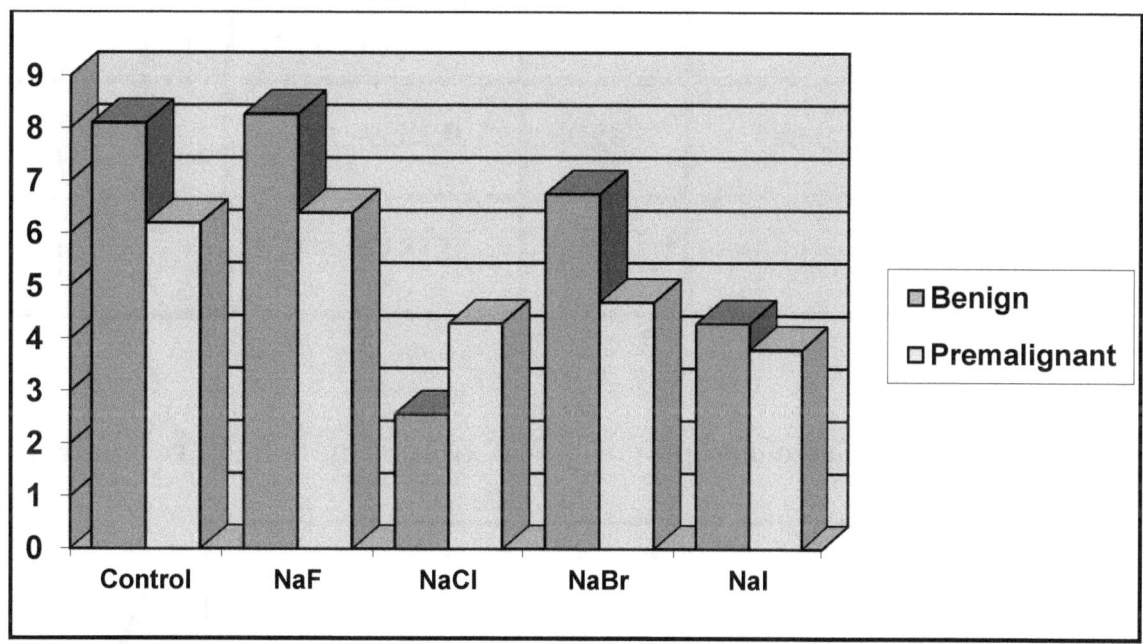

Figure (61): Effect of Halides on the binding of ^{125}I-hCG with hCG receptors. (All details are explained in the text).

Determination of the Concentration of hCG-Receptor and the Affinity concentration of ^{125}I-hCG Association with its Receptors in Benign and Malignant Breast Tumor.

The equilibrium association constant (Ka) values were observed for the whole homogenate, plasma membrane fraction and solubilized hCG receptor in many origin. It gives different values depending on the origin kinds and the method.

The concentration of hCG receptors in benign and malignant breast tumor homogenate (B_{max}) and the affinity constant (Ka) of the binding with ^{125}I-hCG has been measured in the optimum condition of the binding, as in figure (62, A and B).

As in table (35), the affinity constant (Ka) was found to be tumor type dependent (i.e, benign or malignant). It is found to be higher in benign ($1ng^{-1}.ml$) than malignant tumor ($0.7ng^{-1}.ml$). The concentration of hCG receptor in benign breast tumor was found to be ($0.2ng.ml^{-1}$), which is higher than malignant breast tumor($0.1838ng.ml^{-1}$).

Table (35): The kinetic parameters of ^{125}I-hCG binding to hCG receptor in breast tumor homogenate.(All other details are explained in the text).

Tumor type	Temp °C	Binding Capacity B_{max} (ng.ml^{-1})	K_a (ng^{-1}.ml)	K_d (ng.ml^{-1})
Benign breast tumor	45 °C	0.2339	1.0029	0.9971
Malignant breast tumor	37 °C	0.1838	0.7747	1.2908

Figure(61): Scatchard plot of ^{125}I-hCG binding to hCG receptor in

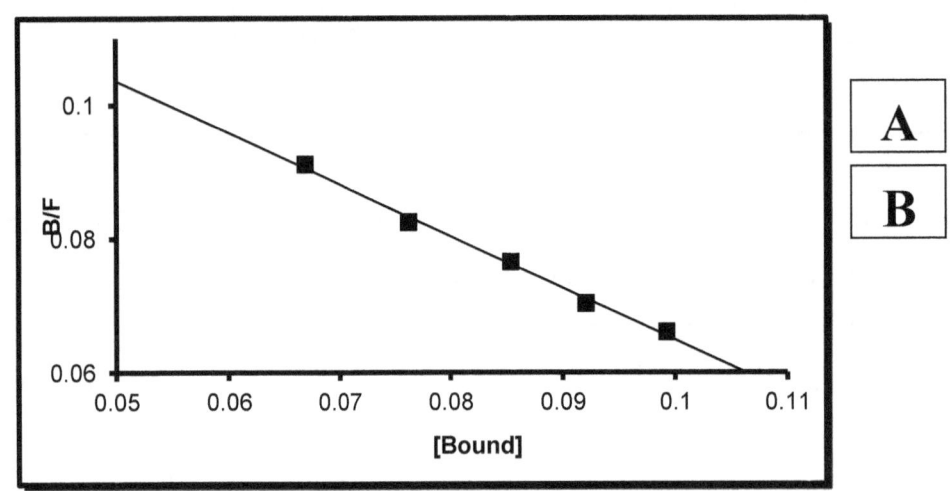

A: benign breast tumor

B: malignant breast tumor

Gel filtration chromatography technique was used for partially purification of hCG

Introduction

Human chorionic gonadotropine is a member of a family of four closely related glycoprotein hormones. In mammals, hCG consists of two subunits. The alpha subunit encoded by a single alpha gene, in contrast, multiple gene exist for the beta subunit of hCG. HCG is produced primarily by the placenta or cells destined to become placental tissue(250). However, It has been demonstrated that the pituitary also produces a small quantity of hCG. It is found in the urine and blood of pregnant women, in pituitary tissue, blood, the urine of postmenopausal women and in the urine from some cancer patients in a variety of forms whose concentration have clinical importance. **So, many efforts to isolate and purification the several forms of hCG and prepare purified standards forms from it was evaluated. These efforts have been used different techniques to purify hCG and its forms, such as, gel filtration, ion and anion exchange chromatography, affinity and immunoaffinity chromatography, hydrophobic interaction chromatography, HPLC chromatographyand PAGE electrophoresis.**

Many reports have been made to characterize the purified hCG by studying its three dimensional structure, the reproductive physiology, receptor interaction, biosynthesis, metabolism, functional and biochemical properties, the relation between biological, immunological activity and the chemical composition of hCG).

These characterization, of hCG composition, have been made by many techniques, such as, proteases analysis, various hCG fragments, enzymatic activities, electrophoresis, SDS-gel electrophoresis, amino-terminal sequence analysis, amino acid analysis, mass spectrometric analysis, immunoassay, bioassay, carbohydrate analysis.

Materials and Methods

- **Patients**

The malignant premenopausal patients tissues were used in the following experiments.

Methods

- **Partially purification of hCG by Gel Filtration Technique**

Gel filtration chromotography technique was used for partially purification of hCG from malignant breast tumor homogenate.

The dimensions of the column were chosen according to the following equation.

$$\text{Diameter} = \sqrt{\frac{m}{10}}$$

Where:

m= amount of protein in mg.

$$L = 30 \times \text{diameter}$$

Where:

L: length of the column

The gel (Sephadex G-100) was allowed to swell in excess of Tris/HCl buffer (0.05M, pH 7.2) containing 5mM EDTA and (0.02%) NaN_3 (20 ml of buffer per gram of gel) and left to stand for three days at room temperature without stirring to equilibrate with the buffer. The buffer was decanted and the gel was resuspended in excess volume of eluent buffer three times. The de-gassed slurry was carefully mixed before pouring into the vertical column which contains 5 ml of eluent buffer using a glass rod attached to the inner surface of the column. After the gel has settled, the column outlet was opened. Packing was continued and column was equilibrating with Tris/HCl buffer for 24 hr until the gel reached a stable bed height 30cm.

- *Void Volume Determination*

The elution volume of blue dextran 2000 is equal to the column void volume (V_o) and it was determined as follows:

A fresh solution of blue dextran (2mg/ml) was prepared in the eluent buffer.555ml of blue dextran solution was carried out with the same buffer, using a flow rate of 10ml/hr. Fractions of 1ml were collected and their absorbence were measured at 600 nm.

Eluent Buffer Preparation

Tris/HCl buffer (0.05 M, pH 7.2) containing 5mM EDTA and (0.02%) NaN_3 was prepared as in partially purification of hCG by gel filtration technique section.

- Sample Addition

A volume of 0.555ml of the tissue homogenate containing approximately 10 mg protein was prepare and then applied to the column equilibrate with Tris/HCl buffer.

The fractions were eluted with same flow rate (10ml/hr), 1 ml for each fraction, gel filtration was carried out at 10 °C. The fractions contained hCG were identified by the assay method.

The binding of each fraction was calculated and plotted against the elution volume. The fractions that gave maximum binding was poled together and used in the next experiment. The degree of isolation (folds) of hCG was calculated from the following formula.

$$\text{Yeild\%} = \frac{\text{Total protein content of partially purified hCG}}{\text{Total protein content of crude hCG}} \times 100$$

Then yield % was determined as follows:

$$\text{Purification fold of hCG} = \frac{\text{Specific binding of partially purified hCG}}{\text{Specific binding of crude hCG}}$$

Calculations

1. The radioactivity (c.p.m) and the absorbency were plotted against the fraction number.

2. The fractions under each peak were poled and the absorption spectrum was measured in the (280nm) using a 1cm cuvette against Tris/HCl buffer (0.05 M, pH 7.2), in reference beam.

- Standandard hCG Addition

A volume of 0.555ml of stander hCG containing approximately 5 mg protein was prepare and then applied to the column equilibrate with Tris/HCl buffer.

The fractions were eluted with same flow rate (10ml/hr), 1 ml for each fraction, gel filtration was carried out at 10°C. The fractions contained hCG were identified by the assay method.

The binding of each fraction was calculated and plotted against the elution volume. The fractions that gave maximum binding was poled together and used in the next experiment.

Calculations

1. The radioactivity (c.p.m) and the absorbency were plotted against the fraction number.
2. The fractions under each peak were pooled and the absorption spectrum was measured in the (280 nm) using a 1cm cuvette against Tris/HCl buffer (0.05 M, pH 7.2), in reference beam.

The Choice of the Appropriate Conditions for the Binding of the Partially purified hCG from malignant premenopausal Breast Tissue to 125I-anti hCG Antibody

- *The Effect of Different Protein Concentration*

1. A volume of 20 µl (0.1176µg.ml^{-1}) of ^{125}I-anti hCG antibody (1470 µg.ml^{-1}) was added to 100ml containing increasing amount (6.8, 13.7, 20.6, 27.4, 34.3, 41.1, 47.9, 54.8, 61.7µg protein) form the poled fractions that containing (685.7 µg.ml^{-1} protein) of breast tumor homogenate in a final volume of 250µl complete with Tris/HCl buffer (0.05M, pH 7.2) containing 0.1% BSA.

1. One additional tube containing only 20 µl of ^{125}I-anti hCG antibody was set aside until counting for total activity computation.
2. The tubes were incubated at 25°C for 60 min.
3. After incubation, 500µl of PEG 6000 (10%) were added to the tubes and incubated again for 15 min at 4°C.

4. After incubation, the tubes were centrifuged at 1500xg for 30 minute at 4°C.

5. The supernatant was discarded by decanting the assay tubes, then the tubes were inverted on a filter paper for 10min.

6. The rims of the tubes were swabbed with a cotton piece and the amounts of bound radioactivity were counted in a gamma counter.

Calculations

1. The bound fraction (B) represents the counted radioactivity in each tube, expressed in c.p.m i.e (^{125}I-anti hCG antibody/hCG) complex.

2. Total activity (T) represents the counted radioactivity in the tubes containing ^{125}I-anti hCG antibody only.

3. The (B/T)% ratio for each tubes was counted as follows:

4. $(B/T)\% = \dfrac{\text{Sample mean counts (B)}}{\text{Total activity mean counts (T)}} * 100$

5. The percent of binding value (B/T)% were plotted against the increasing amount of proteins.

- **The Effect of Different Concentrations of ^{125}I-anti hCG Antibody**

1. One hundred microliters containing the optimum protein amount (34.3μg) form the poled fractions that containing (685.7 μg.ml^{-1} protein) of malignant premenopausal breast tumor homogenate was added to increasing volumes (10, 15, 20, 25, 30, 35 and 40) μl; (0.0882-0.2058μg.ml^{-1}) of ^{125}I-anti hCG antibody (1470μg.ml^{-1}) then the volume were made up to 250μl with Tris/HCl buffer (0.05M, pH 7.2) containing 0.1% BSA.

2. A set of tubes containing only the same increasing volumes of ^{125}I-anti hCG antibody (10, 15, 20, 25, 30, 35 and 40) μl were set a side until counting for total activity computation.

3. The tubes were incubated at 25°C for 60 min.

4. After incubation, the (^{125}I-anti hCG antibody/hCG) complex was estimated by following the steps 4, 5, 6 and 7 in the experiment of protein effect.

Calculations

1. **The (B/T) percent values were determined.**

2. The percent of binding values (B/T)% were plotted against the concentration of ^{125}I-anti hCG antibody.

- **The Effect of Different pH on the Binding**

1. One hundred microliters containing the optimum protein amount (34.3μg) form the poled fractions that containing (685.7 μg.ml^{-1} protein) of malignant premenopausal breast tumor homogenate was added to 20μl (0.1176μg.ml^{-1}) of ^{125}I-anti hCG antibody, the mixtures volumes were made up to 250μl with Tris/HCl buffer (0.05M) containing 0.1% BSA of different pH (6.8, 7.0, 7.2, 7.4, 7.6, 7.8 and 8.0).

2. One additional tube containing only 20μl of the ^{125}I-anti hCG antibody were set aside until counting for total activity computation.

3. The tubes were incubated at 25°C for 60 min.

4. After incubation, the (^{125}I-anti hCG antibody/hCG) complex was estimated by following the steps 4, 5, 6 and 7 in the experiment of protein effect.

Calculations

1. **The values of (B/T)% were determined**

2. The percent of binding value (B/T)% were plotted against their corresponding pH values.

- **Time Course of the Binding**

1. One hundred microliters containing the optimum protein amount (34.3μg) form the poled fractions that containing (685.7 μg.ml^{-1} protein) of malignant premenopausal breast tumor homogenate was added to 20μl

of ^{125}I-anti hCG antibody (1470µg.ml^{-1}), the mixture volume was made up to 250µl with Tris/HCl buffer (0.05M, pH 7.2) containing 0.1% BSA.

2. One additional tube containing only 20µl of the ^{125}I-anti hCG antibody were set aside until counting for total activity computation.

3. All tubes were incubated at 25°C at different time interval (30, 60, 90, 120, 150 and 180) min.

4. After incubation, the (^{125}I-anti hCG antibody/hCG) complex was estimated by following the steps 4, 5, 6 and 7 in the experiment of protein effect.

5. To determine the time course of HCG binding to ^{125}I –anti hCG antibody at different temperatures. Steps 1, 2 and 3 in the same experiment were repeated at different temperatures (5, 37, 45°C).

Calculations

1. **The (B/T) percent values were determined.**.

2. The values of (B/T)% were plotted against the incubation time.

The Kinetic and The Thermodynamic Studies

- **Determination of Affinity Constant (Ka) and the Maximal Binding Capacity (B$_{max}$) of hCG in Partially purification hCG from malignant premenopausal Breast Tissue with ^{125}I-anti hCG Antibody**

1. One hundred microliters containing the optimum protein amount (34.3µg) form the poled fractions that containing (685.7 µg.ml^{-1} protein) of malignant premenopausal breast tumor homogenate was incubated with increasing volumes (6, 8, 10, 12, 14, 16, 18 and 20 µl) of ^{125}I-anti hCG antibody (0.035-0.1176 µg.ml^{-1}). The final volumes were made up to 250 µl with Tris/HCl buffer (0.05M, pH 7.2, 0.1% BSA).

2. Two additional tubes containing only increasing volumes (6, 8, 10, 12, 14, 16, 18 and 20 µl) of ^{125}I-anti hCG antibody were set aside until counting for total activity computation.

3. The tubes were incubated for 60 min at 45°C.

4. After incubation, the steps 4, 5, 6 and 7 of the experiment of protein effect. were repeated.

5. The previous steps were performed at different temperature (5, 25 and 37°C), the time of incubation needed to get the equilibrium state were (60 min at 5 °C, 90 min at 25°C and 90 min at 37°C).

Calculations

1. The values B/F ratio were determined where:-

   B: Is the bound radioactivity (mean of counts c.p.m), which represents the (^{125}I-anti hCG antibody/hCG) complex.

   F: Is the free radioactivity (mean of the counts c.p.m), which represented the non-bound ^{125}I-anti hCG antibody.

   T: Is the total radioactivity mean of the counts.

   F = Total counts (T) – Bound radioactivity (B)

2. The concentration of the (^{125}I-anti hCG antibody/hCG) complex in (ug/ml) that formed after time (t) was calculated from the following equation:-

$$B(mg/ml) = \frac{B(c.p.m)}{T(c.p.m)} \times \text{Concentration of } ^{125}\text{I-anti hCG antibody in the incubation medium } (\mu g/ml)$$

3. The affinity constant and the maximal binding capacity were determined according to Scatchard equation [212,259]: -

$$\frac{B}{F} = \frac{1}{Kd}(B_{max} - B)$$

$$Ka = \frac{1}{Kd}$$

Where:

Ka = Affinity constant

Kd = Dissociation constant

B_{max} = Maximal binding capacity

4. (μg/ml), gives a linear relationship. The values of the affinity constant of the binding (Ka) at each temperature can be calculated from the slop of the straight line, while the value of the total concentration of hCG (B_{max}) in breast tumor tissue was calculated from the intercept with the x-axis.

- **Thermodynamic Studies of the Binding of hCG in Premenopausal Patients with Breast Tumor to ^{125}I-anti hCG Antibody**

According to the steps of the two experiments explained in time course of the binding and determination of kinetic parameters the thermodynamic parameters were calculated.

Calculations

1. The thermodynamic parameters of standard state ($\Delta H°$, $\Delta G°$, $\Delta S°$) were obtained from Van't Hoff plot, the values of the natural logarithm of equilibrium constant (affinity constant Ka) obtained at different temperature were plotted against the reciprocal values of absolute temperatures in kelvin (1/T) was calculated according to the following equation: -

$$\ln Ka = \frac{\Delta S°}{R} - \frac{\Delta H°}{RT}$$

Where :-

$\Delta H°$: The enthalpy change of the standard state .

$\Delta S°$: The entropy change of the standard state .

R: The gas constant (8.31441 J.K^{-1} mol^{-1})

$\Delta H°$ value obtaied from the linear relationship of the plot . The change in Gibbs free energy of the standard state ($\Delta G°$) was obtained from the following equation :-

$$\Delta G° = -RT\ln Ka$$

While the standard state entropy change was obtained from

$$\Delta S° = \frac{\Delta H° - \Delta G°}{T}$$

2. The thermodynamic parameters of the transition state were obtained from Arrhenius plot of $\ln k_{+1}$ values against 1/T values, that gives a linear relationship according to the following equation:-

$$\ln K_{+1} = \ln A - \left(\frac{Ea}{RT}\right)$$

Where :-

A : Arrhenius constant .

Ea: Apparent energy of activation .

T: Absolute temperature in kelvin .

The value of E_a of the binding reaction can be determined from the slop of the straight line.

The enthalpy of transition state (ΔH^*) was obtained from:-

$\Delta H^* = E_a - RT$

The free energy change of the transition state (ΔG^*) was calculated by using the following equation:-

$$\Delta G^* = -RT\ln K_{+1} + RT\ln\left(\frac{KT}{h}\right)$$

Where:-

K: Boltzmann constant = 1.38×10^{-23} J.deg^{-1}

h: Plank's constant = 0.662×10^{-33} J.S^{-1}

The change in entropy of the trasition state (ΔS^*) was calculated from the following equation:-

$$\Delta S^* = \frac{\Delta H^* - \Delta G^*}{T}$$

Result and Discussion
Partially purification of hCG by Gel Filtration Technique

To partially purified hCG from malignant breast tumor, gel filtration technique was preformed by using Sephadex G 100. By this technique hCG was separated from aggregates and other protein having differ molecular weight, as shown in figure (63, A). This experiment revealed the presence of two different eluted components eluted with different elution volume corresponding to their different molecular weights. All fractions that have been collected, 1 ml for each fraction, were tested for hCG presence by assay method, then the fractions that have a higher binding activity were collected, pooled and determined there total protein. It was found that fractions containing hCG were begin within the void volume (V_o), fraction number 10, until the fraction number 16. The isolation of hCG from malignant permenopausal breast tumors group on Sephadex G-100 showed 4.6 folds of purification; as illustrated in table (38). The same isolation procedure was preformed for standard hCG as a comparison with hCG from malignant breast tumor. From this experiment one peak was obtained and as mention above all collected fractions, 1ml for each fraction, were tested for hCG presence by assay method. It was found that fractions containing hCG also begin with void volume fraction until the fraction number 16, as shown in figure (66,B). From these findings it seems that hCG which obtained from malignant breast tumor homogenate is in the same range of molecular weight of the standard hCG.

Table (36): Partial isolation of hCG by gel filtration. (All other details are explained in the text).

HCG Source	Total protein mg. ml^{-1}	Specifically bound of ^{125}I-anti hCG antibody to hCG	Specifically bound ^{125}I-anti hCG antibody/mg protein	Purification fold
Crude extract	0.6	46.2	0.077	1.00
Gel filtration on sehpadex G-100	0.0343	12.1	0.353	4.6

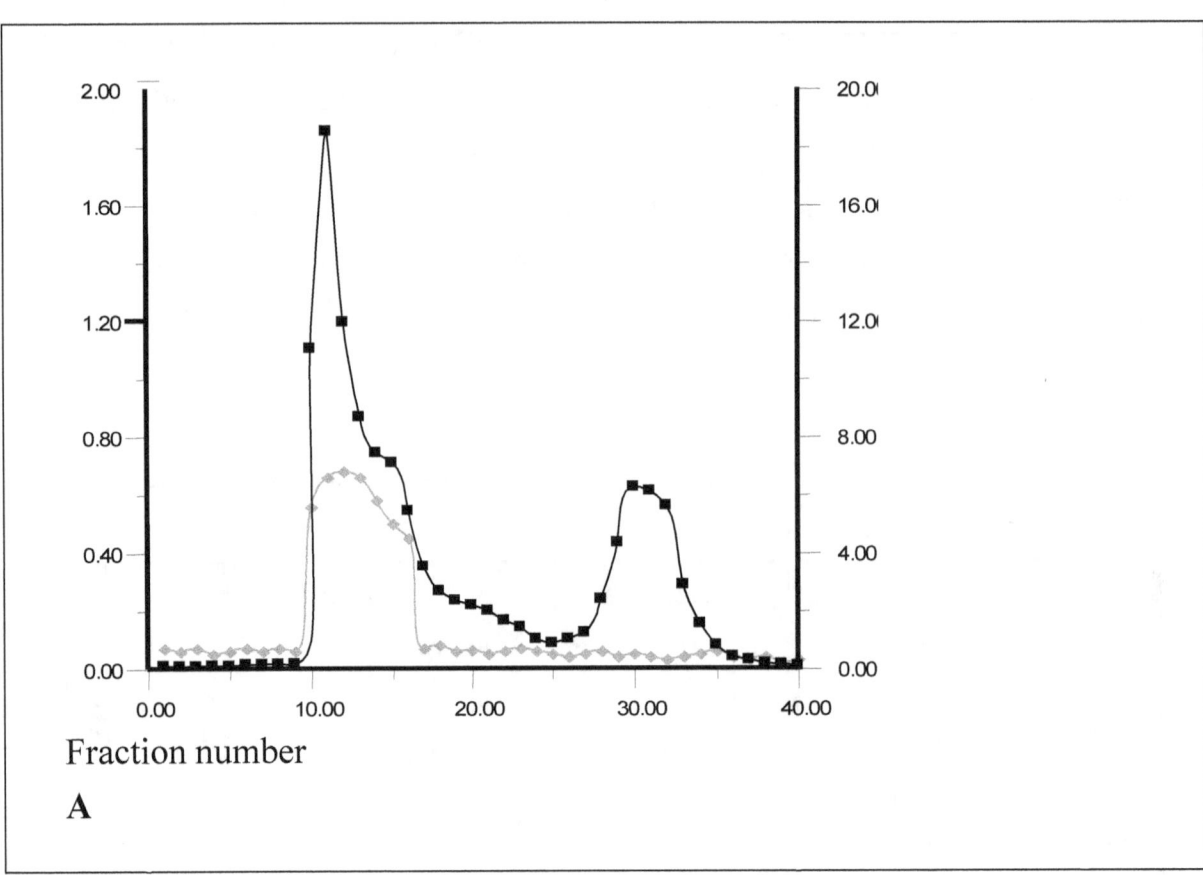

Figure (63): The elution profile of hCG from, A: Pre-malignant breast tumor.

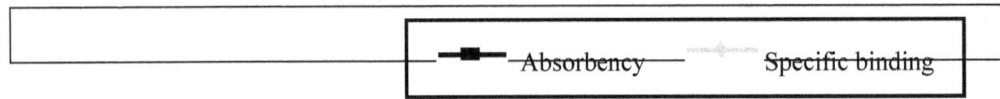

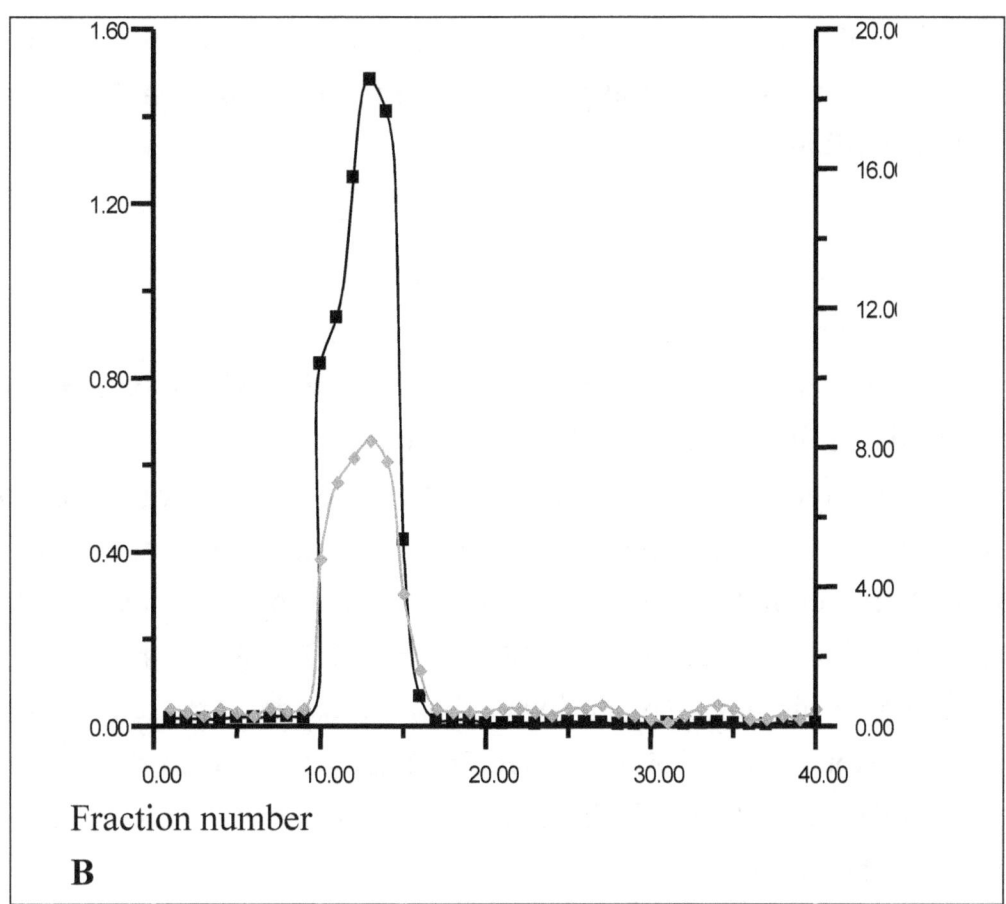

Figure (63): The elution profile of hCG from B: Standard hCG. (All other details are explained in the text).

The Choice of the Appropriate Conditions for the Binding of the Partially Purified hCG from malignant premenopausal Breast Tissue to ^{125}I-anti hCG Antibody

The Effect of Different Protein Concentration

The effect of increasing amount of partially purified hCG to a fixed amount of ^{125}I-anti hCG antibody to produce (^{125}I-anti hCG antibody/hCG) complex was shown in figure (64). At the beginning, the formation of small complex was occurred; then by increasing of the added hCG, large complex was preformed until reach maximum binding (8.4%). Excess addition of hCG make large complex less probable; causing solublization of the complex. The amount of partially purified hCG (34.3 µg) that needed for reaching maximum binding is much lower than the crude homogenate used in the previous experiment of protein effect in chapter two. According to the results, (34.3 µg) was used in all the subsequent experiments since it gives maximum value of binding.

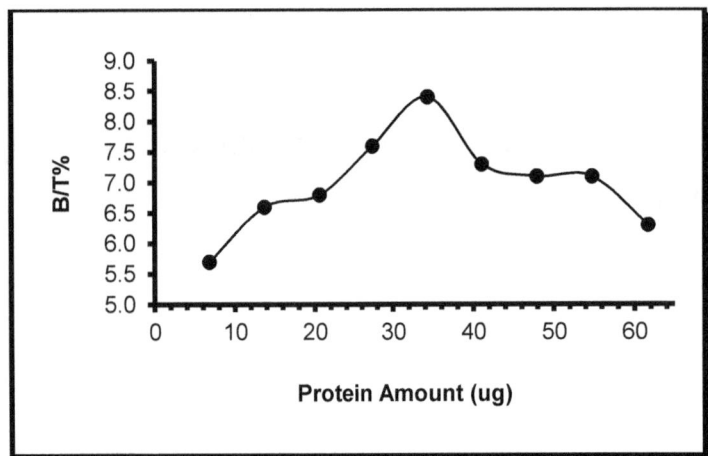

Figure (64): Influence of protein amount on the binding of ^{125}I-anti hCG antibody with partially purified hCG from malignant premenopausal breast tumor. (All other details are explained in the text).

The Effect of Different Concentrations of ^{125}I-anti hCG antibody

Effect of ^{125}I-anti hCG antibody concentration on the (^{125}I-anti hCG antibody/hCG) complex formation was investigated by incubated fixed amount of partially purified hCG with increasing concentration ^{125}I-anti hCG antibody for 60 mint at 25°C. Figure (65) is representative this binding and revealed that percent of binding increased by the amount of ^{125}I-anti hCG antibody added until it reaches the maximum binding (8.4%) at (0.1176 µg.ml^{-1}). This concentration is nearby the concentration needed for reaching the maximum binding with crude homogenate. From these observation it found the tracer concentration that needed for reaching the maximum binding is almost constant while the difference was in the amount of added protein. This is due to that crude homogenate is containing number of proteins; hCG is one of its contents and it seems that its concentration in the crude homogenate is very low as comparison for the needed protein from the crude and partially purified hCG. ^{125}I-anti hCG antibody (0.1176 µg.ml^{-1}) was used in all subsequent experiments since it gives maximum value of binding.

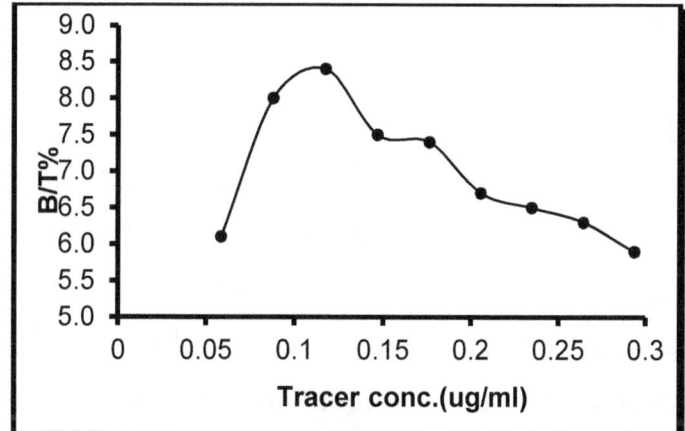

Figure (65): Influence of ^{125}I-anti hCG antibody concentration on its binding with partially purified hCG from malignant premenopausal breast tumor. (All other details are explained in the text).

The Effect of Different pH on the Binding

The influences of pH on the binding of partially purified hCG with ^{125}I-anti hCG antibody is stated in figure (66). In this figure, optimum pH, giving maximum binding, was found to be 7.2 with decreasing in the binding at pH higher or lower than the optimum one. This result indicated that the induction of protonation-deprotonation process occurring within the ionizable groups of the amino acids present in the binding domain of hCG did not effected or changed by the isolation hCG from the crude homogenate media$^)$. According to the results obtained, the pH of the buffer used in all subsequent experiments were adjusted to pH 7.2.

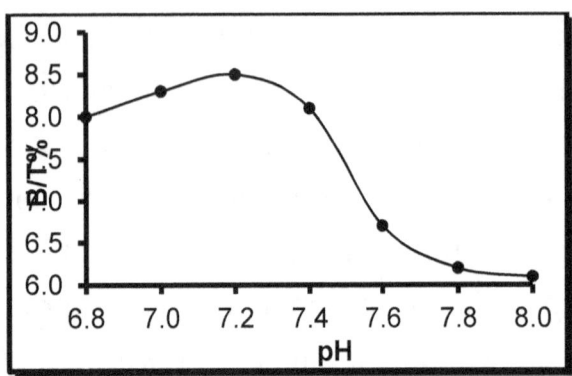

Figure (66): pH effect on the binding of ^{125}I-anti hCG antibody with partially purified hCG from malignant premenopausal breast tumor. (All other details are explained in the text).

Time Course of the Binding

The binding of partially purified hCG and ^{125}I-anti hCG antibody was carried out at different incubation time in four temperatures (5,25,37,45), as shown in figure (67). In this figure the present of binding shows conformance relation with elevation of temperature showing maximum binding (12.1%) at 45°C in 60 mint. These results are in an agreement with the previous experiment for malignant premenopausal breast tumor crude homogenate in chapter two. The binding in the two experiments shows it maximum binding at 45°C with shorter time for the partially purified hCG binding.

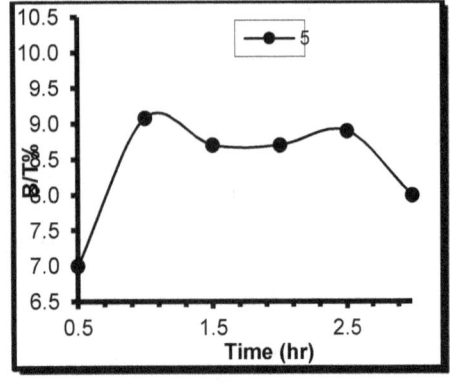
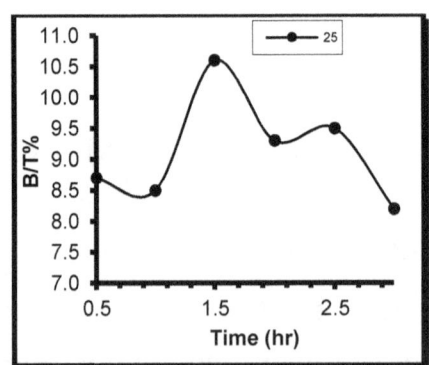

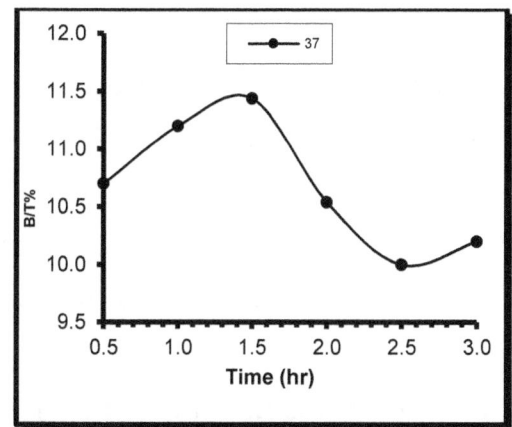
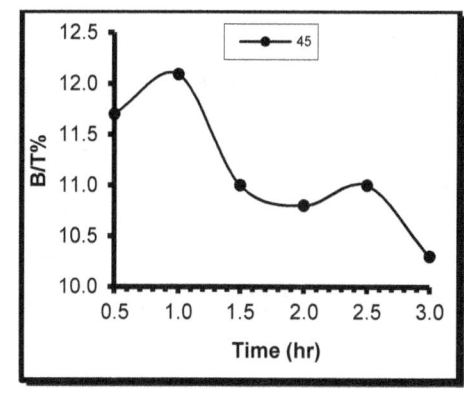

Figure (67): Time-Course of ^{125}I-anti hCG antibody binding to partially isolated hCG from malignant premenopausal breast tumor. (All other details are explained in the text).

Determination of Kinetic Parameters of ^{125}I-anti hCG antibody Binding with partially Purified hCG from malignant breast tumor

The time course of ^{125}I-anti hCG antibody binding with hCG from pre-menopausal malignant breast tumor was carried out to describe kinetic parameters of the binding. The simplest proposed model representing the binding of ^{125}I-anti hCG antibody with hCG could be expressed by the following equation:

$$^{125}\text{I-anti hCG Antibody} + \text{hCG} \underset{k_{-1}}{\overset{k_{+1}}{\rightleftharpoons}} [^{125}\text{I-anti hCG Antibody/hCG}]$$

Where-:

K_{+1}: is the rate of the association of ^{125}I-anti hCG antibody with hCG.

K_{-1}: is the rate of the reverse reaction of the dissociation of the complex formed under the same condition.

At equilibrium:

$$K_a = \frac{[^{125}\text{I-anti hCG Antibody/hCG}]}{[^{125}\text{I-anti hCG Antibody}][\text{hCG}]} \quad \text{......(2)}$$

$$K_d = \frac{[^{125}\text{I-anti hCG Antibody}][\text{hCG}]}{[^{125}\text{I-anti hCG Antibody/hCG}]} \quad \text{......(3)}$$

Thus:

$$K_a = \frac{1}{K_d} = \frac{K_{+1}}{K_{-1}} \quad \text{......(4)}$$

Where-:

Ka: is the equilibrium constant of the association (affinity constant.

Kd: is the equilibrium constant of the dissociation ^{125}I-anti hCGantibody/hCG) complex.

The values of Ka and maximal binding capacity (B_{max}) were calculated from Scatchard plots at five different temperatures as shown in figure (68) and table (39).

As in table (40), the K_a and B_{max} values are temperatures depended. The K_a increased with increased temperature, in the following order 5>25>37>45 °C, to reach it highest value at 45°C (7.756 µg^{-1}.ml), while the concentration of partially purified hCG was determined to be (0.0143 µg^{-1}.ml) at 45 °C.

The value of dissociation constant (K_d) was calculated by using equation (4), which indicate that K_d values are decrease with increasing temperature to reach it lowest value (0.1289 µg^{-1}.ml) at 45 °C.

Table (37): The kinetic parameter of ^{125}I-anti hCG antibody binding to partially purified hCG in pre-menopausal malignant breast tumor homogenate.(All other details are explained in the text).

Temp °C	Binding Capacity B_{max}(µg.ml^{-1})	K_a (µg^{-1}.ml)	K_d (µg.ml^{-1})
5	0.0121	6.4672	0.1546
25	0.0122	6.872	0.1455
37	0.0131	7.066	0.1415
45	0.0143	7.756	0.1289

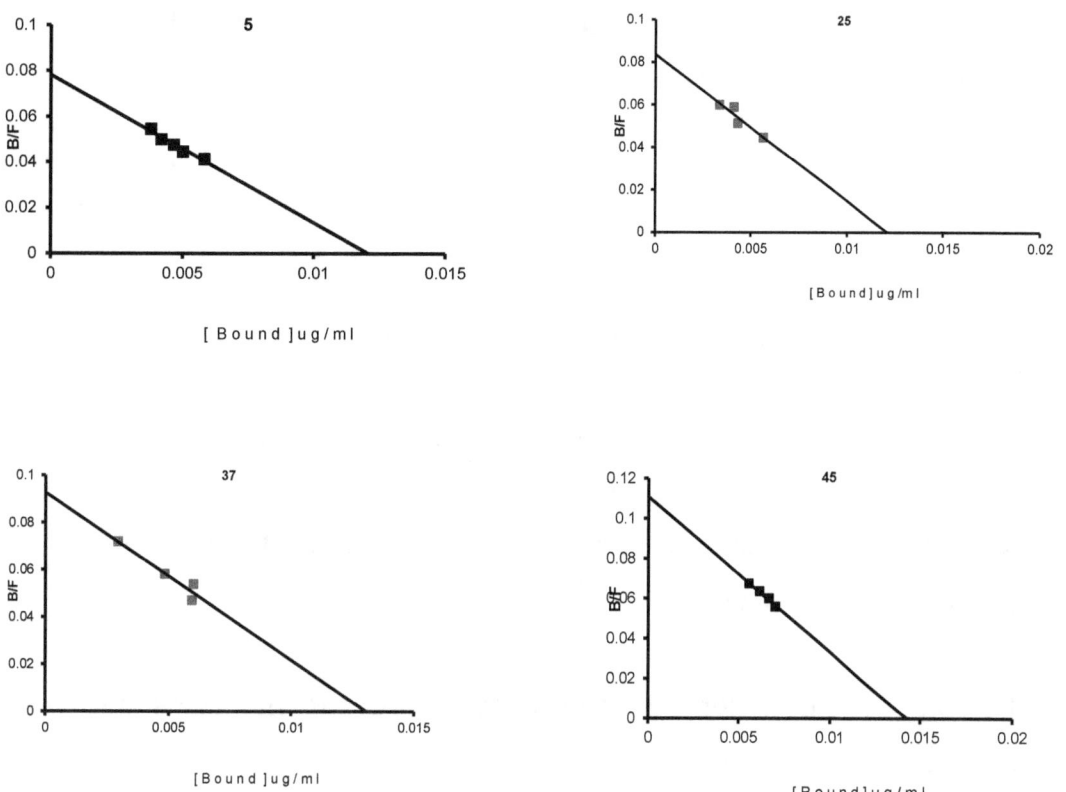

Figure (68): Scatchard plot of ^{125}**I-anti hCG antibody binding to the partially purified hCG in malignant breast tumor at four different temperatures. (All other details are explained in the text).**

Time course data in figure(69) were used to determine the reaction order of hCG binding to its specifically ^{125}I-anti hCG antibody. Because the binding is small and the most labeled antibody remains free and only small fraction binds even at equilibrium, i.e, $[Ab]_t \gg [AbAg]_e$

Thus:

$$[Ab]_t \gg \frac{[AbAg]_t \, [AbAg]_e}{[Ag]_t}$$

$[AbAg]_e$: is the concentration of (^{125}I-anti hCG/hCG)complex formed at equilibrium.

$[AbAg]_t$: is the concentration of (^{125}I-anti hCG/hCG) complex after time (t).

$[Ab]_t$: is the total concentration of ^{125}I-anti hCG antibody in µg. ml^{-1}.

$[Ag]_t$: is the total concentration of hCG in µg. ml^{-1}.

So the following equation could be used in order to fit the pseudo- first order kinetics-:

$$Ln \frac{[AbAg]_e}{[AbAg]_e - [AbAg]_t} = K_{+1} t \frac{[Ab]_t \, [Ag]_t}{[AbAg]_e} \quad \dots\dots\dots(5)$$

Figure (69) shows the plot of $\ln \frac{[AbAg]_e}{[AbAg]_e - [AbAg]_t}$ against time in malignant breast tumor, which gives a straight line with a slope, equal to the observed value of first rat constant K_{obs} in min^{-1}. The rate constant (K_{+1}) in µg^{-1}.ml was calculated at four different temperatures by using the following equation[261]:

$$K_{obs} = K_{+1} \frac{[^{125}\text{I-anti hCG Antibody}]_t \, [\text{hCG}]_t}{[^{125}\text{I-anti hCG Antibody/hCG}]_e} \quad \dots\dots(6)$$

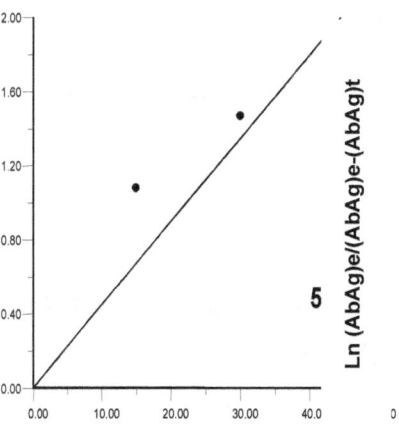
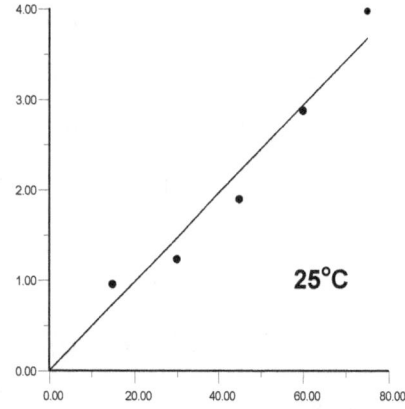
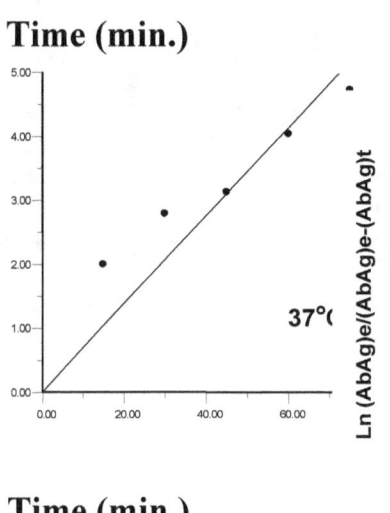
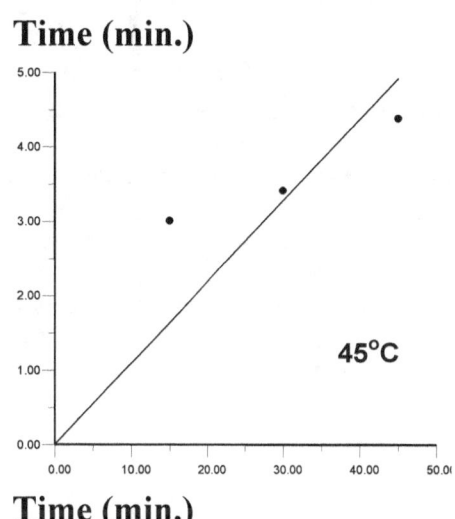

Figure (69): Kinetics of ^{125}I-anti hCG antibody binding to patrially purified hCG in malignant breast tumor at four different temperatures. (All other details are explained in the text).

The values of K_{-1} at four different temperatures were calculated by using equation (4). Half-life time of association $(t\,½)_{ass.}$, Which represented the time needed for the formation of half amount of the complex at equilibrium was determined from the concentration of the complex at equilibrium and the time-course curve. The half-life time of dissociation $(t\,½)_{diss}$, was calculated from the following relation:

$$(t_{1/2})_{diss.} = \frac{\ln 2}{k_{-1}} = \frac{0.693}{k_{-1}}$$

Table (38) summarized the values of K_{obs}, K_{+1}, K_{-1}, $(t_{1/2})_{ass.}$ and $(t_{1/2})_{diss.}$ at four different temperature. These values indicate that highest rate for the association reaction K_{+1} occurs at 45°C and the dependence of association and dissociation reaction rate on temperature.

Table (38): The effect of temperature on the kinetic parameter of ^{125}I-anti hCG antibody binding to partially purified hCG in pre-menopausal malignant breast tumor homogenate. (All other details are explained in the text).

Temp °C	k_{obs} x 10^{-2} (min^{-1})	k_{+1} x 10^{-1} µg^{-1}.ml.min^{-1}	k_{-1} x 10^{-2} min^{-1}	$(t_{1/2})_{ass}$ (hr)	$(t_{1/2})_{diss}$ (hr)
5	4.48	3.37	5.21	3.69	13.3
25	4.89	4.26	6.20	2.5	11.1
37	6.87	5.97	8.45	1.66	8.2
45	10.92	8.66	11.1	1.08	6.2

The Thermodynamic Studies of ^{125}I-anti hCG antibody to the Partially Purified hCG in malignant breast tumor

The dependence of the equilibrium binding constant (affinity constant) for the binding of ^{125}I-anti hCG antibody with partially purified hCG on temperature (Van`t Hoff blote) was clarified in figure (70). Also table (39) clarify the values of thermodynamic parameters of Standard State of partially isolated hCG in malignant breast tumor.

The results indicate that ΔH°, in general, had very small value; indicating a favorable interaction between ^{125}I-anti hCG antibody and partially purified hCG. Also, the positive sign a certain that the reaction was nearly endothermic.

The negative values of ΔG° reflects the stability of the complex hence, the high affinity of the reactants. So this system is characterized by the sole contribution of ΔS° to the stability of the complex formed, while ΔH° has little or no effect. The positive values of ΔS° suggest that the reaction spontaneity be entropically driven Entropy was the driven force for the occurrence of the binding. This indicates that the hydrophobic interactions played an important role are stabilizing the complex.

These include the non-covalent interaction which are fundamentally electrostatic in nature such as charge-charge, charge-dipole, dipole-dipole, charge-induced dipole, dipole-induced dipole interactions and hydrogen bonds. The sum of these types of interactions can yield some stabilization to the folded structure of the complex.

So the negative values of ΔG° showed that the overall reaction energetically favorable in the direction of complex formation.

Table (39): Thermodynamic parameters at standard state of ^{125}I-anti hCG antibody to the partially purified hCG in pre-malignant breast tumor homogenate.(All other details are explained in the text).

Temp°C	ΔH° KJ.mol^{-1}	ΔG° KJ.mol^{-1}	ΔS° J.mol^{-1}.K^{-1}
5	2.952	-4.314	26.14

25	2.952	-4.775	24.93
37	2.952	-5.039	25.78
45	2.952	-5.415	26.31

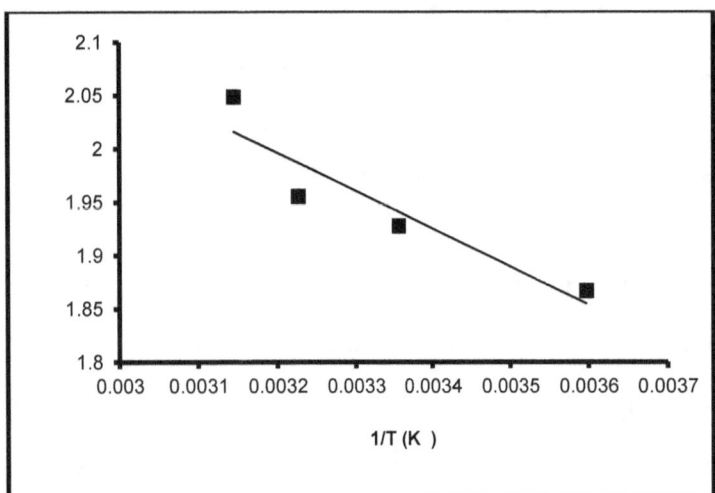

Figure (70): Van`t Hoff lpot for the binding of ^{125}I-anti hCG antibody to the partially purified hCG in pre-malignant breast tumor homogenate.(All other details are explained in the text).

Thermodynamic Parameters of Transition State

According to the transition state theory, the interaction of hCG with it ^{125}I-anti hCG antibody lead to the formation of an activated complex (transition states) then the formation of the final product:-

$$^{125}\text{I-anti hCG} + \text{hCG} \longrightarrow [^{125}\text{I-anti hCG/hCG}] \longrightarrow [^{125}\text{I-anti hCG/hCG}]$$

State(A)   An Activated Complex   Final Product
           Transition State   State(B)

The thermodynamic parameters of the transition state) $\Delta H^*, \Delta G^*, \Delta S^*$ and Ea) could be determined from Arrhenius equation and the kinetic constant.

Figure (71) shows the Arrhenius plot of $\ln K_{+1}$ against $1/T$ values. Table (40) shows the values of thermodynamic parameters of transition state (Ea, ΔH^*, ΔG^*, ΔS^*). The value of Ea that determined from Arrhenius plot represents the apparent energy activation of the binding reaction and the required energy to overcome the energy barrier of the transition state for the formation of (^{125}I-anti hCG antibody/hCG) complex.. The positive value of ΔH^* shows that content of activated complex is more than that of purified species$^)$. The high positive value of ΔG^* indicated that the formation of activated complex was a non spontaneous process and required a lot of energy(equal to Ea) to overcome the transition energy barrier and giving the final product. Also, the ΔG^* positive values is mainly attributed to the decrease in entropy of the transition state ($\Delta S^* < 0$). The high negative ΔS^* revealed that activated complex had more ordered structure than the reactant species($\Delta S^* < 0$).

The values of the thermodynamic parameters of the binding reaction, gave an overall idea about the nature of forces that regulate the formation of complex. It is proposed that the formation of a protein-ligand complex, occurs in two steps. The first stabilization of the complex by hydrophobic interactions and the second is the stabilization by short range interaction, such as electrostatic interaction, hydrogen bonding and vander wall's interactions. Hydrophobic interactions contribute to the complex stability via high positive entropy change ($\Delta S^* > 0$), while electrostatic interactions, hydrogen bonding and vander waal's interactions contribute to the stability of the complex via negative entropy change($\Delta S^* < 0$). The thermodynamic data from this study indicate that the binding of ^{125}I- anti hCG antibody with hCG are entropically driven and in agreement with the concept that hydrophobic interaction play an important role in ^{125}I-anti hCG/hCG) interactions.

Table (40): Thermodynamic parameters at transition state of ^{125}I-anti hCG antibody to the partially purified hCG in pre-malignant breast tumor homogenate.(All other details are explained in the text).

Temp° C	Ea KJ.mol^{-1}	ΔH^* KJ.mol^{-1}	ΔG^* KJ.mol^{-1}	ΔS^* J.mol^{-1}.K^{-1}
5	16.279	13.968	67.950	-194.18
25	16.279	13.802	75.095	-205.68

| 37 | 16.279 | 13.702 | 77.351 | -205.32 |
| 45 | 16.279 | 13.635 | 78.433 | -203.76 |

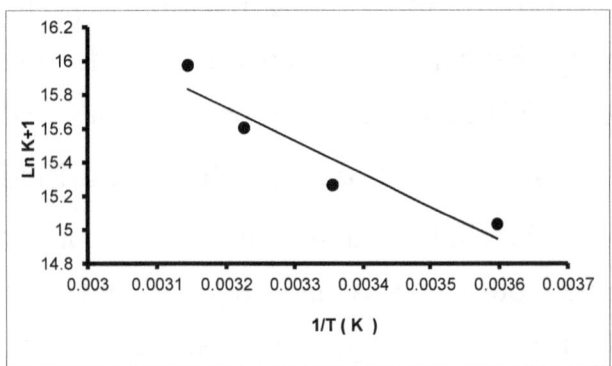

Figure (71): Arrhenius plot for the binding of 125I-anti hCG antibody to the partially purified hCG in pre-malignant breast tumor homogenate.(All other details are explained

Gel filtration technique was used to separate free 125I-anti hCG antibody from 125I-anti hCG antibody/hCG complexes on Sephadex G

Introduction

The protein is a complex molecules that several methods, such as ultraviolet spectra, have been developed to determined it structure. In this spectrum, even limited information about protein molecular is obtained, it can absorb radiation over a wide range of spectra giving a wide range of studies.

Although several immunochemical, biochemical and biophysical studies was carried out to characterize protein-protein interaction, which is operative at almost every level of cell function, UV spectral method remain one of the most important spectral methods in immunology because it provides a sensitive, quantitative methodology for the study of antibody structure and specific ligand binding.

Many studies have been taken, under the supervising of Dr.Almudufer, to investigate many and different kind of proteins and steroids by studying there purified structure, their interaction and binding, studying the effect of different factor that might affect their absorbance . These studies reached to a conclusion that UV studies are useful for quantitative determination of complex formation and for studying these molecules which have a characteristics spectrum for each kind. hCG and its receptor of the proteins, which had, have their share of studying by UV spectra. In this chapter hCG studying have been take as a complementary study for the previous studies.

Materials and Methods

- **Patients**

The same benign and malignant premenopausal patient's tissues mentioned in chapter two were used in the following experiments.

Methods

- ***Gel Filtration Technique for Separation of Free and Bound 125I - anti hCG Antibody***

For Preparation of the Column, the dimensions (0.8x22 cm) were chosen according to the equation in the experiment of partially purification of hCG by gel filtration technique in chapter four. The Sephadex G-150 was used to separate free and bound ^{125}I-anti hCG antibody and eluent buffer and reagents was prepared as mentioned in the experiment of partially purification of hCG by gel filtration technique in chapter four. The void volume was determined and found to be 7 ml.

Separation Procedure of (125I-anti hCG Antibody/hCG) Complex

- **hCG from Benign (Fibrocyst) Breast Tumor Homogenate and its ^{125}I -anti hCG Antibody**

1. Benign breast tumor homogenate, 140μl, containing (1.400μg. ml^{-1}) was incubated with 6μl of ^{125}I-anti hCG antibody (8.82μg. ml^{-1}) and complete the reaction to a final volume of 500μl with Tris/HCl buffer 0.05M pH 7.2. The tubes were incubated for 150 min at 4°C.

2. At the end of incubation, the mixture was applied to the surface of a Sephadex G-150 (0.8x22 cm) equilibrated with Tris/HCl buffer 0.05M, pH 7.2 containing 5mM EDTA and (0.02%) NaN$_3$. Elution was carried out using the same buffer to separate hCG bound to ^{125}I-anti hCG antibody from unbound (Free) ^{125}I-anti hCG antibody with a flow rate (12ml/hr.), and fraction volumes of 1 ml were collected.

3. Gamma counter counted the radioactivity of each fraction.

Calculations

1. Radioactivity (c.p.m) of each eluted fraction was plotted against the fraction number.

2. The percent radioactivity was calculated by dividing the sum of the radioactivity of the fractions under each peak by the sum of radioactivity of all peaks appeared in the profile:

3. Percent radioactivity of each peak = $\dfrac{\text{Radioactivity per peak (c.p.m)}}{\text{Sum of radioactivity of all peaks (c.p.m.)}} \times 100$

- **hCG from Premenopausal Malignant (IDC) Breast Tumors Homogenate and Its ^{125}I-anti hCG Antibody**

1. Premenopausal malignant breast tumors, 66.6µl, containing (1.200µg.ml^{-1} protein) was incubated with 50µl of ^{125}I-anti hCG antibody (8.82µg.ml^{-1}) in a final volume 500µl with Tris/HCl buffer 0.05M, pH 7.2. The tubes were then incubated for 120 min at 45°C.

2. Steps 2 and 3 in experiment of hCG complex separation from benign tumor were repeated.

Calculation

The same calculation that mentioned in experiment of hCG complex separation from benign tumor were repeated. was used to calculate the radioactivity; the absorbence and the percent of radioactivity of each eluted fraction.

- **Standard hCG and Its ^{125}I-anti hCG Antibody Reagents**

1. Standard hCG, 12.5µl, containing (1218.75 µg. ml^{-1} protein) was incubated with 50µl of ^{125}I-anti hCG antibody (73.5µg.ml^{-1}) in a final volume 500µl with Tris/HCl buffer 0.05M, pH 7.2. The tubes were then incubated for 60 min at 25°C.

2. Steps 2, and 3 in experiment of hCG complex separation from benign tumor were repeated.

Calculation

The same calculation that mentioned in experiment of hCG complex separation from benign tumor were repeated. was used to calculate the radioactivity; the absorbence and the percent of radioactivity of each eluted fraction.

- **^{125}I-anti hCG Antibody Reagent**

1. Fifty microliters of ^{125}I-anti hCG antibody (73.5µg.ml^{-1}) was completed to 500µl with Tris/HCl buffer 0.05M, pH 7.2, then this volume was injected to the column as mentioned in step2, then steps 3 were repeated.

2. Gamma counter counted the radioactivity of each fraction.

Calculation

The same calculation that mentioned in experiment of hCG complex separation from benign tumor were repeated. was used to calculate the radioactivity; the absorbence and the percent of radioactivity of each eluted fraction.

Partially purification of hCG by Gel Filtration Technique

- *Preparation of the Column, Gel and Determination of Void Volume*

The Sephadex G-100 was used to isolate hCG and was prepared as mentioned in the experiment of partially purification of hCG by gel filtration technique, the void volume was determined and found to be 10ml.

- **Samples Addition**

The same standard hCG and malignant premenopausal patient's tissues sample and procedure mentioned in the experiment sample and standard addition in chapter four were used in this experiment.

Calculations

The same Calculations mentioned in the experiment sample addition in chapter four were used in this experiment.

- **. The UV Spectrum**

- **The UV Spectrum of (^{125}I-Anti hCG Antibody / hCG) Complex of Benign Breast Tumor**

The gel filtration profile in experiment of hCG complex separation from benign tumor gave two peaks. The fractions under first peak were pooled and the absorption spectrum was scanned in UV Region (200-350nm) against the appropriate blank in the reference beam.

- **The UV Spectrum of (^{125}I-Anti hCG Antibody / hCG) Complex of Malignant Breast Tumor**

The gel filtration profile in experiment of hCG complex separation from malignant tumor gave two peaks. The fractions under first peak were pooled and the absorption spectrum was scanned in UV Region (200-350nm) against the appropriate blank in the reference beam.

- **The UV Spectrum of (^{125}I-Anti hCG Antibody / hCG) Standard hCG Complex**

The gel filtration profile in experiment of hCG complex separation from standard tumor gave two peaks. The fractions under first peak were pooled and the absorption spectrum was scanned in UV Region (200-350nm) against the appropriate blank in the reference beam.

- **The UV Spectrum of ^{125}I-Anti hCG Antibody**

Twenty microliter of ^{125}I-Anti hCG Antibody was completed to 1ml with eluant buffer and the absorption spectrum was scanned in UV Region (200-350nm) against the appropriate blank in the reference beam.

- **The UV Spectrum of Partially purified hCG from malignant premenopausal Breast tumor tissue**

The gel filtration profile in experiment of sample addition in chapter five gave two peaks. The fractions that gave higher radioactivity under the first peak were pooled and the absorption spectrum was scanned in UV region (200-350nm) against the appropriate blank in the reference beam.

- **The UV Spectrum of the Standard HCG**

The gel filtration profile in experiment of hCG complex separation from benign tumor gave one peak. The fractions under the peak were pooled and the absorption spectrum was scanned in UV region (200-350nm) against the appropriate blank in the reference beam.

Factors Affecting the Absorption Properties of Benign, Malignant, Standard (^{125}I-Anti hCG Antibody/hCG) Complex, standard and partially purified hCG.

- **pH Effect**

Two hundred microliters of pooled fractions under the first peak of each group, which represents the (^{125}I-anti hCG antibody/hCG) complex was completed to 1 ml with different buffers at different pH values (4, 7.2, 11) then each of which was placed in 1cm cuvette in the sample beam and the buffer at the adjusted pH in the reference beam the absorption spectrum was measured in the area of (200-350nm).

- **Effect of Solvents Polarity**

1. Two hundred microliters of pooled fractions under the first peak of each group, which represents the (^{125}I-anti hCG antibody/hCG) complex was completed to 1ml with The eluent buffer in the presence of 20% ethanol.

2. The mixture was placed in the sample beam using 1cm cuvette against 20% ethanol prepared in the same buffer in the reference beam. The absorption spectrum was measured in the area of (200-350nm).

3. The experiment was repeated in the presence of 20% Glycerol, Ethylene glycol and Sucrose prepared in the same buffer.

- **Effect of Urea, KCl and (Urea, KCl) Mixture**

Two hundred microliters of pooled fractions under the first peak of each group, which represents the (^{125}I-anti hCG antibody/hCG) complex were pipetted in a set of three tubes. The volume was completed to 1ml with The eluent buffer contains (0.03M KCl, 8M urea and mixture 1:1 of both 0.03M KCl and 8M Urea) respectively, then each sample was placed in 1cm cuvette in the sample beam and the buffer at the same pH in the reference beam. The absorption spectrum was measured in the area of (200-350nm).

- *Spectrophotometric pH Titration of the Complex*

Two hundred microliters of pooled fractions under the first peak of each group, which represents the (^{125}I-anti hCG antibody/hCG) complex was completed to 1ml with different eluent buffers pH values at ranging from 8 to 11. The maximum absorbency of each sample was measured at 295nm; the absorbency of λ_{max} at each pH value was plotted versus the corresponding pH. Other Two hundred microliters of pooled fractions under the first peak of each group was completed to 1ml with different buffer pH values at ranging from 4 to 8. The maximum absorbency of each sample was measured at 211nm. The absorbency of λ_{max} at each pH value was plotted against the corresponding pH.

Results and Discussion

Gel Filtration Technique for Separation of Free and Bound ^{125}I-anti hCG Antibody

Gel filtration technique was used to separate (^{125}I-anti hCG Antibody/HCG) complexes from unbound ^{125}I-anti hCG Antibody for benign, malignant breast tumor and standard hCG, as shown in figures (72,A-C).

Filtration technique was preformed on G-150, yielding two peaks. First peak represents (^{125}I-anti hCG Antibody/HCG) complex, while the second peak represents the unbound (free) ^{125}I-anti hCG Antibody.

Separation process was depending upon the difference in the molecular weight of the compounds, so the first peak was presenting the complex and the second was presenting the unbound ^{125}I-anti hCG Antibody, due to that the complex has greater molecular weight than the unbound ^{125}I-anti hCG Antibody.

To insure this observation, unbound ^{125}I-anti hCG Antibody was applied alone to the same column as a comparison, as shown in figure (72,D). This experiment appeared only one peak in the same position of the second peak of the figures (72,A-C) which represent the unbound ^{125}I-anti hCG Antibody.

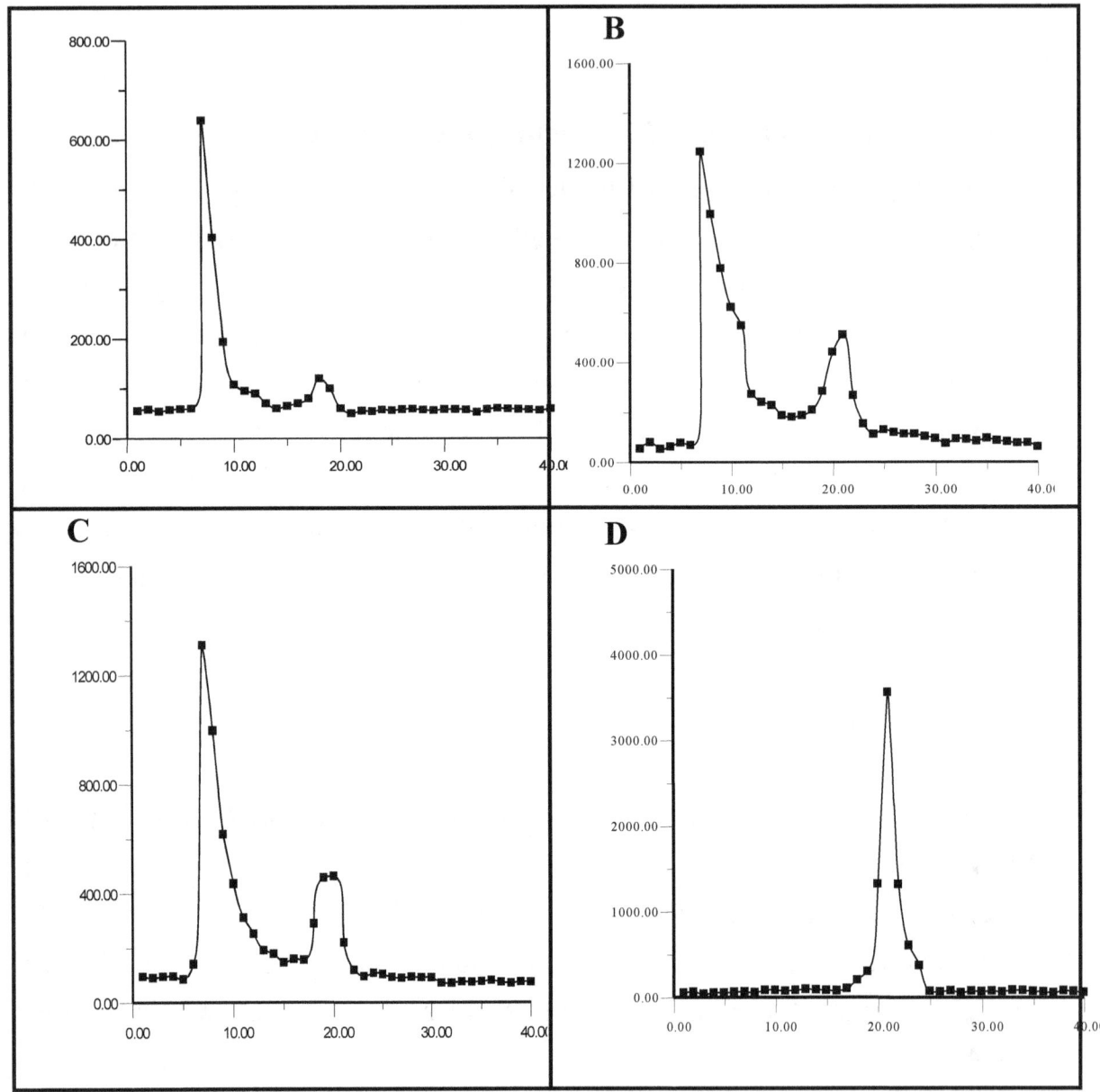

Figure (72): The elution profile of hCG from, A: benign breast tumor, B: pre-malignant breast tumor, C: standard hCG and the elution of ^{125}I-anti hCG Antibody (D). (All other detailes are explained in the text).

The UV spectra of (^{125}I-anti hCG Antibody/HCG) complexes, standard, partially purified hCG and ^{125}I-anti hCG Antibody

The UV spectra of (^{125}I-anti hCG Antibody/HCG) complexes at pH 7.2 shows one maximum wavelength at 222nm, 220nm and 220nm for standard hCG, benign and malignant breast tumor respectively, as shown in figure (73,A-C). This wavelength is assigned to the peptide bonds of the complex molecule), as shown in table (41).

The UV spectra of ^{125}I-anti hCG Antibody at pH 7.2 shows one maximum wavelength at 227nm, as in table (42), which could be a signed to the side chain chromophore of the amino acid residues, as in figure (73,D).

Figure (74,E and F) shows the UV spectra of standard and partially purified hCG at pH 7.2. It consisted of one maximum wavelength at 279nm for standard hCG, which shows a slit difference from the wavelength of partially purified hCG at 271nm. These wavelengths are assigned to the side chain chromophore of amino acids residues).

Table (41): The λ_{max} values of (^{125}I-anti hCG Antibody/HCG) complexes, ^{125}I-anti hCG Antibody, standard hCG and partially purified hCG. (All other details are explained in the text).

Fractions	λ_{max}
Standard (^{125}I-Anti hCG Antibody/ hCG) Complex	222nm
Benign (^{125}I-Anti hCG Antibody/ hCG) Complex	220nm
Malignant (^{125}I-Anti hCG Antibody/ hCG) Complex	220nm
^{125}I-Anti hCG Antibody	277nm
Standard hCG	279nm
Partially purified hCG	271nm

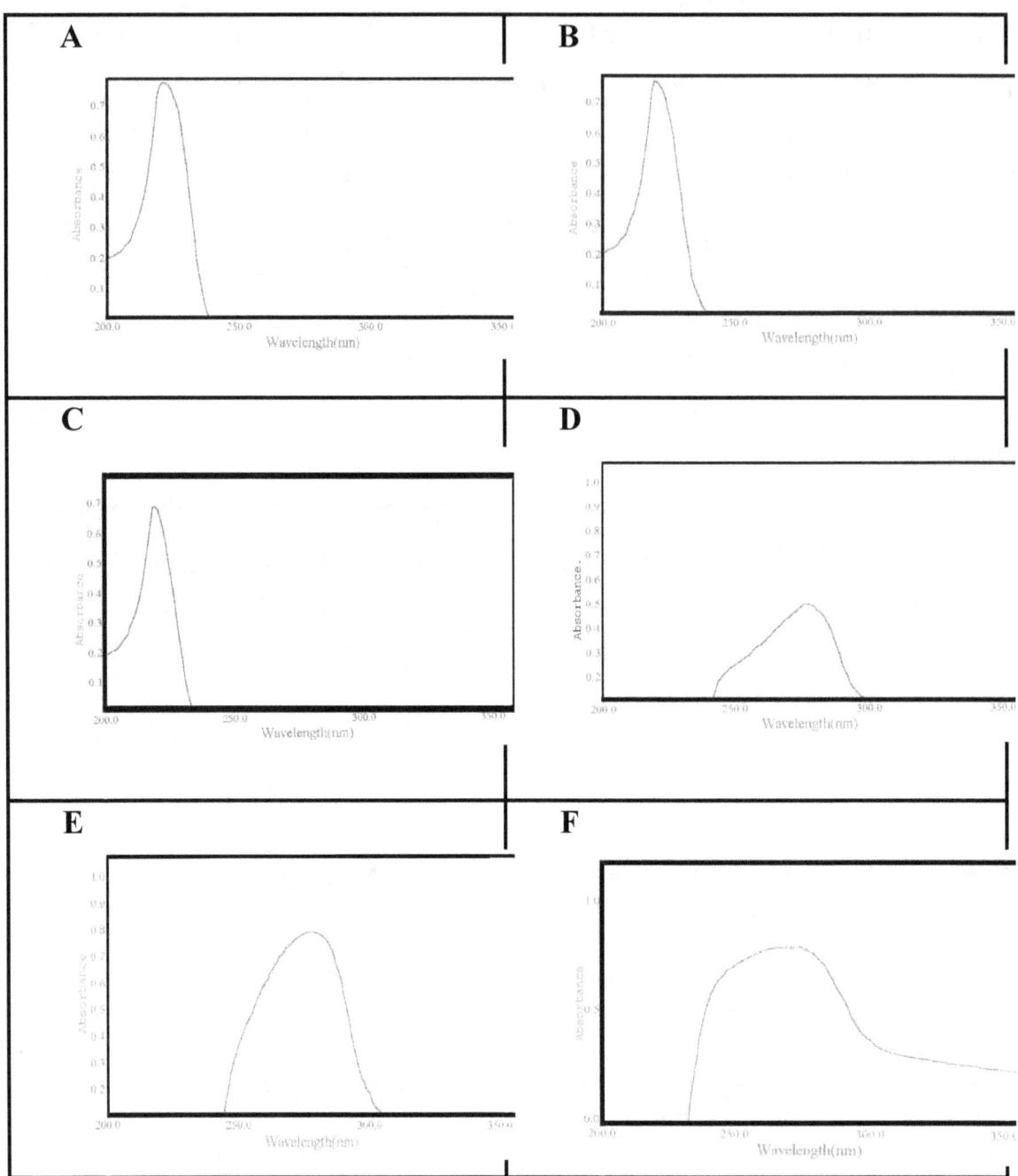

Figure (73): The UV spectrum at pH 7.2 for A: Standard complex, B: Benign complex, C: Malignant complex, D: ^{125}I-Anti hCG Antibody, E: Standard hCG and F: Partially purified hCG.

Factors Affecting the Absorption Properties of Standard, Benign, Malignant, (^{125}I-Anti hCG Antibody/hCG) Complex, standard and partially purified hCG.

Large number of environmental factors produces detectable change in λ_{max} and absorbency of protein molecules. Such factors are pH, polarity of the solvent).

pH Effect

The ionization state of ionizable chromophores in the protein molecular is determine by the pH of the solvents. So, the λ_{max} values of UV spectrum for the three hCG complexes, standard and partially purified hCG were determined at three different pH (4, 7.2, 11), as shown in table (43). In acidic region, pH 4, one maximum wavelength was obtained at 226nm, 226nm, 225nm for standard hCG, benign and malignant breast tumor respectively. With increasing of pH value of the complexes medium from 4 to 7.2, the λ_{max} were slightly shift to a shorter wavelengths for the three complexes 222nm, 220nm and 220nm respectively, while, no λ_{max} was observed for the three complexes at pH 11. Standard and partially purified hCG show two maximum wavelength at 236nm, 279nm and 227nm, 271nm respectively, while at pH 7.2 only one maximum wavelength was obtained at 279nm and 271nm respectively. The disappearance of the λ_{max} of the polypeptide bound could be due to a conformation change in the protein[274]. When the pH value was increased from 7.2 to 11, two maximum wavelengths were obtained again at 240nm, 279nm and 220nm, 274nm for standard and partially purified hCG respectively. The spectrum shifts of protein produced by pH cannot be simply attributed to the inductive effect at vicinal charges, such spectral changes must therefore be attributed mainly to rearranged of secondary and tertiary structure, although the possibility of field effects due to unusually close conjunction of charges aromatic groups is not excluded.

Table (43): The effect of different pH on λ_{max} values of (^{125}I-anti hCG Antibody/HCG) complexes, standard hCG and partially purified hCG. (All other details are explained in the text).

Fractions	pH	λ_{max}
Standard (^{125}I-Anti hCG Antibody/ hCG) Complex	4	226nm
	7.2	222nm
	11	-
Benign (^{125}I-Anti hCG Antibody/ hCG) Complex	4	226nm
	7.2	220nm
	11	-

Malignant (^{125}I-Anti hCG Antibody/ hCG) Complex	4	225nm
	7.2	220
	11	-
Standard hCG	4	236nm 279nm
	7.2	279nm
	11	240nm 279nm
Partially purified hCG	4	227nm 271nm
	7.2	271nm
	11	220nm 274nm

Effect of Solvents Polarity

Significant solvent effects can be induced by use of a mixture of water and a substance of a reduced polarity such as ethanol, ethylene glycol, glycerol and sucrose. Several spectra changes were obtained in the presence of these perturbants, like the alteration of λ_{max} positions and intensities of protein spectrum and the appearance of new chromophores on the surface of the protein. When no effect on the maximum absorbency is observed, this indicated no interaction or any change that happened between the solvent and the molecules. When one band was observed, may be attributed to the amino acids buried in the internal region of the protein and surrounded by non-polar amino acids. A blue shift in the wavelength in some solvents may be attributed to n $\rightarrow \pi^*$ transition and also may be attributed to hydrogen bonding of the solvent to the amino acid. The polar chromophores show a red shift if their

hydrogen bonding to solvent molecules increases in the excited state, but a blue shift if their hydrogen bonding to solvent molecules decreases in the excited state). Solvent may produce a blue shift by hydrogen bonding to the oxygen atom in the amino acids and withdrawing electrons from benzene chromophore). The λ_{max} values under the effect of 20% of different solvents at pH 7.2 are shown in table (45). For (^{125}I-Anti hCG Antibody/ hCG) Complex of standard hCG, the λ_{max} for the amide groups of polypeptide bond showed longer wavelength than the λ_{max} original in the presence of 20% ethanol (225nm), ethylene glycol (232nm), glycerol (235nm) and sucrose (231nm). For benign and malignant breast tumor complexes, Ethanol has showed no effect on the λ_{max} values of two complexes, as comparing these values to the λ_{max} in the table (46) While 20% ethylene glycol, glycerol, and sucrose caused in red shift in the λ_{max} of both benign and malignant complexes. For standard hCG, it seems that 20 % ethanol have no effect on the position of the λ_{max}. While for partially purified hCG there was a slight red shift toward longer wavelength as comparing these values to the λ_{max} in the table (46). In 20% ethylene glycol, stander and partially isolated hCG, was a significant red shift in the λ_{max} at 291nm and 317nm respectively with appearance of a new λ_{max} at 229nm for the amide groups of the polypeptide bond of partially purified hCG. When 20% glycerol was used, a slight blue shift in the λ_{max} at 277nm with appearance of a new λ_{max} at 236nm observed for standard hCG. While for partially purified hCG, the original λ_{max} was disappeared with appearance of a new λ_{max} at 233nm. Even 20% sucrose has no effect on the original λ_{max} of standard hCG but it caused in the presence of a new λ_{max} at 236nm. The original λ_{max} of partially purified hCG showed a slight blue shift at 269nm and the appearance of a new λ_{max} at 219nm.

Table (44): The effect of 20% of ethanol, ethylene glycol, glycerol and sucrose on λ_{max} values of (^{125}I-anti hCG Antibody/hCG) complexes, standard and partially purified hCG. (All other details are explained in the text).

Fractions	λ_{max}			
	20% Ethanol	20% Ethylene glycol	20% Glycerol	20% Sucrose
Standard (^{125}I-Anti hCG Antibody/ hCG) Complex	225nm	232nm	235nm	231nm
Benign (^{125}I-Anti hCG Antibody/ hCG) Complex	220nm	229nm	233nm	227nm
Malignant (^{125}I-Anti hCG Antibody/ hCG) Complex	220nm	227nm	233nm	228nm
Standard hCG	279nm	291nm	277nm 236nm	279nm 236nm
Partially purified hCG	247nm	317nm 229nm	233nm	269nm 219nm

Effect of Urea, KCl and (Urea, KCl) Mixture

In the presence of urea (8M), as shown in table (45), The λ_{max} of the polypeptide bond of hCG complexes showed significant red shift (15-23) for both standard and benign complex, while a slight red shift in the λ_{max} at 224nm for malignant complex was observed. These changes were concomitant with the appearance of a new λ_{max} at 279nm, 278nm and 278nm for the three complexes respectively.

A slight degrease in λ_{max} of standard hCG was obtained at 278nm, while a slight red shift than the original λ_{max} at 273nm and a new λ_{max} at 229nm was obtained for polypeptide bond of the partially purified hCG.

The blue shift could be due to that the protein is unfolded in the high concentration of urea, the chromophores buried in the interior are transformed into the solvent. This transfer produces a blue shift in the absorption of these chromophores, giving rise to a moderate decrease in the absorption at this wavelength. Also this transfer leads to a new λ_{max} to be appear.

Adding 0.03M KCl caused red shift for the λ_{max} of benign complex at 229nm with appearance of a new wavelength for the three complexes at 279nm, 278nm and 278nm respectively. When 0.03M KCl was added, no alteration in the position of the λ_{max} of standard hCG was detected. For partially purified hCG a slight red shift was observed for the λ_{max} at 272nm with a new λ_{max} at 230nm.

A slight blue shift was observed for the λ_{max} of both standard and malignant complexes at 219nm and 219nm respectively. While. Such a blue or a red shift can arise by introducing positive (K^+) or negative (Cl^-) charges near the chromophore (the amide group) which might interact directly with the π-electron system of the amide group.

When 8M urea was mixed with 0.03M KCl, there was no alteration in the λ_{max} position of the standard hCG, while there was a slight red shift (1nm) for the original wavelength at 272nm, like the same shift when 8M urea was used, with the appearance of a new wavelength at 233nm. For the three hCG complexes, a slight red shift was observed for standard and benign breast tumor complexes, while blue shift for malignant breast tumor complex was observed. It seems that the shift in the case of the mixture may be due to the effect of the 8M urea and 0.03M KCl. As seen, the changes in absorption near 230nm were larger than those near 278nm, which have been noted by Glazer who noted that solvent perturbation or denaturation the protein.

Table (45): The effect of 8M Urea, 0.03M KCl and mixture 8M Urea and 0.03M KCl on λ_{max} values of (^{125}I-anti hCG Antibody/HCG) complexes, standard and partially purified hCG. (All other details are explained in the text).

Fractions	λ_{max}		
	8M Urea	0.03M KCl	Mixture 8M Urea and 0.03M KCl
Standard (^{125}I-Anti hCG Antibody/ hCG) Complex	237nm 279nm	219nm 279nm	224nm
Benign (^{125}I-Anti hCG Antibody/ hCG) Complex	243nm 278nm	229nm 278nm	229nm
Malignant (^{125}I-Anti hCG Antibody/ hCG) Complex	224nm 278nm	219nm 278nm	219nm
Standard hCG	278nm	279nm	279nm
Partially purified hCG	229nm 272nm	230nm 272nm	233nm 272nm

Spectrophotometric pH Titration of the Complex

Spectrophotometric pH Titration is the following of the change in absorbance of the chromophore with increasing pH. The study of protein structure require the determination of pka values for proton dissociation from ionizable amino acide chains, because these values give an indication of the location of the amino acid in the protein. This can often be done spectrophotometrically becouse dissociation often changes the spectrum of one of the chromophores, the observation of tyrosine dissociation was performed by measuring the absorption at 295nm (λ_{max} for the ionized form of tyrosine), and the observation of histidine dissociation was carried out by measuring the absorption at 211nm[283].

Figure (74) shows the pH titration curve of standard hCG and partially purified hCG for histidine and tyrosine respectively. Figure (47,A) shows that pka values for tyrosine are (3) for standard hCG and (2.8) for partially purified hCG. From the same curve it seems that about (40%) histidine residues of standard hCG are internal and a large arise in the absorbance at very high pH was observed. While, for the partially purified hCG about (35%) histidine residues are internal. The pka value from figure (47,B) appears to be about (40%) for tyrosyl residues are internal for standard hCG and (28%) tyrosine residue are internal for purified hCG, indicating the high content of histidine compared to the low content of tyrosine in the standard hCG and the partially purified hCG

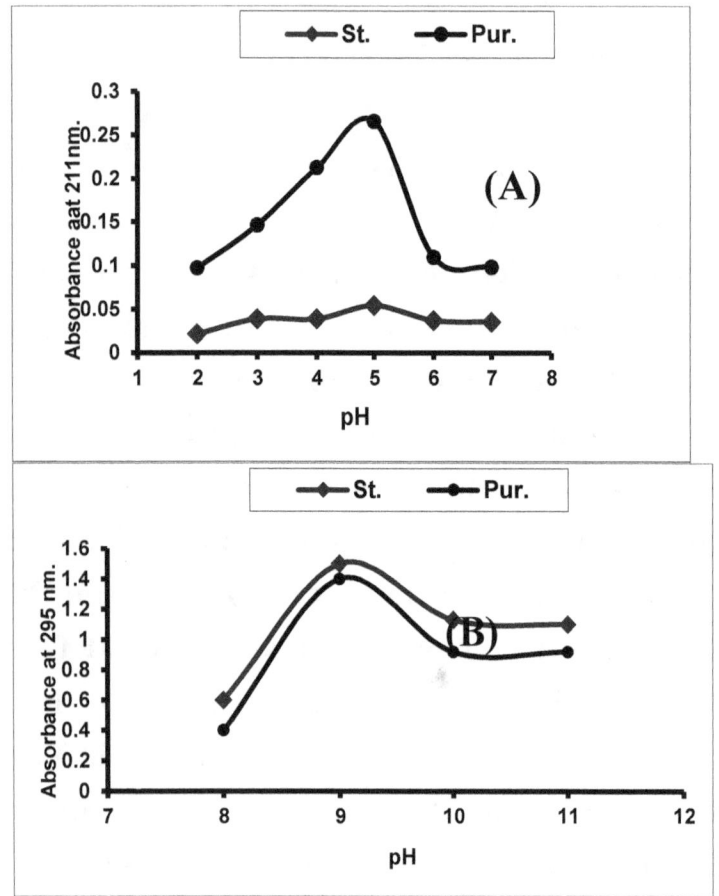

Figure (74): Spectrophotometric pH titration of standard hCG and partially purified hCG: (A) for histidine, (B) for tyrosine.

(St): Standard hCG, (Pur): partially purified hCG. (All other details are explained in the text).

Chapter Five

Isolation of β_2M Receptors From Breast Tumors by Gel Filtration Chromatography

Introduction

Column chromatography represents an extremely efficient family of techniques for the isolation and identification of proteins in biological extracts where they exploit the differences in physical and chemical properties.

The purification of proteins is the most common use of gel filtration chromatography due to the ability of a gel to fractionate molecules on the basis of size and shape .

The efficiency of the isolation method is given by the purification factor by which the specific activity of the preparation has increased .

Unfortunately, no similar studies are available dealing with the receptors for β_2M. However, gel filtration chromatography has been employed to isolate the β_2M receptor from murine cells. The isolated receptor has been found to comprise a 48 KD glycoprotein. Occasionally, variable quantities of non-glycosylated 25 KD component was also present .

The scope of this chapter is to retain as much of the β_2M receptors as possible while getting rid of as much of the other proteins as possible and to definitely establish the optimum conditions of their binding reaction.

The receptors for β_2M were isolated from a malignant breast tumor homogenate by using gel filtration chromatography technique.

Gel filtration chromatography technique was also used in the isolation of (^{125}I-β_2M/receptor) complexes, which were prepared from crude and isolated receptors. The results revealed the presence of two forms of β_2M receptors (protein I and II) and also, two complexes appeared for those forms.

Gel filtration chromatography technique was also employed in the determination of molecular weight of the isolated β_2M receptors using a calibration kit containing highly purified low molecular weight proteins (albumin, ovalbumin, chymotrypsinogen and ribonuclease).

The molecular weight of protein I and II was found to be 52.8 KD and 29 KD respectively. The isolation of the receptors showed a purification fold of 6 for protein I, and 4.5 for protein II.

Tissue Homogenate

A breast tumor homogenate from a post-menopausal patient with infiltrative ductal carcinoma was used in the experiments of this chapter.

Methods

Preparation of the Column

A sephadex G-100 column was used to isolate β_2M receptors, the dimensions of the column were chosen according to the following equation:

$$\text{Diameter} = \sqrt{\frac{m}{10}} \quad \text{and} \quad L = 30 \times \text{diameter}$$

Where:

m: amount of protein in milligrams.

L: length of the column.

Preparation of the Gel

The gel was prepared by allowing the pre-swollen gel to swell again in tris buffer solution (0.05 M) pH 7.4, then it was left to settle and the excess of the buffer was decanted. This step was repeated several times. The gel was degassed using evacuating pump and slurry was left for 24 hours to be equilibrated with buffer.

The swollen gel was suspended and carefully poured into a vertical glass column (0.9 × 30 cm.) down the wall using a glass rod. After the gel has settled, the column was equilibrated with tris buffer for 24 hours.

Void Volume Determination

The void volume of the column was determined by using blue dextran 2000 at a concentration of 2 mg/ml dissolved in tris-buffer pH7.4, then the elution was carried out with the same buffer at a flow rate of 10ml/hr. Fractions of 1ml were collected and their absorbencies were measured at 600 nm.

Gel Filtration Chromatographic Studies

Isolation of β_2M Receptors by Sephadex G-100 Column

1. The sample of tissue homogenate (555 μl) containing approximately 10 mg protein was applied to the surface of the gel, and then equilibrated with 0.05M tris buffer pH 7.4.

2. The sample was eluted by using the same tris buffer with a flow rate of 10 ml /hr. Fractions of 1ml volume were collected, and the gel filtration was carried out at 10°C.

3. The protein content of each fraction was determined according to Lowry method; the absorbance of each fraction was measured at 280 nm.

4. The assay method was carried out using the collected fractions in order to identify the one that contains the β_2M receptors.

Calculations

1. The percentage of specific binding (SB%) was calculated ..

2. The SB% values were plotted against the fraction number.

3. The purification fold of β_2M receptors was calculated by the equation:

$$\text{Purification Fold} = \frac{\text{Bound of purified receptors/mg Protein}}{\text{Bound of crude receptors/mg Protein}} \times 100$$

Isolation of (^{125}I-β_2M / Receptor) Complex

Procedure

1. ^{125}I-β_2M was reacted with its receptors in the crude sample of breast tumor homogenate at their optimum conditions.

2. At the end of the reaction, 555 ml of the reaction mixture were applied to the surface of the gel, equilibrated with tris buffer pH 7.4, a flow rate of 10 ml /hr was adjusted.

3. Fractions of 1 ml were collected. The radioactivity of each fraction was then counted.

4. Unreacted ^{125}I-β_2M was poured into the column, and the radioactivity of each fraction was counted.

Calculations

The counted radioactivity as (cpm) was plotted against the corresponding fraction number.

Determination of Molecular Weight for the Isolated Receptors

The molecular weight of the isolated receptors was determined using pharmacia calibration kit. This kit contains highly purified low molecular weight proteins (albumin, ovalbumin, chymotrypsinogen and ribonuclease).

Procedure

1. At the first run, 555 μl of the mixture (albumin and chymotrypsinogen) were applied to the sephadex G-100 column.
2. Fractions of 1ml volume were collected at a flow rate of 10 ml /hr.
3. At the second run, 555 μl of the mixture (ribonuclease and ovalbumin) were applied to the sephadex G-100 column. Then step 2 was repeated.

Calculations

The K_{av} values for standard proteins and isolated receptors were calculated using the formula:

$$k_{av} = \frac{V_e - V_o}{V_t - V_o}$$

Where:

V_o : void volume.

V_e : elution volume of each protein.

V_t : total gel bed volume which was calculated as follows:

$$V_t = \left(\frac{d}{2}\right)^2 \times 3.14 \times h$$

d=0.9 cm.
h=30 cm.

Determination of the Optimum Conditions for ^{125}I-$\beta_2 M$ Binding to the Isolated Receptors

The optimum conditions of ^{125}I-$\beta_2 M$ binding to its isolated receptors including protein concentration, tracer concentration, pH, temperature and time were determined using the same experiments previously mentioned.

Results and Discussion

Isolation of $\beta_2 M$ Receptors by Sephadex G-100 Column

The isolation of $\beta_2 M$ receptors was achieved using gel filtration chromatography technique in which the protein content of the malignant post-menopausal breast tumor homogenate was separated according to the differences in molecular weight. Figure (74) shows the elution profile of blue dextran 2000 that was used to determine the void volume of sephadex G-100 column employed in these experiments. The void volume was found to be 10 milliliters.

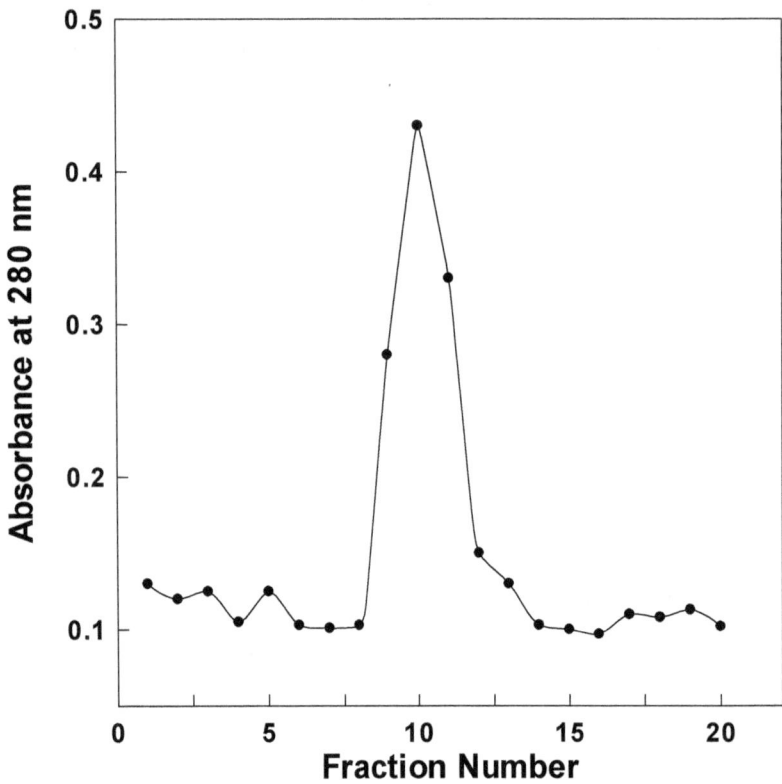

Figure (74). The elution profile of blue dextran, using Sephadex G-100 gel, 10ml/hr flow rate, tris-buffer pH 7.4 and at 25°C. All details are described in section (3.2.4).

The elution profile of the breast tumor homogenate, figure (75), revealed two main peaks at fraction number 11 and 26 when the collected fractions were measured at 280 nm. representing the protein content of the homogenate.

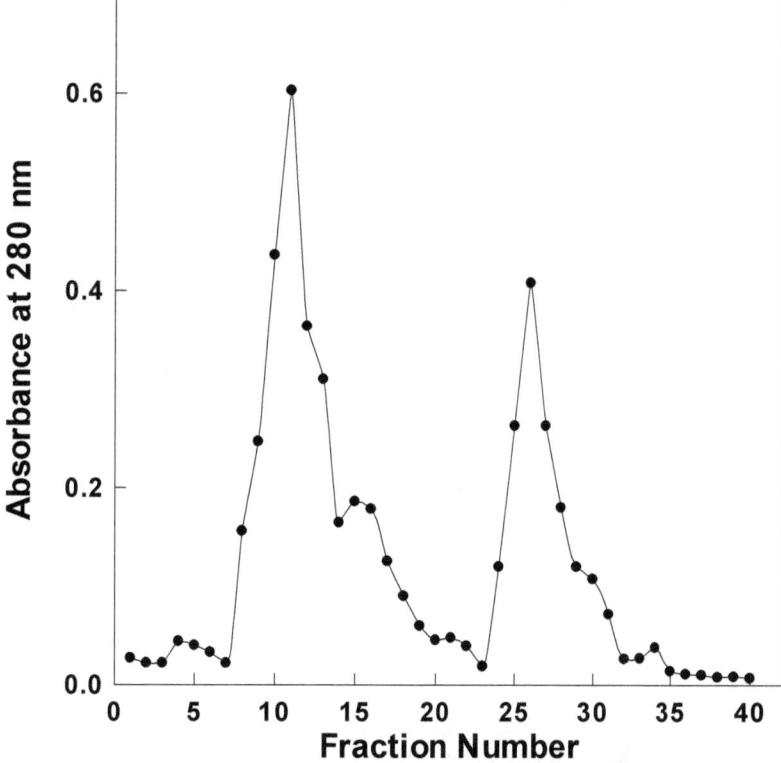

Figure (75). The elution profile of the breast tumor homogenate, using Sephadex G-100 gel, 10ml/hr flow rate, tris-buffer pH 7.4 and at 25°C. All details are described in section (3.3.1).

The collected fractions were reacted with ^{125}I-β_2M at their optimum conditions previously determined, the binding assays gave rise to figure (76) in which two peaks can be seen at fraction number 14 and 25 indicating the presence of β₂M receptors in those fractions.

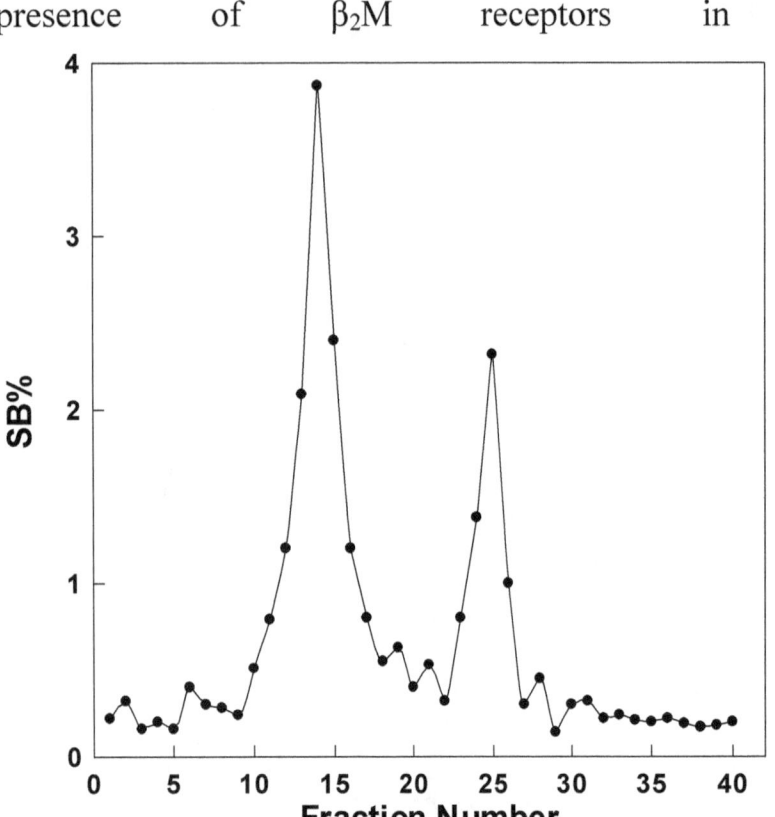

Figure (76). The binding of the eluted fraction of the breast tumor homogenate, using Sephadex G-100 gel, 10ml/hr flow rate, tris-buffer pH 7.4 and at 25°C.

Isolation of (^{125}I-β₂M/Receptor) Complex

^{125}I-β₂M and its receptors in crude breast tumor homogenate were incubated at their optimum conditions; the resulting complex was isolated from the free ^{125}I-β₂M by its application to the gel filtration column. The elution profile shown in figure (77) reveals two peaks related to two complexes and a third peak at fraction number 36 related to the free ^{125}I-β₂M.

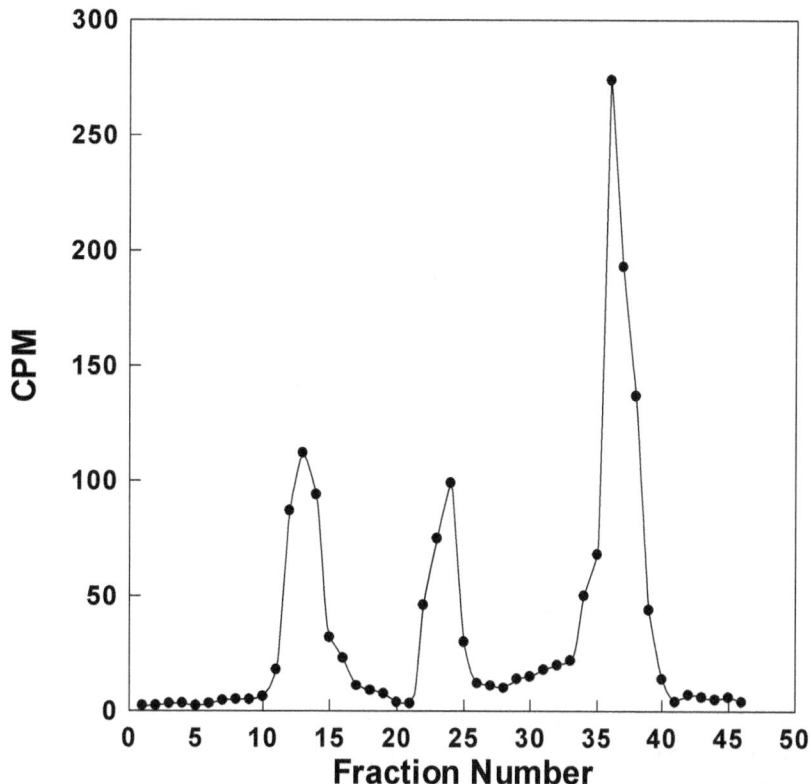

Figure (77). The elution profile of the complex (^{125}I-β_2M/receptor) prepared from the crude sample, using Sephadex G-100 column, 10ml/hr flow rate, tris-buffer pH 7.4 and at 25°C.

The fractions forming the two peaks in figure (78) were pooled separately and reacted with ^{125}I-β_2M at the optimum conditions. Each complex resulted from the binding reaction was separated from the free ^{125}I-β_2M by gel filtration using the same column. Figure (79) shows the elution profile of the two complexes prepared in this experiment with their peaks appeared at fraction number 14 and 25. These results are in agreement with those obtained from the experiment of the isolation of β_2M receptors from crude sample, figure (80), they are both confirm the presence of two protein components that are able to be bound to β_2M.

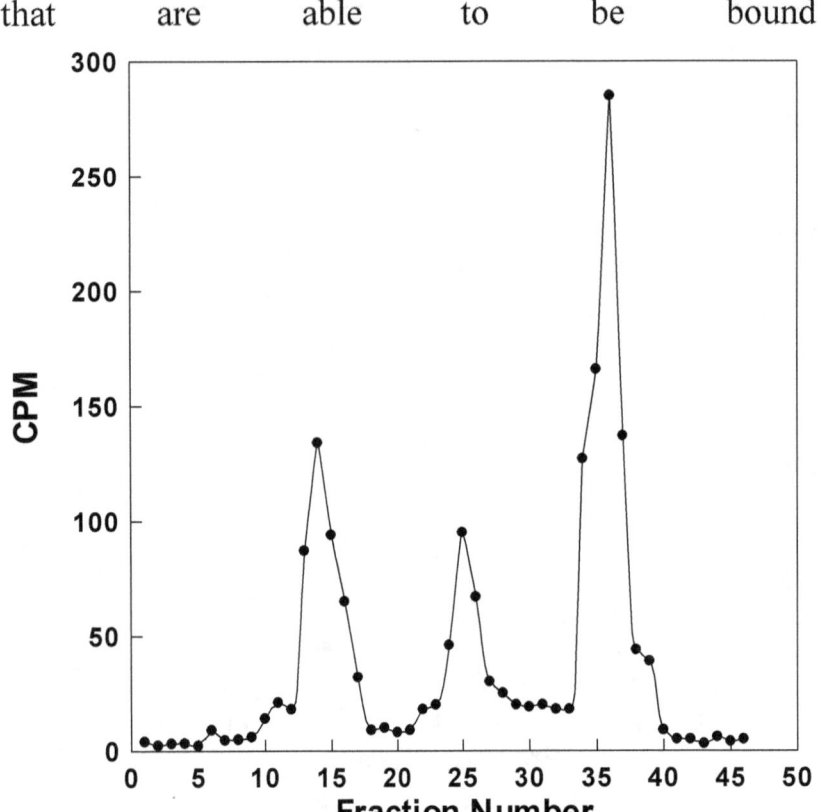

Figure (78). The elution profile of the complex (^{125}I-β_2M/receptor) prepared from the pooled peaks, using Sephadex G-100 column, 10ml/hr flow rate, tris-buffer pH 7.4 and at 25°C. All details are described in the text.

To ascertain the position of the free tracer peak, ^{125}I-β_2M was applied to the gel filtration column. The results gave rise single peak appeared at fraction number 37.

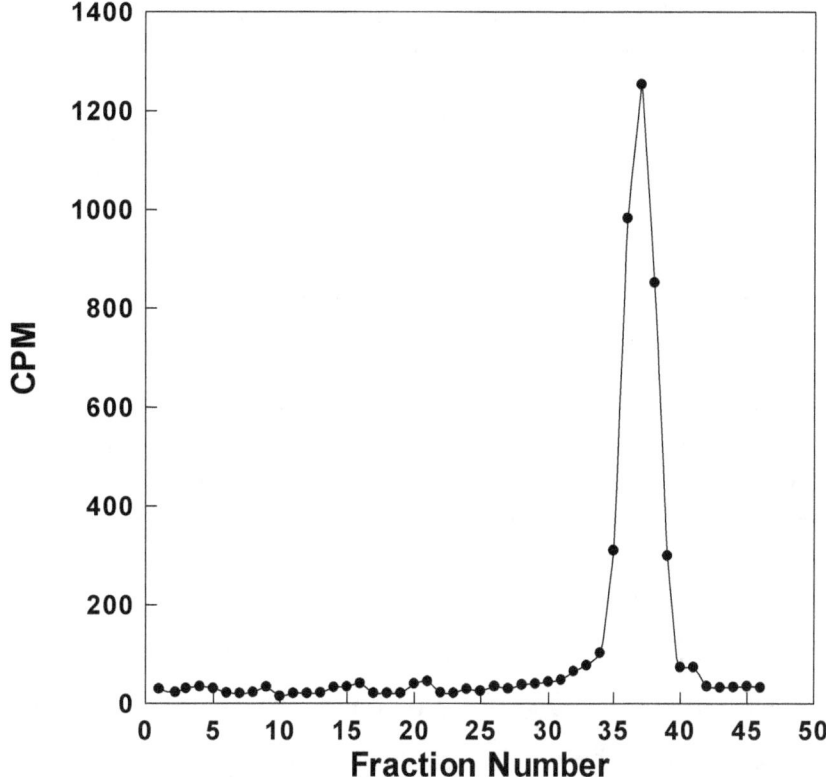

Figure (79). The elution profile of the free ^{125}I-β_2M, using Sephadex G-100 column, 10ml/hr flow rate, tris-buffer pH 7.4 and at 25°C

Gel filtration chromatographic studies indicate clearly that ^{125}I-β_2M could bind to two protein components (receptors) having a certain difference in molecular weight. To explain further these results, it could be said that the low molecular weight component might be a degradation product of the high molecular weight one. Another possibility, the obtained results may indicate that β_2M receptors exist in two isoforms. Whatever the matter, the specific binding per milligram protein has increased, table (46).

It seems that the presence of the receptors with many other proteins in the crude homogenate could markedly attenuate their binding affinity towards ^{125}I-β_2M, the gel filtration process act to decrease the number of proteins in the reaction mixture and hence increase the probability of ^{125}I-β_2M to bind its receptors.

Table (46). Purification parameters of β_2M receptors isolated from a post-menopausal breast tumor homogenate (ThI) using gel filtration chromatography.

Receptor	Protein (mg)	Specific binding	Specific binding /mg protein	Purification fold
ThI	0.40	12.5	31.25	1.0
Protein I	0.04	7.4	185.00	6.0
Protein II	0.04	5.6	140.00	4.5

Determination of Molecular Weight

Figure (80) shows the elution profile of the standard proteins used in this experiments. The K_{av} value of each protein was plotted against logarithm of molecular weight of the corresponding protein. The straight line equation generated from the resulting plot, figure (81), was used to determine the molecular weight of the isolated β_2M receptors. This experiment showed that the first receptor has a molecular weight of 52.8 KD while the second of 29 KD.

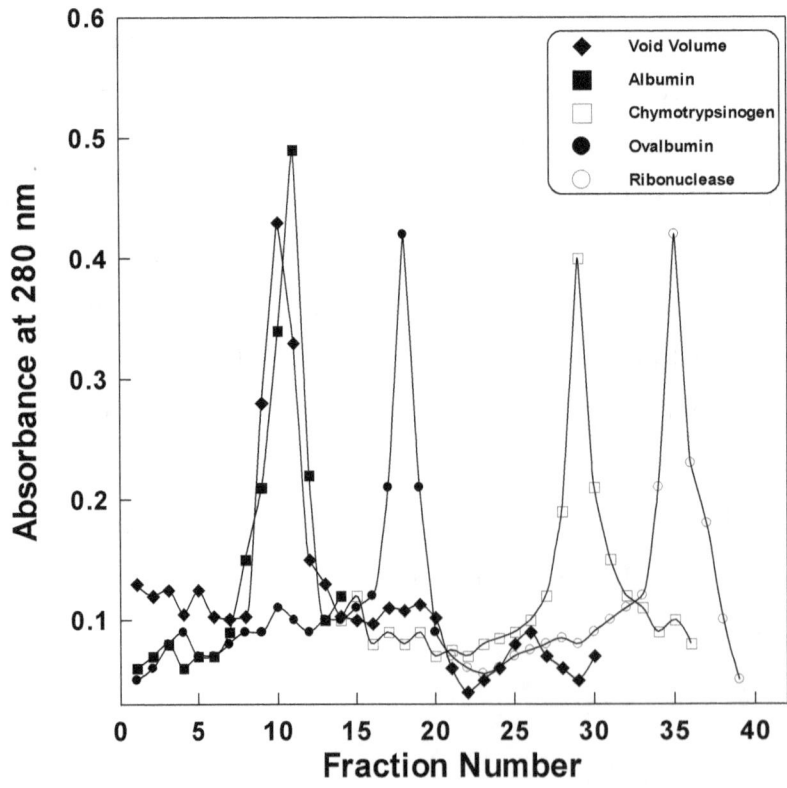

Figure (80). The elution profile of standard proteins used in molecular weight determination, using sephadex G-100 column, 10 ml/hr flow rate, tris-buffer pH 7.4 and at 25°C.

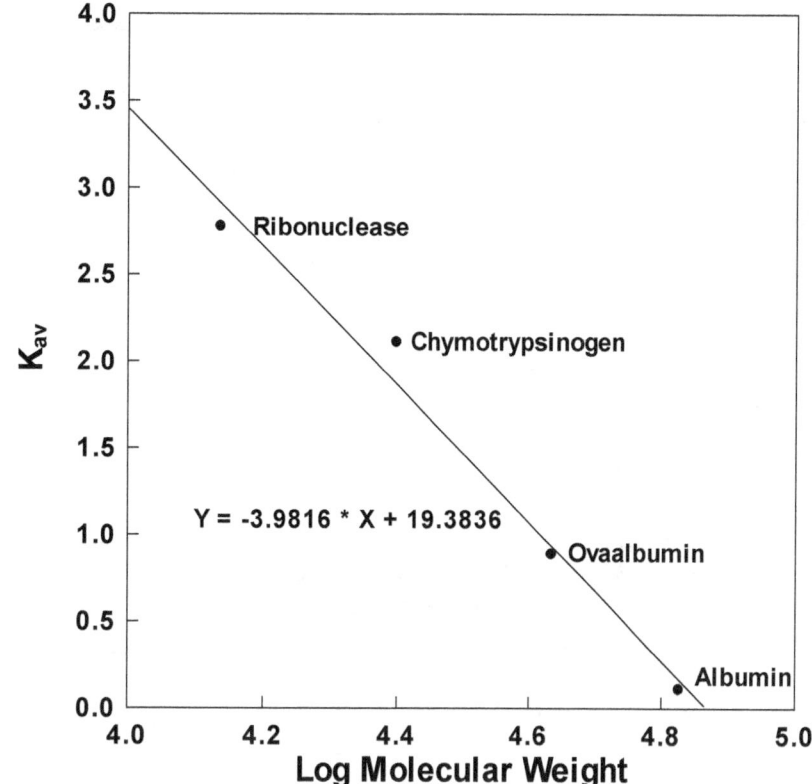

Figure (81). Calibration curve for molecular weight determination by gel filtration chromatography using low molecular weight Pharmacia Calibration Kit.

Determination of the Optimum Conditions for ^{125}I-β_2M Binding to the Isolated Receptors

Determination of Optimum Receptor Concentration

The effect of increasing amounts of isolated β_2M receptors (protein I and II) was investigated, figure (82) shows an increase in the specific binding (SB%) with increasing receptor concentration (for the two proteins) until reaching the saturation region where the binding remains unchanged whatever the concentration is.

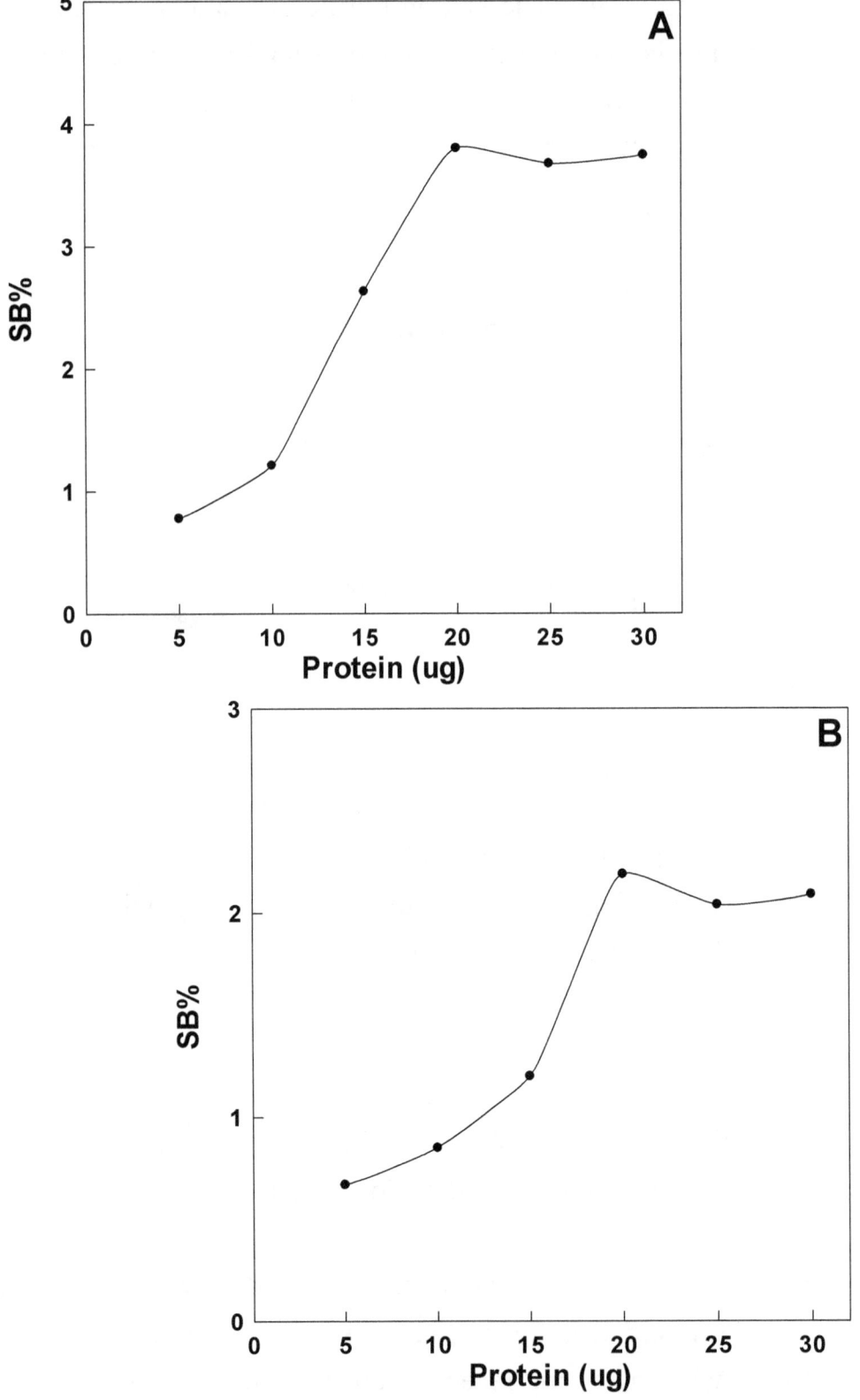

Figure (82). The effect of protein concentration on ^{125}I-β_2M binding to the isolated receptors, A) Protein I, B) Protein II.

Determination of Optimum ^{125}I-β_2M Concentration

Different concentrations ^{125}I-β_2M were used in the reaching mixture to investigate its effect on the binding. Figure (83) shows a behavior similar to that obtained with the crude homogenate in which a biphasic response curve may indicate a probable multivalent character of β_2M.

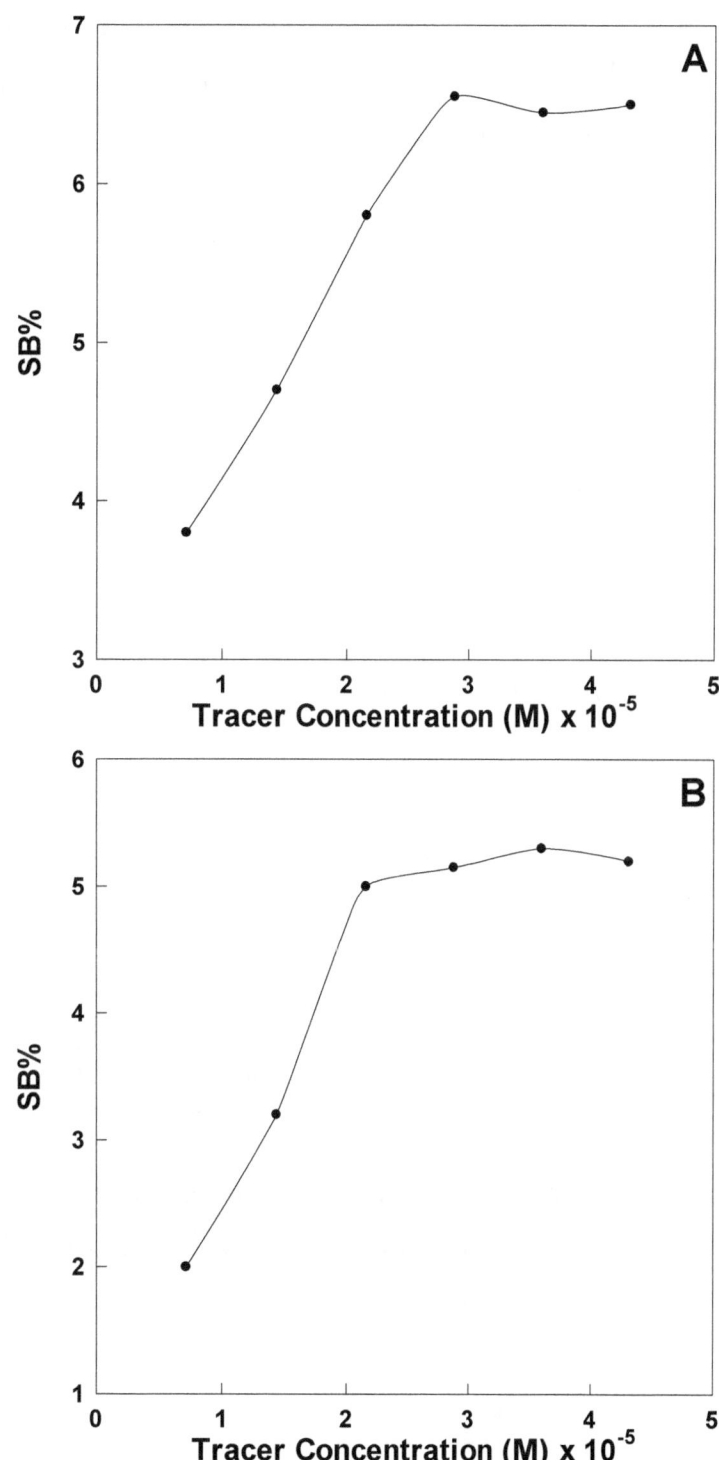

Figure (83). The effect of ^{125}I-β_2M concentration on the binding with the isolated receptors, A) Protein I, B) Protein II.

Determination of the Optimum pH

Figure (84) demonstrates the effect of medium pH on ^{125}I-β_2M binding to the partially purified receptors. It is noticeable that the percentages of specific binding decline above and below the optimum value of pH 7.4. Generally, changing the environment pH can lead to a dramatic changes in protein conformation through the induction of protonation-deprotonation process within the ionizable groups resulting in formation of improper ionic forms.

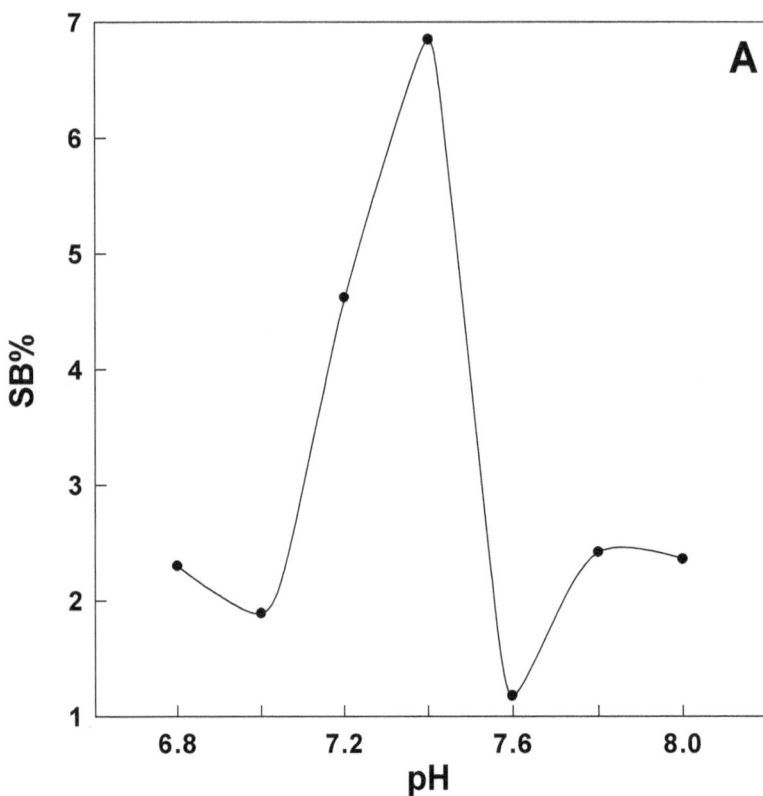

Figure (84). The effect of medium pH on ^{125}I-β_2M binding to the isolated receptors, A) Protein I, B) Protein II.

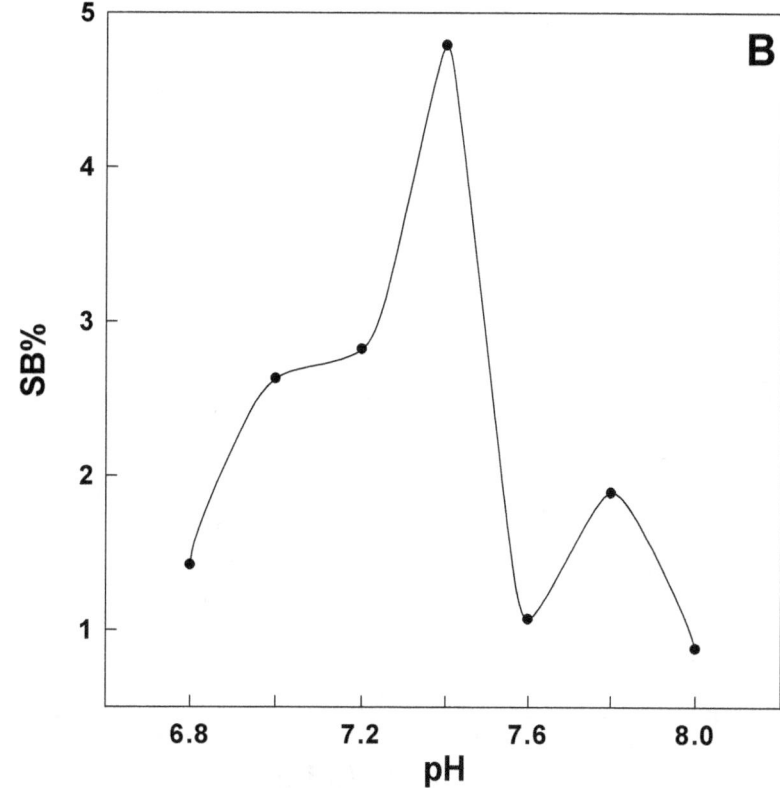

Figure (84). Continued.

Time Course of ^{125}I-β_2M Binding to its Isolated Receptors

The results of time course experiments shown in figure (85) indicate that the binding reaction is a time and temperature dependent process. It was found that the optimum reaction occurs at 37°C after incubation for 90 minutes. These results are in agreement with those obtained when the crude receptors were used.

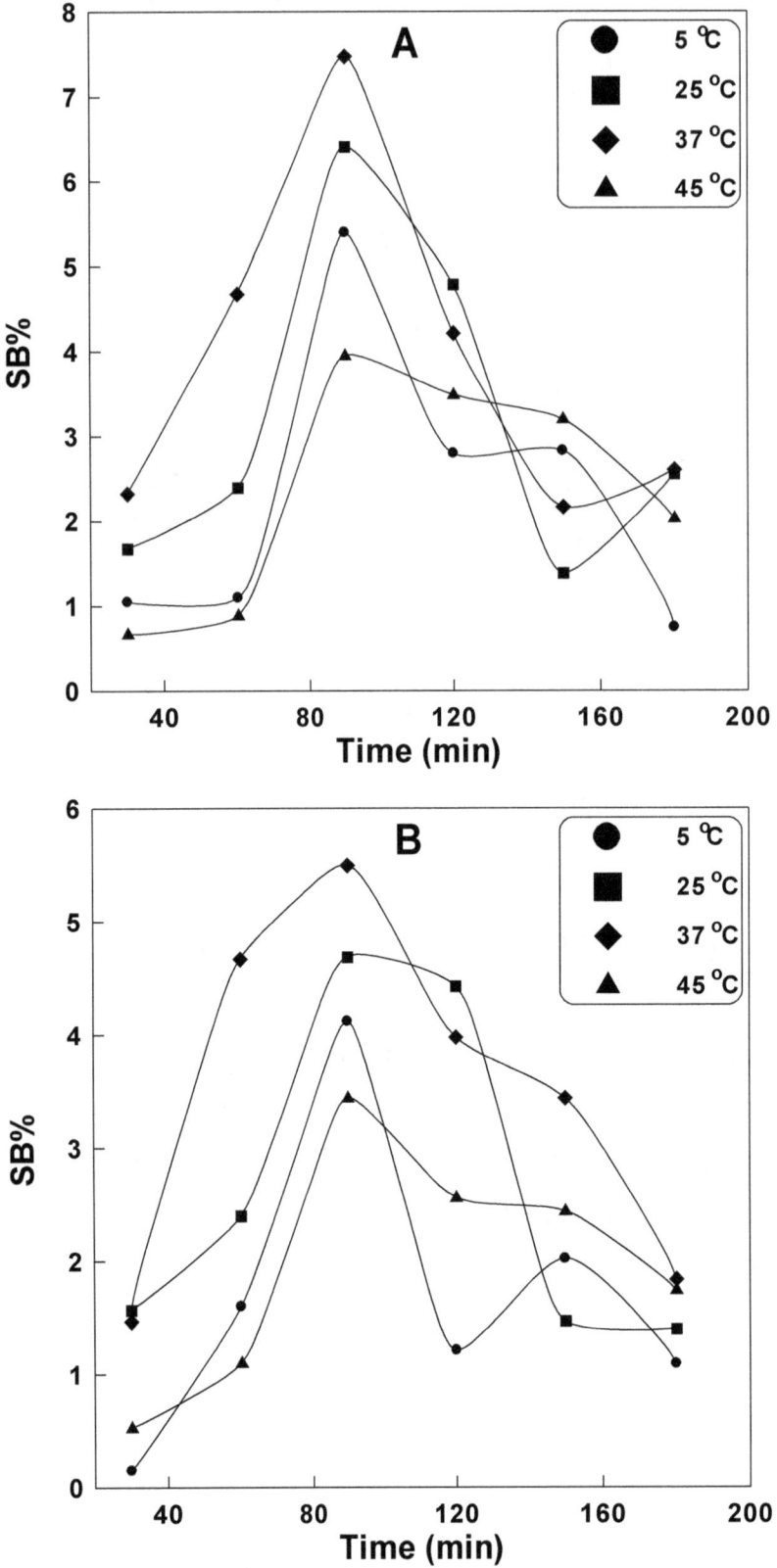

Figure (85). Time course of ^{125}I-β_2M binding to the isolated receptors, A) Protein I, B) Protein II.

Chapter six
Biochemical studies of Sialic Acids in Breast Tumors

Introduction

The Sialic Acids:

The term sialic acids represent a group of nine-carbon sugar collectively called neuraminic acid. Sialic acids, are the N-acetyl or N-glycolyl derivatives of the parent D-neuraminic acid (Fig86). They contain carboxyl, acetamide, ketone, deoxy group, as well as hydroxyl group.

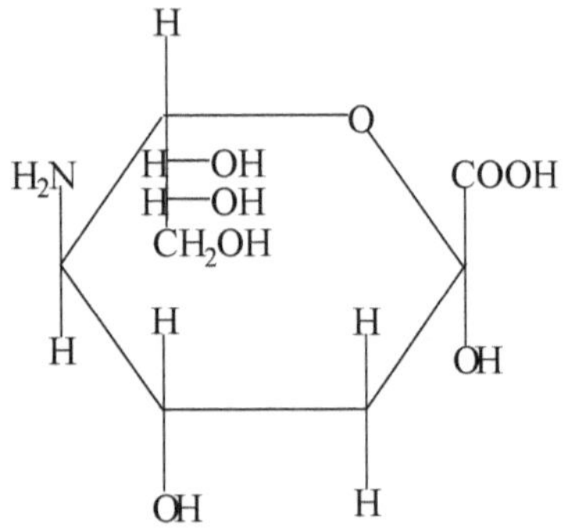

($C_9H_{17}O_8N$) D-Neuraminic acid

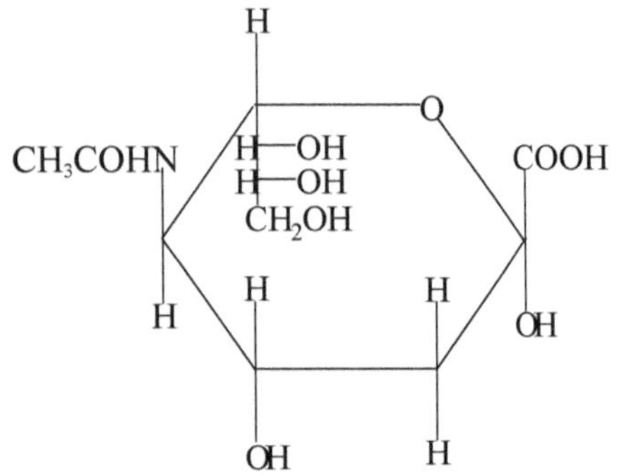

($C_{11}H_{19}O_9N$) N-Acetyl Neuraminic acid

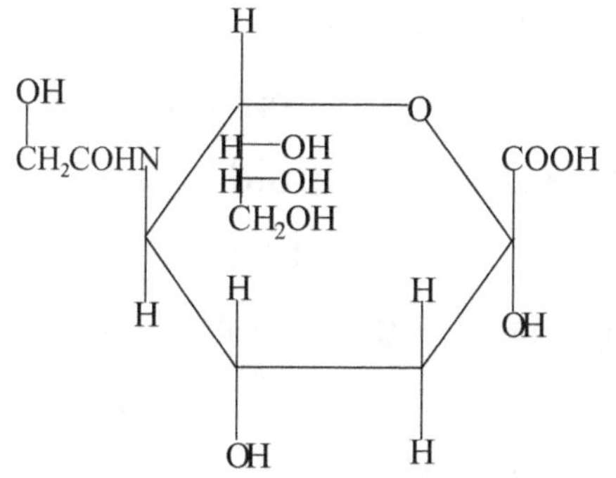

($C_{11}H_{19}O_{10}N$) N-glycolyl Neuraminic acid

Fig. (87): Structure of the Sialic acids

Some of these sialic acids are known to carry o-acetyl substituents located at carbon No. 4 (C_4) and/or at the various position of the polyhydroxy side chain which are carbon No. 7, 8 and 9 (C_7, C_8, and C_9), (Fig. 88).

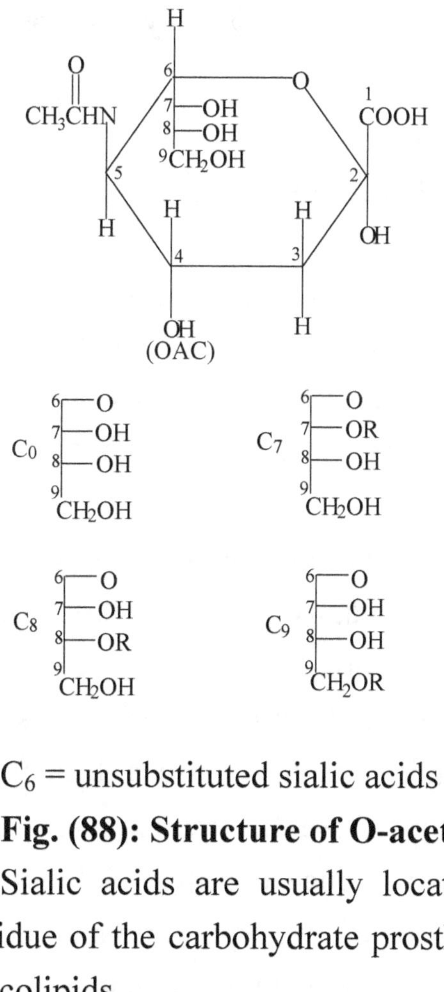

C_6 = unsubstituted sialic acids

Fig. (88): Structure of O-acetyl substitute sialic acids.

Sialic acids are usually located terminally and they are non-reducing residue of the carbohydrate prosthetic group of numerous glycoproteins and glycolipids.

Types of Sialic Acids:

Ten sialic acids were isolated, they differ in having various degrees of O-acetylation and N-glycolyation. (Nana) which is N-acetyl neuraminic acid is the most common in mucous glycoproteins and in glycolipids. The major derivatives of neuraminic acid as components of glycoproteins and glycolipids are N-acetylated or N-glycoloylated and frequently also O-acetylated. O-acetyl groups were found at C-4, C-7, C-8 and most frequently at C-9 of Neu5Ac or Neu5Gc; in some cases O-lactoyl groups are also present at C-9. Mono-and oligo-O-acetylated sialic acids were isolated in pure form e.g. from glycoproteins of submandibular glands and horse erythrocyte membranes. In rat, rabbit and mouse erythrocyte membranes 9-O-acetylated sialic acids, and in horse erythrocytes 4-O-acetylated neuraminic acid derivatives were identified. It is difficult to identify the particular type of sialic acids present in certain tissue because, the two common analytical methods used, the Erlich and the Resorcinol method, are carried out in strong acids and under such condition the acyl groups are removed.

Physical and Chemical Properties of Sialic Acids:

Purified sialic acids are colourless; it is only after heating at 100°C for several hours that they turn yellowish. They do not melt, but decompose with discoloration over a range of several degrees, discoloration usually preceding decomposition. The sialic acids are rather strong acids (α-keto-acids) with a pK_a value of 2.6.

The sialic acids and methoxyneuraminic acid are easily soluble in water. Bovine O,N-diacetylneuraminic acid dissolves readily in methanol, the other acids are only sparingly soluble in methanol. All sialic acids and methoxyneuraminic acid are insoluble in ether and light petroleum. The sialic acids are very unstable to both acid and alkali; aqueous solutions decompose already when kept for some time at room temperature due to their acidity. All sialic acids, but not methoxyneuraminic acid, reduce Fehling's solution on heating. The sialic acids consume in the oxidation by hypoiodite an amount of iodine varying with time between 33 and 90% of that calculated for the presence of one aldehyde group per molecule.

Methods for Determination of Sialic Acid:

A variety of procedures have been used for the measurement of total sialic acid. These can be broadly classified as colorimetric, fluorometric, enzymatic and highly sensitive high performance liquid chromatographic (HPLC) procedures.

Colorimetric procedures:

Two classical procedures have stood the test of time. One uses resorcinol and the other uses periodic and thiobarbituric acids. The resorcinol based assay uses heat and strong acid to hydrolyze glycosidic bonds. The released free sialic acids are reacted with resorcinol and copper ions to give a colored compound, which is extracted and measured at 580 nm. While the procedure described by Warren is typical of the periodic and thiobarbituric acid procedure, which measures only free sialic acid that is released after an initial hydrolysis step. In this procedure, formyl pyruvic acid formed as a result of periodic acid oxidation of free sialic acid is reacted with thiobarbituric acid to yield a red color, which is measured at 549 nm.

Fluorometric procedures:

In a typical and more specific assay formaldehyde that is formed upon oxidation of free sialic acid by periodic acid is reacted with acetyl acetone. The yellow product is excited at 410 nm and the resulting fluorescence is measured at 510 nm.

Enzymatic procedures:

Enzymatic assays are based on conversion of free sialic acids released by the enzyme neuraminidase to **pyruvate and acetyl monnosamine with the aid of the enzyme acetyl neuraminic acid pyruvatelyase or** neuraminic acid (NANA) aldolase. The resulting pyruvate can be coupled to the lactate dehydrogenase NADH system to measure the oxidation of NADH to NAD at 340 nm. Alternatively pyruvate can be coupled to pyruvate oxidase, flavine adenine dinucleotide (FAD), and thiamine pyrophosphate (TPP) to form hydrogen peroxide, which in turn is coupled to peroxidase in presence of 4-

aminoantipyrine and a toluidine derivative to form a red chromogen, which is measured at 550 nm. The reactions associated with the NANA-aldolase-pyruvate oxidase-peroxidase system, as shown below:

- NANA-bound $\xrightarrow{\text{neuraminidase}}$ NANA + aglycone

- NANA $\xrightarrow{\text{NANA-aldolase}}$ N-acetylmannosamine + pyruvate

- Pyruvate + O_2 + $\xrightarrow[\text{FAD, TPP}]{\text{Pyruvate oxidase}}$ Acetylphosphate + CO_2 + H_2O_2

- $2H_2O_2$ + 4-Aminoantipyrine + N-2-hydroxyethyl toluidine $\xrightarrow{\text{Peroxidase}}$

Red dye 550 nm

Reactions involved in a typical enzymatic assay for the measurement of sialic acid.

High performance liquid chromatographic (HPLC) procedures:

The HPLC procedures provide the ultimate sensitivity. In one such procedure, sialic acid released from the sample by acid hydrolysis is converted to highly fluorescent derivatives by reacting with a fluorogenic agent for alpha-keto acids such as 1,2-diamino 4,5-methylene dioxybenzene in dilute sulfuric acid. The fluorescent derivatives are separated on an octadecyl (C18) bonded silica column using a reverse phase solvent system. The chromatographic step takes only 12 minutes allowing detection of levels as low as 25 femtomoles (F.mol) or 7.7 picograms (pg) of N-acetylmeuraminic acid and 23 F.mol or 7.5 (pg) of N-glycolylneuraminic acid, in an injection volume as small as 10 microliter. The procedure is capable of analyzing precisely sialic acids in a 5 μL of serum sample).

Functions of Sialic Acids:

Sialic acids are widely distributed in the human body and they perform important functions. The plasma glycoproteins are soluble and they have a hydrophilic character, it is believed that these two characters are due to the presence of sialic acids as a terminal sugar residues in these plasma glycoproteins . In addition to the role of sialic acids in protection of the plasma glycoproteins from splitting by proteolytic enzymes, they also play a role in the turn over of some plasma glyscoproteins; for example ceruloplasmin, which its removal from the circulation is believed to occur when sialic acids has been removed from its molecule; so it will be recognized by the liver cell plasma membrane .

The human red blood cell is studied with nearly 20 million molecules of sialic acid on the outer cell membrane which contributes to its electronegative charge, and by cell to cell repulsion prevents red blood cell from aggregating. Owing to its negative charge, sialic acid can bind positively charged molecules and thus play a role in the transport of such molecules . RBC aging process is due to the removal of sialic acids from its plasma membrane, so RBC will be taken up by the endothelial system to be destroyed. Similarly, the injection into rabbits of a desialylated preparation of ceruloplasmin resulted in the removal of the desialoglycoprotein from the circulation within a few minutes after injection, in contrast to the intact glycoprotein, which exhibited a normal survival time . Sialic acids are believed to decrease the immunogenicity of the glycoproteins, for example, human orosomucoid (which contain 11.8% NANA) was found to be a poor immunogenic in the rabbit but the removal of NANA by neuraminidase enhance the immunogenic properties of this glycoprotein at least five folds . In the same way the removal sialic acids from feutin enhances its combination with the antibodies. In RBC, the removal of sialic acids from its surface enhances the binding of influenza viruses to the RBC. Thus, sialic acid seems to protect self protein from self immune response and unfortunately it also seems to mask a foreign protein from rejection caused by host immunogenicity .

The presence of sialic acids in the sera of patients with various tumors inhibit the attack of immune lymphocytes to the tumor, so they cause masking to the tumor cells. However, the increased amount of sialic acid on the tumor cell surface can, by increasing adhesiveness, contribute to the formation of larger tumor emboli. Metastatic spread is also facilitated by sialic acid molecules increasing the adherence of tumor cells to vascular endothelium at secondary sites of implantation and by increasing the ability to aggregate platelets.

In glycohormones sialic acids perform important function, and their biological activity affected by the removal of sialic acids, example of these hormones are Human Chorionic Gonadotropin (HCG) and Follicular-Stimulating Hormone (FSH), this effect is presumably because sialic acids removal destroys the ability of these hormones to reach the target cells, so sialic acids seem to posses a hormone-receptor recognition function.

In mucin, sialic acids known to give the mucin it's viscosity and protect it from proteolytic splitting. Colonic mucin are predominantly neuraminidase resistant and contain O-acetyl substituent in the polyhydroxy side chain of the sialic acids, and it was found that the reduction in this substituent associated with diseases of the colon like ulcerative colitis and Crohn's disease.

Role of Sialic Acids in Disease Conditions:

McNeil et al, stated that the determination of sialic acids in serum is a sensitive, accurate method of estimating the glycoproteins content of the blood. Thus, sialic acids level increase in the same diseases that cause glycoproteins level to increase. Carter and Martin demonstrated an increase in the level of sialic acids in serum of the patients with rheumatic arthritis, bacterial infection, cirrhosis, myeloma and macroglobulinanemia. They suggested that this increase is due to increased production of abnormal glycoproteins with normal sialic acids.

Elevated serum sialic acids has been shown to increase in patients with variety of malignant diseases including, melanoma, breast cancer and lung cancer. Furthermore, patients with metastatic cancer had significantly higher sialic acids level than cancer patients without metastatic. Increase in the level of serum sialic acids is usually due to bound sialic acids, which is bound to lipids or proteins, causing increase concentration of total serum sialic acids.

Brozmanova and Skrovina, studied serum sialic acids level in patients with bon tumors. They observed significant elevation among sarcoma patients as compared to those with benign tumors, therefore sialic acids level was suggested to be useful for comparison and diagnosis of bone tumors.

Serial determination of serum sialic acids has found to be useful monitor of tumor burden, and it had been shown to be proportional to the tumor stage.

Again the relation between sialic acids, tumor stage and clinical course indicates that serum sialic acids analysis could prove clinically important as a monitor of tumor burden in diseased patients, specially as sialic acids correlate both with disease progression and remission, and it may provide the earliest indication of tumor response to drugs. One of the interesting facts about sialic acids is that it tend to increase in smokers and with age. The study of side chain O-acetyl substitution of sialic acids is important, it has a value in identifying mucin producing metastasis arising from carcinoma of the colon, important in studies of perianal Paget's diseases, in distinguishing between carcinoma arising from the anal gland and those arising from the rectal epithelium.

In addition, reduction in the side chain O-acetyl substitution pattern of these sialic acids has been found to be associated with colonic malignancy, Crohn's disease and ulcerative colitis. However, detailed studies about the biological significance of O-acetyl groups in sialic acids have shown that the degree of O-acetylation and the position of O-acetyl groups in sialic acids play a significant role in the action of sialidases. O-Acetyl groups at the sialic acid side chain reduce the rate of enzymatic hydrolysis, while a corresponding residue at C-4 prevents the action of all sialidases tested so far. It was furthermore observed that the degree of 9-O-acetylation exerts an effect on immunological and complement reactions. Therefore, the distribution of N-acetyl and/or N-glycolyl, with or without O-acetyl substituent on different tissue or secretion differ appreciably and may express the special function of these sialic acids in that tissue, for example the presence of O-acetylated sialic acids confers some protection to the epithelial surface of colon against the fecal stream (Bacteria and Enzymes) and it was suggested that the substitution with O-acetyl at C-4 position is responsible for sialic acids vibrio cholera neuraminidase resistance.

Sialic Acid and Cancer:

There has been much discussion of the relation between sialic acid and cancer. The elevation of serum sialic acid levels has been reported in cancer patients . It has also been reported that, in melanoma patients serum sialic acid levels are greater than those in normal donors and the levels are in proportion to tumor burden . The increase of sialyated carbohydrates, such as sialyl Lewis ^{a} (CA19-9) and sialyl Lewis^x (CSLEX-1), on the cell surface is generally observed with malignant or transformed cells . Plasma from cancer patients shows an almost uniform elevation in sialyltransferase activity. The elevated sialylated carbohydrate level in the plasma of cancer patients may contribute to the pathological immunodepression by blocking leukocyte interaction with endothelial cell leukocyte adhesion molecule-1 . These sialylated carbohydrates that are increased in cancer patients have a basic oligosaccharide structure in common, with blood group antigens (ABH, Lewis, etc.); thus, it is considered that the former is induced by modification of the latter with cancerization. It is reported that oligosaccharides with the structure of ABH antigenic determinants are present in normal human urine . However, there has been no investigation of the nature of oligosaccharides in the urine of cancer patients. LASA levels have been reported to be useful in monitoring patients with malignant melanoma. In one study when tumor recurrence was correlated with elevated LASA levels, the increased level was found as early as 9.3 months (median value) prior to recurrence . Higher levels of TSA and LASA have been reported in leukemia patients compared to patients with anemia. The TSA levels were significantly higher in acute myeloid leukemia compared to chronic myeloid leukemia and acute lymphatic leukemia patients. The LASA levels were significantly elevated in acute myeloid leukemia patients as compared to other leukemic patients. The sensitivity of sialic acid as a marker for leukemia is high with the sensitivity of LASA approaching 85 percent. The TSA levels in patients with oral and maxilla facial malignancy were reported to be significantly higher in patients with stage III and IV cancer, when compared to patients with stage I and II cancer. During follow-up of response to treatment while TSA levels declined during remission of disease, they became elevated with recurrence and metastasis .

Proteins Containing Sialic Acid:

Glycoproteins:

Glycoproteins can be best defined as "conjugated proteins containing as prosthetic group one or more heterosaccharides, the latter is usually branched, lacking repeating units and bound covalently to the peptide chain".

Chemical characteristics and structure:

There are great variations in chemical and physical properties of glycoproteins according to their location and function [102]. Molecular weight of glycoproteins may vary from 15,000 to over 2 million Dalton, usually contain 15 or fewer sugar units which are attached to the protein back-bone.

Glycoproteins are isolated from most organisms including: plants, bacteria, fungi, viruses and animals, and their chemical structure differ in the different organs, for example: the submandibular salivary mucin has a simple composition, it is almost exclusively N-acetylneuraminic acid and N-acetylgalactosamine ; while the intestinal glycoproteins are more complex, they have no unique amino acid composition but they do contain a characteristic group of sugars that include D-galactose, L-fucose, N-acetylglucosamine, N-acetylgalactosamine, and furthermore the chemical structure of glycoproteins may differ even in the different parts of the same organ in the same individual, for example, mucin in small intestine contain both neutral fucomucin and acid non-sulphated mucin however, in large intestine acid-sulphated and acid non-sulphated mucin are present . In addition to the variation in glycoproteins structures in the same individual, variation among different strains of the same species has been reported . In general, mucus glycoproteins from various sources share the following features: (i) very high molecular weight, usually in excess of 10^6 Dalton: (ii) approximately 75% carbohydrate and 20% protein core, with the remainder consisting of variable amounts of sulphate and water: and (iii) isoelectric point below 4, due to charged sialic acids and sulphated groups .

Structure of the Oligosaccharides attached to Glycoproteins:

Nine different sugar residues are generally found in the oligosaccharides chain attached to the protein core, these sugars differ according to their site and function, for example; glucose is found only in collagen, while galactose and mannose are more common and widely distributed. The two most frequently found hexoses are N-acetylgalactosamine and N-acetylglucosamine. Fucose, which is 6-deoxygalactose, is a common constituent, and frequently located at the terminal site in the neutral glycoproteins.

Two pentoses; arabinose and xylose were isolated from different tissues like dermatine. The ninth sugar is the sialic acids, of which N-acetylneuraminic acid (NANA) is an example, these acids are terminally located non-reducing residue of the carbohydrate prosthetic groups in the acidic glycoproteins.

On the other hand the N-acetylhexosamine are the most common sugars present at the proximal end of the oligosaccharide side chain, attaching to the protein back-bone, through the amino acids asparagine, serine, therionine, hydroxylysine or hydroxyproline (the first three usually present in mucus glycoproteins, while the last two present in connective tissue glycoproteins).

There are two types of linkage between carbohydrate of the polysaccharides side chain and the above amino acids, these are (i) the O-glycosidic linkage; and (ii) the N-glycosidic linkage (Fig. 89).

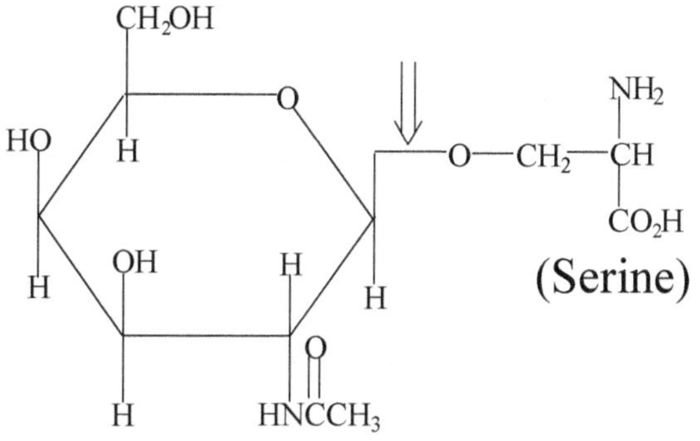

(N-acetylgalactosamine) — (Serine)

(i)

(N-acetylglucosamine) — (Asparagine)

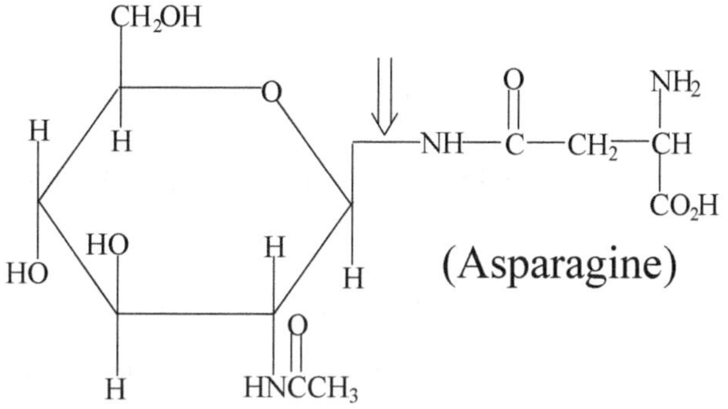

(ii)

Fig. (89): The O-glycosidic linkage (i) and the N-glycosidic linkage (ii).

Although it is possible to dissociate the polysaccharides chain complex by changes in pH or ionic strength, yet it is not possible to separate carbohydrate from peptide portion of the glycoproteins without directly degrading the entire molecule, therefore carbohydrates are integral part of the glycoproteins

Functions of glycoproteins:

Glycoproteins are thought to participate in many important functions. Glycoproteins may serve as a structural molecules in the cell, the major portion of the glycoprotein in the animal cell is associated with the cell surface, and approximately 70% of the sialic acid-containing glycoproteins are found in the surface membrane.

Glycoproteins also serve as a structural molecules in collagen, elastine, fibrine, and bone matrix. Concerning collagen Herp and Pigmen in 1958 found that rat skin contained an insoluble component; after the collagen fraction had been removed by intensive treatment of the collagen with hot alkali; this component showed to contain the sugar characters of glycoproteins . The other important function of glycoproteins is that they act as lubricants and protective agents, one of the well known examples is mucin; mucin is a glycoproteins containing viscous fluid being continuously secreted by the wet mucosa like respiratory, gastrointestinal and urinary systems. An example of the protective function of the glycoproteins is that of the human fetus, in which the surface epithelium of the stomach secretes neuraminidase resistant acid-mucin to protect the stomach wall from digestion by hydrochloric acid and the digestive enzymes .

Not only gastrointestinal mucin have this lubricant and protective function, but other mucin also performs similar function, for example respiratory mucin which protect the respiratory epithelium from the external environment by providing a barrier to the epithelium.

Furthermore, the immunoglobulin content of the respiratory mucin; which is recognized as a first line of defense against infection; is a glycoprotein .

Other important function of glycoproteins is to serve as a transport molecules for vitamins, lipids, minerals, trace elements and hormones. Example of those glycoproteins are transcobalamin (bind vitamin B_{12}), prealbumin and transferrin. Prealbumin, which is of a special biological interest because it is responsible for the transport of a hormone and a vitamin together, it appears to contain one binding site for one molecule of thyroxine and it also transports retinol (vitamin A), which is bound indirectly by the retinol-binding protein that forms a protein-portion complex with prealbumin . The other example of glycoprotein as a transport molecule is transferrin, which transport iron, it contains 8-9% carbohydrate in its molecule.

Glycoproteins could serve as a defense mechanism molecule in the body for example immunoglobulins, histocompatability antigen, complement and interferon. Immunoglobulines form a set of glycoproteins that have the ability to bind other molecules with a high degree of specificity. These molecules are foreign bodies or non-self and they are called antigens . Furthermore, the carbohydrate moieties of glycoproteins which are displayed on the cell surface act as immunogenic determinants (antigenic behaviour of the molecule).

Example of this immunogenic determinant is the blood group antigen; i.e. the four different types of the blood group (A, B, AB and O); which are determined by the sequences and arrangements of the sugar residues and the glycosidic bond. So the type of terminal sugars and their arrangement in the oligosaccharides side chain of glycoproteins, which are present in the RBC membrane, will determine the blood group, and it is specificity of the individual.

The other and important function of glycoproteins, is that same hormones are glycoproteins (glycohormones) for example Human Chorionic Gonadotropin (HCG) and Thyroid-Stimulating Hormone (TSH). Human chorionic gonadotropin appears in the urine and serum of women in significant quantities during the first trimester of pregnancy. Thus, pregnancy test based on the estimation of the level of this glycoprotien in urine and serum of pregnant women.

Not only hormones, but glycoproteins may be a part of an enzyme (glycoenzymes); these are widely distributed among animals, plants and microorganisms; and they are biologically active molecules for example proteases, nucleases and clotting factors. The last are essential glycoenzymes present in the blood, they are important in haemostasis (stop bleeding). Some of these clotting factors may be deficit in certain diseases causing defect in haemostasis, an example of such a disease is haemophilia, which is a congenital disease . Glycoproteins may also act as cell attachment and recognition sites, for example cell-cell, virus-cell, bacterial-cell and hormone-receptors. In 1953 Coman, D.R., suggested that malignant cells were less adhesive than their normal counterparts. Adhesion is defined as the attachment of the cell to each other, leading to the formation of tumor or, under normal conditions, to organs. Concerning recognition, the presence of galactose on the serum glycoprotein was necessary for the latter to be taken up by liver cell membrane, which in turn require sialic acids for this function. On the other hand, the circulating glycoproteins rapidly leave the portal system after desialization .

In hormone-receptor recognition as a functions of glycoproteins, insulin and glucagon are excellent examples. The receptors of these two hormones have been recognized as glycoprotein . One of the interesting facts in glycoproteins function, is that they act as antifreeze in Antarctic fishes.

Role of Glycoproteins in Disease Conditions:

Considerable interest has been focused toward the plasma glycoproteins in the past few years, centered largely around demonstration of increased level of glycoproteins in the plasma of individuals suffering from various disease conditions.

Most pathological conditions involve glycoproteins are due to defective degradation of glycoproteins rather than synthesis.

In a previous study , it was demonstrated that patients suffering from various malignant conditions postulated that, the increase in glycoproteins arises as a result of depolymerization of the ground substance adjacent to the tumor; with subsequent release of the glycoproteins to the circulation, on the other hand, many other workers suggested that this increase is due to tumor destruction, proliferation and repair .

However, other workers suggested that the major possibilities of the increased level of glycoproteins in malignant diseases are due to; (i) increase in carbohydrate content of normal glycoproteins, (ii) increase glycoproteins production by the tumor itself, (iii) increase glycoproteins synthesized by the liver or by the lymphoreticular tissue.

Glycoproteins, which increase in malignant diseases), are also known to increase in the acute phase of diseases (acute phase protein or reactant), infections and inflammatory diseases like rheumatic arthritis .

It was found that continuous estimation of glycoproteins level in plasma may give a clue about the progress of the disease. The other interesting fact about glycoproteins was that their level is more in smoker than in non-smoker persons.

It has been found that, glycoproteins level decreased in vitamin A deficiency, and drugs like Glutamine also cause reduction in glycoproteins level because both affect glycoproteins synthesis .

In several congenital diseases, including mucopolysaccharidoses and sphingolipidoses, glycoproteins have been blamed. These diseases are due to defect in glycoproteins catabolism, causing abnormal storage of glycoproteins and increase their accumulation. These diseases are called inbora error of metabolism and affect mostly serum and membrane glycoproteins .

It is well known that one of the glycoproteins functions is to determine the blood group of the individual. Furthermore, there are some relation between blood group and susceptibility to certain malignant diseases, for example blood group A individuals are more likely to have salivary gland tumors and carcinoma of the stomach than are individuals who are O, B or AB blood group. This susceptibility is of unknown etiology but it is believed to be due to the alteration in glycoproteins secreted by salivary gland or stomach of those people.

Deficiency in certain glycoproteins cause a disease condition, for example deficiency in the clotting factors (which are plasma glycoproteins) cause a defect in clotting mechanism of the affected patient with defect in haemostasis.

The tow leading diagnostic markers in human cancer, carcinoembryonic antigen (CEA) and α-fetoprotein are both glycoproteins, they increase in malignant diseases and their level may return to normal after treatment (with surgery or radiotherapy), but their level starts to increase again if the tumor starts regrowth. From this point, it seems that their level is beneficial to estimate the progress and follow up of the disease if estimated in many occasions.

The Lectins:

Lectins are divalent or multivalent carbohydrate-binding proteins with the ability to agglutinate erythrocyte (RBC of certain blood group), bacteria and other normal and malignant cells.

A great variety of carbohydrate-binding proteins (lectins) have been isolated from plants and have proved to be very useful in investigations of glycoproteins, glycolipids and polysaccharides and in studies of cells.

Many of these plant lectins are able to agglutinate erythrocytes or other cells and they have therefore often been termed phytohaemagglutinins. Lentil lectin is the haemagglutinaing lectine isolated from common lentil lens culinaris and shows a specific binding affinity towards α-D-dlucose and α-D-mannose residues. The ability of lectins to interact with soluble glycoproteins has been used to isolate and fractionate glycoproteins, this ability is due to binding of lectin to the oligosaccharides moieties of glycoproteins. So it is used in chromatographic techniques for glycoproteins fractionation. Most of the soluble lectins isolated from vertebrate tissues bind β-galactoside. The best studied are group of dimeric protein found in many organism including the electric eel, chicken and man.

Another group of vertebrate β-galactoside-binding lectins can be isolated as monomers, other soluble lectins are multimeric. The serum of the eel contains lectin composed or have twelve subunit per molecule.

Biological functions of lectins:

It is reasonable to expect that the known biochemical properties of lectins dictate their endogenous biological function. Some of these properties and the functions that might be inferred from them are summarized in (Table 49).

Table (47): Some common properties of lectins that suggest biological functions.

Property	Function suggested
1. Specific binding sites: a. All of one kind b. Of different kinds	Recognition of complementary oligosaccharide receptors (range of specificities)
2. More than one carbohydrate- binding site	a. Cross-linking glycoproteins or glycolipids in membranes and/or solution b. High affinity (multisite) binding to molecules or a cell surface with multiple receptors
3. Agglutinin	Binding cells together: a. Like cells (promoting adhesion, fusion, etc.) b. Unlike cells (promoting symbiosis, infection, phagocytosis, etc.)
4. Abundant	Structural rather than catalytic function
5. Generally not integrated in membranes	Relative freedom of movement in or between cellular compartments

One major property of lectins is their specific saccharide-binding sites. In the many lectins that are multimers of an identical subuint these binding sites are the same. In contrast, some lectins are composed of subunits with different binding sites. These include the lectin from the red kidney bean, phaseolus vulgaris. It is composed of two different subunits combined into five different forms of noncovalently bound tetramers.

Since the subunits have markedly different specificities for cell surface receptors, each combination could be envisioned to have a different function. For example the homotetramer of one subunit might agglutinate cells with an appropriate receptor, whereas a tetramer that contains only one of these subunits per molecule might inhibit such agglutination.

The common finding of more than one lectin in seeds, slime molds and even vertebrate tissues , raises the possibility of concerted specific reactions due to concurrent display of these proteins on a structure like a cell surface. A highly specific interaction directed by the binding properties of more than one type of lectin molecule could result.

The specificity of the binding sites of the lectins suggests that there are endogenous saccharide receptors in the tissues from which they are derived or on other cells or glycoconjugates with which the lectin is specialized to interact. Unfortunately no endogenous receptor for a lectin has yet been unambiguously identified, despite the fact that their carbohydrate-binding sites may be specialized for association with highly specific complex oligosaccharides . The other properties and functions of lectins included that the fact of lectins have more than one carbohydrate-binding site suggests that they could act to cross-link glycoproteins and glycolipids in membranes of the same cell for various organizational purposes. Agglutination activity suggests functions in binding like or different cells for a variety of purposes ranging from morphogenesis in embryos to phagocytosis of one cell by another. The marked abundance of lectins suggests that they play a structural role rather than an enzymatic or catalytic function. The fact that many lectins are readily isolated as water-soluble materials suggests that if they play a role in membrane function it is by association with oligosaccharides on membranes without the constraints imposed by being integrated within the membrane bilayer. It of course remains possible that some functions of lectins may be mediated by properties that have not yet been discovered .

Applications:

Lectins play an important role in research into a variety of cellular properties and processes.

Their major biological effects, such as cell agglutination and mitogenic stimulation, appear to be mediated initially through interactions at the level of cell surfaces and mimic various physiologically important processes. Since lectins can be obtained in a purified form and show well-defined interaction specificities, they have frequently been used in model systems. Boldt and

Coworkers, used lentil lectin, to study the mitogenic responses of various populations of human lymphocytes. Lectins also provide convenient "markers" in cell surface studies. Lentil lectin was used by Scott and Rosenthal, to characterize plasma membrane vesicles shed by guinea pig macrophages on exposure to sulphydryl-blocking reagents.

This investigation was used to determine the concentration of total sialic acid (TSA), lipid-associated sialic acid (LASA) and total protein (TP) in sera of normal donors (n = 25), patients with chronic non-malignant diseases (Asthma disease n = 20) as a pathological control, patients with breast cancer (premenopausal n = 40, postmenopausal n = 35 and benign patientsn = 25) and patients with other cancer (endometerial carcinoma n = 15 and sarcoma n = 5).

Data analysis show a significant increase ($P < 0.001$) in the mean values ($\pm$ SD) of TSA and TSA/TP in sera of cancer patients when compared to normal healthy individuals and pathological control (Asthma disease). However, the mean values ($\pm$ SD) of total protein (TP) in sera of breast cancer patients did not show significant differences with respect to normal controls and patients with chronic non-tumoral disease (Asthma disease) except for other cancer patients, which show a significant differences ($P < 0.001$). Also, from results analysis there is a statistically significant rise in the mean values ($\pm$ SD) of LASA in sera of cancer patients when compared to the mean values ($\pm$ SD) with normal controls and Asthma disease.

glycoproteins

Many glycoproteins and glycolipids from malignant cells differ in carbohydrate composition from those found in normal cells. Since many of these glycoconjugates contain sialic acid, which can be shed into the circulation, total sialic acid (TSA) is of great interest as a marker of malignancy, although it has not been demonstrated to be specific for any type of cancer. There are many controversies regarding the quantitative changes in TSA occurring in cancer patients. It can only be said that increase in TSA is associated with certain diseases, including cancer, and is roughly related to tumor size.

Some authors have based their work upon the fact that alterations in glycolipid metabolism are well documented in many tumors, including human cancers. These authors have reported raised levels of lipid sialic acid (LSA) in sera of patients with various neoplasms, but others have found raised LSA values in sera of patients with acute inflammatory disease.

This has led to the conclusion that the high levels of the so-called LSA in the sera of patients with cancer probably emerge from associated inflammation and, hence, the acute phase reactant glycoproteins mainly build up the LSA fraction. Moreover, Dnistrian et al and Katopodis et al reported that excessive sialic acid levels are found frequently in carcinomas of the breast, pancreas, colon, ovaries and other organs. In actuality, some investigators believe that high levels of sialic acid are markers for neoplasms of these organs, but they are not thought to be very reliable.

In a previous study, it was demonstrated that patients with breast cancer had lower neuraminidase levels compared with subjects without a personal or family history for breast cancer. Consequently, it was suggested that inadequate activity of neuraminidase enzyme may be a marker for breast cancer.

The objective of this part was to investigate the diagnostic utility of serum levels of total sialic acid (TSA) and lipid-associated sialic acid (LASA) in breast and other cancer patients.

Materials and Methods:

Patients and Blood Samples:

Twenty five samples of blood were taken from physically normal volunteers, these samples were used as a control, the control volunteers aged between 25-40 years, and consisted of 20 females and 5 males, who gave no history of previous diseases.

Twenty samples of blood were taken from patients with Asthma diseases were used as a pathological controls.

Twenty samples of blood were taken from patients with other cancer also were used as a pathological controls.

Two groups of breast cancer patients and one group pf patients with benign breast tumors were included in this study.

Group 1 contained 40 premenopausal patients with breast cancer. Group 2 consisted of 35 postmenopausal patients with breast cancer. Group 3 comprised 25 patients with benign breast tumors.

All patients were admitted for treatment to AL-Karama teaching hospital, Saddam Medical City and AL-Saddoon Hospital. Patients suffered from any disease that may interfere with our study were excluded.

The blood samples were collected from these patients, left for 30 minutes at room temperature; blood clots were separated at 3000 RPM for 10 minutes by using centrifuge. Sera were aspirated and stored in caped sterilized tubes at -20°C until time of analysis.

The host information of all patients and normal healthy subjects is summarized in Table (48).

Table (48): The host information, which are used in this study

Groups	Number	Type of tumor	Metastases	Age (range)
Controls: Normal healthy	25	-	-	25-40
Non-tumoral	20	-	-	20-60

disease: Chronic Asthma disease				
Breast cancer: **Premenopausal**	40	38 Infiltrative ductal carcinoma 2 Infiltrative labular carcinoma	Liver and axillary lymphnode	30-42
Postmenopausal	35	34 Infiltrative ductal carcinoma 1 Infiltrative lobular carcinoma	Liver and axillary lymphnode and bone	48-68
Benign	25	24 Fibroadenomas 1 Fibrocystic disease		21-45
Other cancers	20			30-50

Estimation of Total Serum Proteins:

The method of Lowry et al was used to estimate serum total proteins, using bovine serum albumin as standard. Protein, concentrations are expressed as g/dL of sera. Fig. (90) shows the standard curve of protein, which was constructed by plotting the absorbance at 600 nm against standard protein concentration.

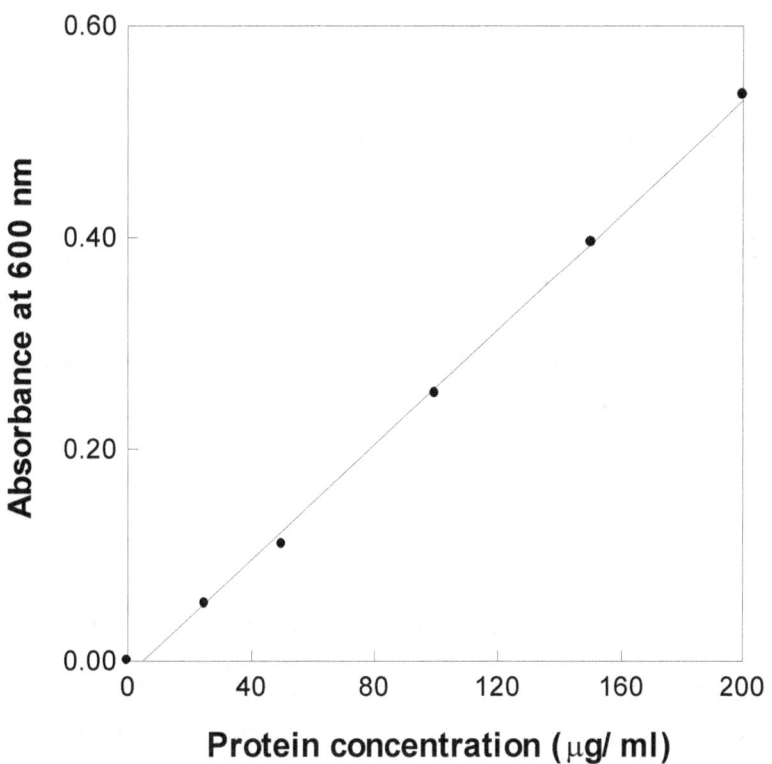

Fig. (90): Standard curve of protein determination in human sera by Lowry method.

Estimation of Total Serum Sialic Acid and Lipid-Associated Sialic Acid:

Serum LASA was estimated spectrophotometry by the method of Katopodis et al, with slight modification in the volume of sample used. Fifty microliters of serum was placed in test tube (triplicate) with 150 µL deionized water, then vortexed for five seconds and placed on ice. Three ml of cold (4°C) chloroform: methanol (2:1 v/v) was added, then again vortexed and 0.5 ml of cold deionized water was added, then also vortexed after that, centrifuged for 5 minutes at 3000 rpm, at room temperature. One ml of the resulting upper layer was transferred to another tube and 50 µL of phosphotungstic acid (1g/ml) was added, then vortexed and allowed to stand at room temperature for 5 minutes, then centrifuged for 5 minutes at 3000 rpm. The supernatant fluid was removed and the remaining precipitate was dissolved in one ml deionized water.

TSA levels were determined as follows:

Twenty microliters of sera and 980 µL of deionized water were placed in test tube, vortexed and placed on ice.

To each assay tube for LASA and TSA one ml of resorecinol reagent was added, and placed in a 100°C (boiling water bath) for exactly 15 minutes, followed by 10 minutes on a ice bath, two ml of butyl acetate: methanol (85:15 v/v) was added to each tube, then vortexed and centrifuged for 10 minutes at 3000 rpm. The extracted chromophore was read at 580 nm against distilled water.

Standard sialic acid: The standard sialic acid solutions with different concentrations (5, 10, 15, 20, 25, 30, 35, 40, 45) µg/ml were prepared by serial dilutions from a stock standard solution of sialic acid (1000 µg/ml). Fig. (91) shows the standard curve of sialic acid which was constructed by plotting the absorbance at 580 nm against the corresponding concentration of standard sialic acid solution, and was used to determine the TSA and LASA levels in the serum samples.

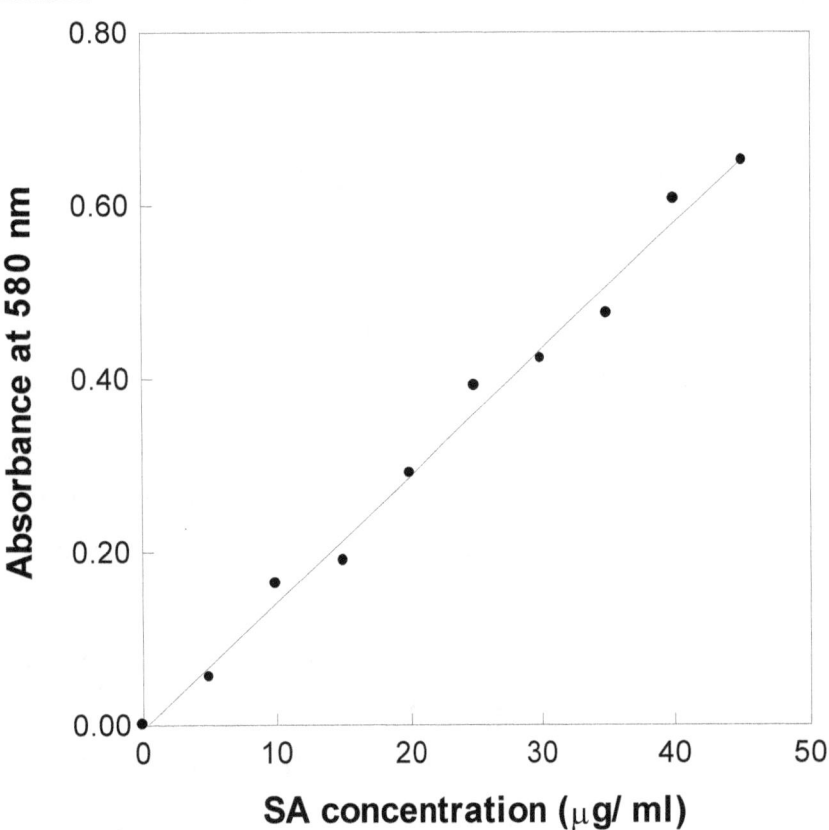

Fig. (91): Standard curve for determination of sialic acid concentration in human serum.

1. 2.6. STATISTICAL METHODS:

The results for TSA, LASA, TP and TSA/TP ratio were analyzed statistically and values were expressed as mean ±SD. The level of significance was determined by student's t-test.

Results and Discussion

Determination of Total Sialic Acid (TSA) and Total Protein (TP) Concentrations in Sera of Patients With Breast Cancer, other Cancers and Non-C Individuals:

The individual and mean serum concentrations of TSA and TP for the normal healthy controls, pathological controls (chronic Asthma disease) and patients with breast cancer and other cancers are summarized in table (49). This table presents comparisons of the mean values of sera TSA an TP between different groups above, it shows that the TSA levels in breast and other cancer patients were significantly elevated ($P < 0.001$) as compared to the normal healthy and pathological controls ($P < 0.001$ and $P < 0.01$ respectively). However, comparison of serum TSA for different types of breast cancer was studied, the postmenopausal patients show a high level of TSA as compared to the premenopausal patients.

Table (49): Comparison of mean values for TSA, LASA and TP in sera of normal, pathological controls and in patients with breast cancer and other cancer.

Groups	No.	TSA (mg/dL) (mean ±SD)	LASA (mg/dL) (mean ±SD)	TP (g/dL) (mean ±SD)	TSA/TP (mg/g) (mean ±SD)
Controls: Normal healthy	25	56.7±8.2	17.86±1.59	7.03±0.83	8.27±0.73
Non-tumoral disease: Chronic Asthma disease	20	80.8±1.25	21.57±4.02	7.06±0.91	11.54±0.7

Breast cancer:					
Premenopausal	40	100.29±17.38	32.13±10.01	7.1±1.02	14.23±1.45
Postmenopausal	35	128.25±21.43	30.93±10.16	7.16±1.13	18.01±1.71
Benign	25	85.6±0.16	19.29±1.84	7.08±0.93	12.28±0.8
Other cancer:					
Endometerial carcinoma	15	103.5±11.5	27.3±7.02	7.91±1.2	13.4±1.7
Sarcoma	51	94.6±9.31	23.92±5.88	8.1±1.03	12.7±1.9

On the other hand, the mean values of total protein (TP) in sera of breast cancer patients did not find any significant differences when compared to the mean values with normal healthy and pathological controls, and on the contrary, the mean values of total protein in sera of other cancer patients were observed a significant differences ($P < 0.01$) than that in breast cancer patients, normal healthy and pathological controls (Asthma disease). It is clear from table (50) the mean values of total sialic acid to total protein ratio (TSA/TP) in sera of breast cancer patients (premenopausal and postmenopausal patients) were significantly elevated ($P < 0.01$) as compared to the normal healthy and patients with chronic Asthma disease, whereas, the serum levels of TSA/TP ratio in postmenopausal patients were significantly higher ($P < 0.01$) than those in premenopausal patients with regard to the mean values of TSA/TP ratio in sera of other cancer patients (endometerial carcinoma and sarcoma), show a significantly higher increase than those observed in normal healthy, patients with non-tumoral disease and benign patients ($P < 0.01$), but it is lower than in breast cancer.

Table (50) shows the specificity and sensitivity of TSA test in normal controls and in patients with breast cancer, other cancer and non-cancer (Asthma disease). The specificity was calculated as % of patients with cancer who had normal TSA values than normal while the sensitivity was calculated as % of patients with cancer who had higher TSA values than normal. Through table (51), the results show that 37 out of the 40 patients with breast cancer (premenopausal patients) 92.5% had elevated values of TSA. Of the groups of 35 patients with breast cancer (postmenopausal), 34 (97.1%) had elevated values for TSA. For benign patients, 22 out of the 25 patients included in group (88%) had elevated values of TSA. But in other cancer, test sensitivity of TSA was varied from 93.3 to 100%. However, in normal controls test specificity of TSA was 84% so that only 16% of those tested were falsely positive. Fig. (96) shows the distribution of the individual values of TSA in sera of cancer patients, non-cancer patients (Asthma disease) and normal healthy.

Table (51): Specificity and sensitivity of TSA in normal, controls and in patients with breast cancer, other cancer and non-cancer.

Groups	No.	TSA<65 mg/dL (normal)	Specificity % true negative	TSA>65 mg/dL (elevated)	Sensitivity % true positive
Normal	25	21	84	4	16
Breast cancer:					
Premenopausal	40	3	7.5	37	92.5
Postmenopausal	35	1	2.9	34	97.1
Benign	25	3	12	22	88
Other cancer:					
Endometerial carcinoma	15	1	6.7	14	93.3
Sarcoma	5	0	0	5	100
Non-cancer:					

| (Asthma disease) | 20 | 4 | 20 | 16 | 80 |

Note: Sensitivity and specificity of TSA measuring as follows: Sensitivity was calculated as % of patients with cancer who had higher TSA values than controls (normal). Specificity was calculated as % of patients with cancer who had normal TSA values than controls (normal).

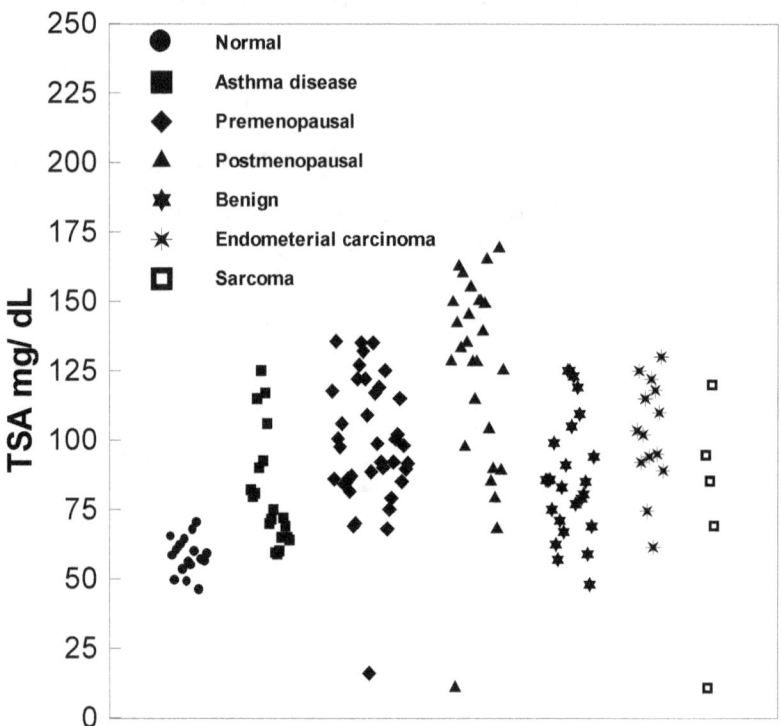

Figure (91): Distribution of the individual values of TSA in sera of breast cancer patients, other cancers, non-cancer (Asthma disease) and normal healthy.

Glycoproteins and glycolipids are present in membrane of malignant cells and differ in carbohydrate composition. This contribute to aberrant cell to cell recognition, cell adhesion, antigenicity and the invasiveness demonstrated by malignant cells . Therefore, alterations in important glycoprotein constituents like sialic acid naturally assume importance in malignancy. Variations in serum TSA levels have been found to be useful for diagnosis, staging, prognosis and treatment monitoring of cancer patients

[148]. TSA levels when normalized for TP variations and expressed as TSA/TP, become more tumor specific. Increased glycolytic activity has been observed in malignant cells. During neoplastic transformation, the carbohydrate chains in glycolipids and glycoproteins are frequently altered. There is a close relationship between the expression of certain carbohydrate antigens and oncogenesis.

In an elegant study examining the significance of the linkage of sialic acid residues in cancer-associated carbohydrate antigens, by using specific monoclonal antibodies, it was demonstrated that not all sialic acids are specific to cancer. Indeed, this study demonstrated that there are significant variations in the cancer specificity depending on the difference in linkage in sialic acid residues. However, the relevance of sialic acid to the tumor cell is apparent from the increased sialylation and sialyltransferase activity observed in many cancer cells. The aberrant glycosylation found in cancer cell membranes is presumably due to the activation of new glycosyl transferases that are characteristic of tumor cells and are absent or present only in small quantities in normal cells. Thus, for instance, a relatively specific sialyltransferase is found to be present by as much as 2.5 to 11 times in greater amounts in transformed cells when compared to control cells.

Sialic acid bound to membrane glycoproteins and glycolipids apparently enters the circulation by either shedding or by cell lysis.

Approximately 98 to 99.5 percent of total sialic acid found in serum or plasma is bound to glycoproteins. Only a very small fraction of sialic acid is bound to lipids, which is mainly in the form of gangliosids. Normal levels of total sialic acid in serum are approximately in the range of 51 to 84 mg/dL. In contrast, the contribution of the pure lipid fraction to total sialic acid level is barely in the range of 0.4 to 0.9 mg/dL.

Since sialic acid is one of the component of glycoprotein and glycolipid, several investigators and our study concentrate on their levels in the sera or plasma or tissues of patients with cancer diseases. On the other hand, most of these studies have been concerned with TSA level.

Total sialic acid (TSA) level is increased in a variety of tumors, and this level is directly related to cancer burden and disease recurrence.

In a recent study on the usefulness of TSA in lung cancer, data obtained in this study show that the mean concentration of TSA was significantly higher in lung cancer patients when compared to benign and normal controls [155].

Erbil et al reported on increased levels of TSA in genitourinary tumors, concluding that serum TSA levels were highly correlated with the stage and grade in patients with advanced urological cancer.

Determination of Lipid-Associated Sialic Acid (LASA) Levels in Sera of Patients with Breast Cancer, Other Cancers and non-Cancer individuals:

Serum lipid-associated sialic acid (LASA) levels were determined in normal healthy persons and in patients with non-cancer (Asthma disease) as a pathological controls and patients with breast cancer, other cancers, using the method of Katopodis et al.

Table (52) shows the individual and mean serum concentrations of LASA in different groups of cancer patients, non-tumoral disease (Asthma disease) and in normal healthy, also this table presents comparisons of the mean values of sera LASA between these groups.

The results in this table reveal an overall elevation in LASA levels for each group of patients with cancer when compared to the normal healthy and pathological controls. This increased LASA values in sera of breast cancer patients except benign patients showed statistically significant differences than those values obtained from sera of normal control, patients with non-malignant diseases and other cancers ($P < 0.001$, $P < 0.01$ and $P < 0.01$) respectively.

It is clear from table (53) the level of LASA in both types of breast cancer patients (premenopausal and postmenopausal) show equivalent value.

Table (52) represents the percentages of LASA test specificity and sensitivity in normal individuals, patients with breast cancer, other cancers and non-cancer (Asthma disease), using the value of 19 mg/dL as the upper limit of normal.

The results presented in this table show that 35 out of the 40 patients with breast cancer (premenopausal) 87.5% had elevated levels of LASA. Of the groups of 35 patients with breast cancer (postmenopausal), 32 had elevated levels of LASA 91.4%.

Table (52): Specificity and sensitivity of LASA in normal controls and in patients with breast cancer, other cancer and non-cancer.

Groups	No.	LASA≤19 mg/dL (normal)	Specificity % true negative	LASA>19 mg/dL (elevated)	Sensitivity % true positive
Normal	25	22	88	3	12
Breast cancer:	40	5	12.5	35	87.5
Premenopausal	35	3	8.57	32	91.4
Postmenopausal					
Benign	25	5	20	20	80
Other cancer:	15	0	0	15	100
Endometerial carcinoma					
Sarcoma	5	0	0	5	100
Non-cancer: (Asthma disease)	20	2	10	18	90

Note: Sensitivity and specificity of LASA measuring as follow: Sensitivity was calculated as % of patients with cancer who had higher LASA values than controls (normal). Specificity was calculated as % of patients with cancer who had normal LASA values than controls (normal).

For benign patients, 20 out of 25 patients included in group (80%) had elevated values of LASA. But in other cancer patients, test sensitivity of LASA had elevated values (100%).

On the other hand, in normal controls test specificity of LASA was 88% sothat only 12% of those tested were falsely positive. Figure (55) shows the distribution of the individuals values of LASA in sera of cancer patients, non-cancer patients (Asthma disease) and normal healthy.

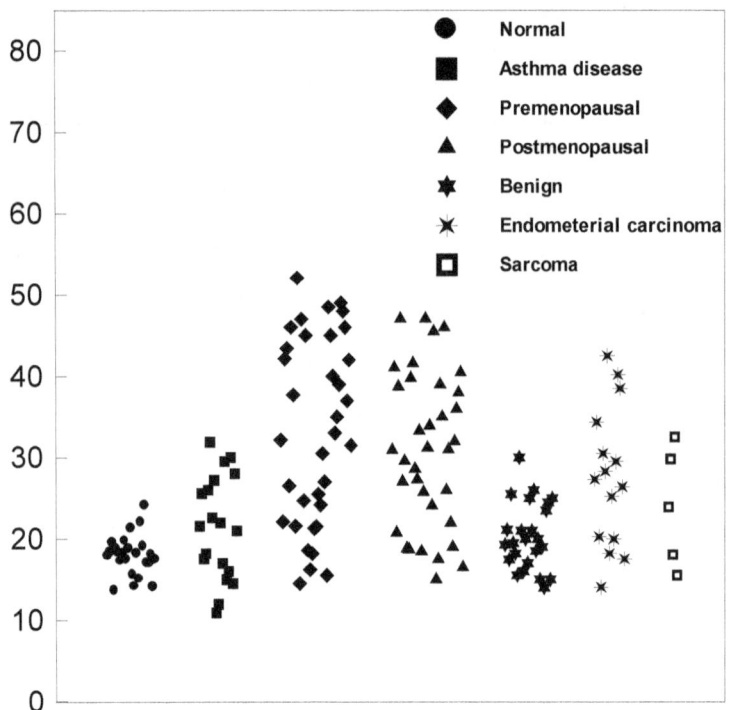

Figure (92): Distribution of the individual values of LASA in sera of breast cancer patients, other cancers, non-cancer (Asthma disease) and normal healthy.

Lipid-associated sialic acid (LASA) level is increased in a variety of tumors, and this level is directly related to cancer burden and disease recurrence.

In another studies LASA levels have been reported to be useful in monitoring patients with malignant melanoma . However, LASA-P is a biomarker useful in a wide range of malignancies. It reflects alteration in the surface membrane of tumor cells. The LASA-P assay measures total gangliosides and glycoproteins. Elevated LASA-P levels in breast cyst fluid have been associated with increased risk of breast cancer . Also, LASA-P levels are higher in women with benign or malignant breast tumors than in controls . Some conditions other than cancer affect LASA-P. Among them, are myocardial infarction , infections, rheumatoid arthritis, and collagen degeneration. Polivkova et al , believe that the determination of LASA levels could be useful not only for cancer diagnosis but also prognosis.

In this investigation, it is established that the data for the TSA and LASA test confirms the previous observations, which have indicated that levels are significantly increased in the sera of breast cancer and other cancer patients, and both tests may prove to be of clinical value.

The increase of TSA are frequently modest in premenopausal patients and high in postmenopausal patients as compared to non-tumoral diseases and normal healthy controls and the demarcation between normal and abnormal levels is sufficiently sharp sothat with a 65 mg/dL cutoff. The specificity and sensitivity data for the results was a favorable when compared to those of the most widely used immunodiagnostic test, CEA.

Sialic acid measurements appear to have a high sensitivity for a wide range of tumors.

However, the specificity of sialic acid measurements, especially the non-specific LASA measurements, is low since

INTRODUCTION:

The measurement of biochemical markers is being increasingly used for early diagnosis and monitoring the progress of cancer. Increased levels of protein-bound carbohydrates have been shown to occur frequently in patients with neoplasms . Glycoproteins play an important role in the cellular phenomena that undergo alterations during cancerours transformation.

Barlow and Dillard demonstrated that serum L-fucose could be helpful as a means of assessing disease status in patients with breast and cervical cancer.

Sugars in glycoproteins, glycolipids, glycosaminoglycans, oligo-and polysaccharides have been identified in a variety of animal tissues . Moreover, various carbohydrate fractions bound to the plasma proteins are elevated in patients suffering from cancer diseases and also certain non-cancer disease state .

Lawrence et al described a group of glycoproteins that are synthesized and released by human breast cancer maintained in organ culture and similar glycoproteins released by a human breast carcinoma cell line (BT-20). The electrophoretic mobility of these glycoproteins on cellulose acetate is consistent with increased glycoproteins staining material present in the α_2 to β-globulin region of serum glycoprotein electropherograms from patients with breast cancer.

However, the protein-bound carbohydrate and seromucoid of plasma have been demonstrated to be elevated in a wide variety of pathological conditions including spontaneous human carcinoma .

The present investigation was carried out to clarify further the possible usefulness of serum protein-bound sugars and seromucoid as for possible identification of the status of advanced breast and other cancers.

MATERIALS AND METHODS:

DETERMINATION OF BIOCHEMICAL CONSTITUENTS IN SERA OF PATIENTS WITH BREAST CANCER, OTHER CANCERS AND NON-CANCER (ASTHMA-DISEASE):

Determination of Seromucoid:

The method of Weimer and Mashin , was used to determine serum seromucoid. This method includes the following steps:

(1) Half milliliter of serum was added to 4.5 ml of 0.85% NaCl, then mixed and 2.5 ml of 1.8 M perchloric acid was added, after mixing by inversion, the assay tubes were allowed to stand at room temperature for 10 minutes, then centrifuged for 15 minutes at 3500 r.p.m. to obtain clear supernatant.

(2) To five milliliter of the supernatant, 1 ml of phosphotungstic acid reagent was added, after mixing the tubes were allowed to stand for 10 minutes.

(3) The tubes were centrifuged, after removing of the supernatant, five milliliter of 95% ethanol was added, then strie, centrifuged and the supernatant was removed.

(4) The resulting precipitate was dissolved in 0.5 ml of (0.1 N) NaOH, this was considered as the unknown.

(5) Set up blank with 0.5 ml of distilled water, and standard using 0.5 ml of working standard.

(6) To each unknown, blank and standard, added 1.25 ml of orcinol reagent and 7.5 ml of (60% v/v) H_2SO_4.

(7) All tubes were placed in a water bath at $(80 \pm 0.5°C)$ for 20 minutes, cooled and read against distilled water at 520 nm.

Calculations:

$$\text{mg seromucoid}/100 \text{ ml} = \frac{A_x - A_b}{A_s - A_b} \times 0.1 \times \frac{100}{0.333}$$

$$= \frac{A_x - A_b}{A_s - A_b} \times 30$$

Where:

A_x = The absorbance of unknown solution at 520 nm.

A_s = The absorbance of standard solution at 520 nm.

A_b = The absorbance of blank solution at 520 nm.

Estimation of Serum Protein-Bound Hexose:

The method includes the following steps:

(1) Hundred microliter of serum was added to 5 ml of 95% v/v ethanol and mixed carefully, then centrifuged for 15 minutes at 3500 r.p.m., after that the supernatant was decanted.

(2) The remaining precipitate was washed with 5 ml of 95% ethanol, then stired, after that centrifugation.

(3) The supernatant was decanted, carried out in the practical part.

Calculations:

$$\text{mg protein bound hexose/100 ml} = \frac{A_x - A_b}{A_s - A_b} \times 0.1 \times \frac{100}{0.1}$$

$$= \frac{A_x - A_b}{A_s - A_b} \times 100$$

Where:

A_x = The absorbance of unknown solution at 520 nm.
A_s = The absorbance of standard solution at 520 nm.
A_b = The absorbance of blank solution at 520 nm.

RESULTS AND DISCUSSION

The level of protein-bound hexose (galactose and mannose) was estimated in the sera of normal volunteers and in patients with breast cancer, other cancers and non-cancer (Asthma disease), using the method of Weimer and Mashin.

The mean concentrations of protein-bound hexoses in sera of all patients and normal healthy are summarized in table (53). Figure (93) shows the distribution of the individual values of protein-bound hexoses in sera of all cancer patients, non-cancer and normal controls.

Table (53): Serum protein-bound hexose (galactose and mannose) in normal controls and in patients with breast cancer, other cancers and non-cancer (Asthma disease) ± standard error.

Groups	Number.	Protein-bound hexose mg/dl±S.E.

Normal	25	87.09±1.82
Breast cancer: Premenopausal	40	111.9±5.55
Postmenopausal	35	103.30±4.48
Benign	25	90.7±5.01
Other cancers: Endometerial carcinoma	15	127.75±3.25
Sarcoma	5	131.5±4.87
Non-cancer: (Asthma disease)	20	101.66±4.1

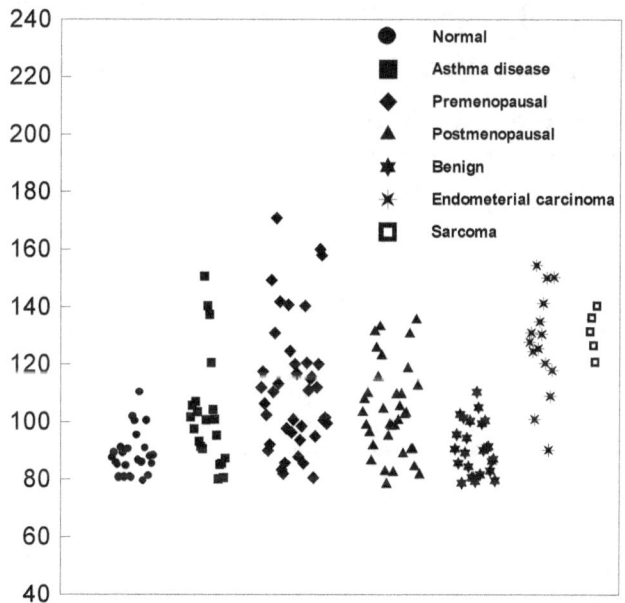

Figure (93): Distribution of the individual values of protein-bound hexoses in sera of patients with breast cancer, other cancers, non-cancer (Asthma disease) and normal healthy.

From the results presented in this table, reveal that the mean values of protein-bound hexose in sera of patients which suffer from breast and other cancers were elevated significantly in comparison to those of normal controls and patients with benign ($P < 0.01$).

The mean concentrations of protein-bound hexose reached to 111.9 ± 5.55 mg/dl in premenopausal patients, 103.3 ± 4.48 mg/dl in postmenopausal patients, whereas in other cancers it was 127.75 ± 3.25 mg/dl in endometerial carcinoma patients, 131.5 ± 4.87 mg/dl in sarcoma patients.

However, the mean level of protein-bound hexose in patients with non-cancer (Asthma disease) was significantly elevated when compared to those observed in normal healthy controls and with benign patients ($P < 0.01$), but the elevation was not significant when compared to the mean values of protein-bound hexose between patients with benign and normal healthy individuals ($P > 0.05$). The serum levels of protein-bound hexose in patients with other cancers were significantly higher ($P < 0.01$) than those in breast cancer patients and non-cancer, and also than those observed in normal healthy controls and benign patients ($P < 0.001$).

Table (54) shows the specificity and sensitivity of the protein-bound hexose in normal controls and in patients with breast cancer, non-cancer.

Table (54): Specificity and sensitivity of protein-bound hexose in normal controls and in patients with breast cancer and non-cancer (Asthma disease).

Groups	No.	Protein-bound hexsoe≤89 mg/dL (normal)	*Specificity % true negative	Protein-bound hexsoe>89 mg/dL (elevated)	**Sensitivity % true positive
Normal	25	6	24	19	76
Breast cancer: Premenopausal Postmenopausal Benign	40 35 52 25	3 4 1	7.5 11.43 4	37 31 24	92.5 88.57 96
Non-cancer: (Asthma	2	0	0.00	20	100

disease)	0				

* Calculated as the number of cases having ≤ 89 mg/dl divided by the total number of cases by 100.

** Calculated as the number of cases having >89 mg/dl divided by the total number of cases by 100.

In this test using 89 mg/dl as the upper limit of normal, test sensitivity in those with breast cancer patients was varied from 88.57% to 92.5%, but this test had elevated in benign patients and non-cancer patients (Asthma disease) 96% and 100% respectively. The test was more sensitive for breast cancer patients and non-tumor disease (Asthma disease), as compared to the normal healthy controls.

Also, the results obtained revealed low specificity in different types of breast cancer.

On the other hand, the level of seromucoid was estimated in sera of normal healthy controls and in patients with breast cancer, other cancers and non-cancer (Asthma dusease).

The data obtained in this study are summarized in table (55). Figure (94) shows the distribution of the individual values of seromucoid in sera of all cancer patients, non-cancer and normal healthy controls.

Table (55): Serum seromucoid in normal controls and in patients with breast cancer, other cancers and non- cancer (Asthma disease) ± standard error.

Groups	Number.	Seromucoid mg/dl±S.E.
Normal	25	10.29±0.29
Breast cancer:	40	17.58±2.08
Premenopausal	35	15.92±1.83
Postmenopausal	25	12.63±0.33
Benign		
Other cancer:		

Endometerial carcinoma	15	18.16±1.3
Sarcoma	5	19.78±1.7
Non-cancer: (Asthma disease)	20	13.24±0.73

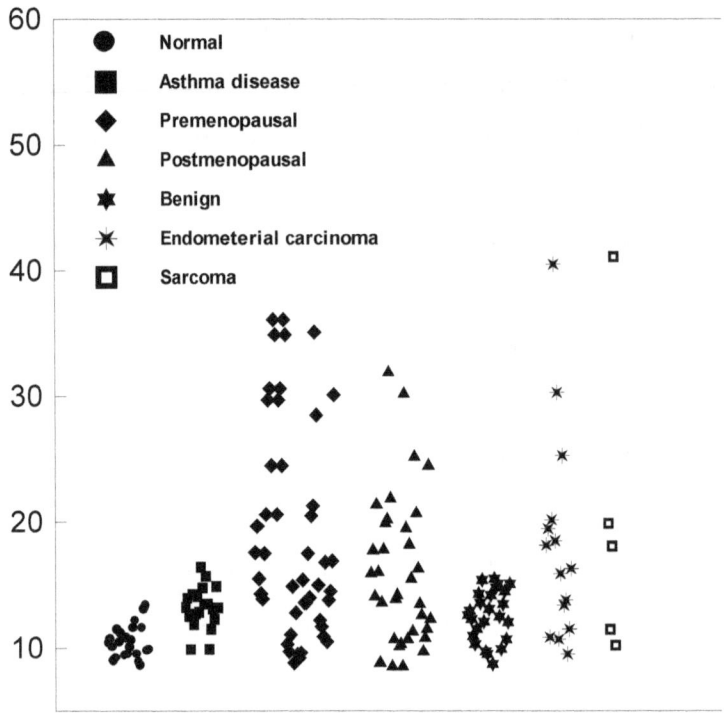

Figure (94): **Distribution of the individual values of seromucoid in sera of patients with breast cancer, other cancers, non-cancer (Asthma disease) and normal healthy.**

From these data it is apparent that the mean values of seromucoid level in hospitalized patients suffering from breast and other cancers were significantly elevated when compared to the benign, normal controls and patients with non-cancer (Asthma disease), ($P < 0.001$).

But there is no statistically significant differences between patients with benign and Asthma disease ($P > 0.05$).

However, the mean values of seromucoid level in other cancer patients show a significantly higher increase than in patients with breast cancer ($P <$

0.01) and also as compared with normal healthy controls, non-cancer and benign patients (P < 0.001).

The specificity and sensitivity of the seromucoid test are considered in table (56), using 11 mg/dl as the upper limit of normal, this test was sensitive for different types of breast cancer as compared to the normal healthy.

Table (56): Specificity and sensitivity of seromucoid test in normal controls and in patients with breast cancer and non-cancer (Asthma disease).

Groups	No.	Seromucoid ≤11 mg/dL (normal)	*Specificity % true negative	Seromucoid >11mg/dL (elevated)	**Sensitivity % true positive
Normal	25	18	72	7	28
Breast cancer:					
Premenopausal	40	7	17.5	33	82.5
Postmenopausal	35	6	17.14	29	82.86
Benign	25	2	8	23	92
Non-cancer: (Asthma disease)	20	1	5	19	95

* Calculated as the number of cases having ≤11 mg/dl divided by the total number of cases by 100.

** Calculated as the number of cases having >11 mg/dl divided by the total number of cases by 100.

Test sensitivity in those with breast cancer patients (premenopausal and postmenopausal) was 82.5% and 82.86% respectively, but in benign patients and non-cancer patients had elevated, 92% and 95% respectively. Test specificity in different types of breast cancer ranged between 17.5% and 17.14%.

From these results, it possible to conclude that it was useful for differential diagnosis and disease monitoring, but not for the early diagnosis of these tumors.

Elevated levels of glycoproteins have been reported in the sera of patients with metastatic breast carcinoma . Alterations in serum glycoproteins have been studied by determining the carbohydrate moieties, viz. fucose, hexose, hexosamine and sialic acid. Variations in glycoproteins in human uterine cervical carcinoma has been reported. These alterations are due to the exponential growth of malignant cells, which results in a rapid rate of membrane glycoprotein turnover and shedding of these excessive glycoproteins into the sera.

Patel, PS. et al has observed significant increase in the levels of the protein-bound hexose and seromucoid in breast carcinoma patients compared with the normal controls, also the differences were significant when compared to the patients with benign breast diseases, it has also been suggested that the measurement of the two parameters be helpful in the diagnosis of breast carcinoma as well as in differentiating between lobular carcinoma and infilterating duct carcinoma patients.

Serum levels of tumor-associated glycoprotein-72 in patients with gynecological malignancies can be used as a clinical marker . The potential role of sialic acid in the mechanism of tumor formation is indicated by the finding that sialic acid-rich glycoconjugates mask the surface of certain tumor cells by interfering with the immune response of the host [172], and that sialic acid content appears to be correlated with metastatic ability in a variety of tumor cells .

On the other hand, Yamamooto et al , reported that the increase in protein-bound hexose arises as a result of depolymerization of the ground substance connective tissue adjacent to cancer with release of these compound into circulation.

Bolmer et al have suggested that the elevation of the plasma glycoprotein reflects merely the occurrence of tissue destruction, while Yaskhiko et al have concluded that tissue proliferation or repair is a more probable etiological factor.

SERUM SUPEROXIDE DISMUTASE ACTIVITY:

INTRODUCTION

An enzyme which catalyzes the dismutation of superoxide free radicals ($O_2^{-\bullet}$) according to the reaction:

$$O_2^{-\bullet} + O_2^{-\bullet} + 2H^+ \rightarrow O_2 + H_2O_2$$

It has been purified by a simple procedure from bovine erythrocytes [177]. This enzyme, called superoxide dismutase, contains 2 equivalent of copper per mole of enzyme. The copper may be reversibly removed, and it is required for activity. Superoxide dimutase has been shown to be identical with the previously described copper-containing erythrocuprein (human) and hemocuprein (bovine). The enzyme has since been detected in a large number of tissues and organisms, and it is thought that it is present to protect the cell from damage by the highly reactive superoxide free radical.

Superoxide is formed by the one-electron reduction of oxygen, and has been identified as a product in a number of biological reactions. It is particularly likely to be formed in the red cell and has been shown to be produced when oxyhemoglobine is outoxidized to methemoglobine.

$$Hb - Fe^{2+} + O_2 \rightarrow Hb - Fe^{3+} + O_2^{-\bullet}$$

Other likely sources include reactions initiated by ionizing radiation.

Stable solutions of the superoxide radical were generated by the electrolytic reduction of O_2 in an aprotic solvent, dimethylformamide.

Slow infusion of such solutions into buffered aqueous media permitted the demonstration that $O_2^{-\bullet}$ can reduce ferricytochrome C and tetranitromethane, and that superoxide dismutase, by competing for the superoxide radicals, can markedly inhibit these reactions.

Superoxide dismutase was used to show that the oxidation of epinephrine to adrenochrome by milk xanthine oxidase is mediated by the superoxide radical

The content and composition of glycoproteins and glycolipids are affected, with an increase in sialic acid on the cell surface membrane and in the serum . Decreased activity of the enzyme superoxide dismutase (SOD) has also been found in all malignant tumors investigated so for . Such changes in the content of glycolipids (LBSA) and the activity of SOD are not found in patients with benign tumors . The present investigation was carried out to compare the content of LBSA and activity of SOD in sera of patients with breast cancer as compared to the normal healthy individuals.

MATERIALS AND METHODS:

PATIENTS AND COLLECTION OF SPECIMENS:

Two groups of breast cancer patients and one group of patients with benign breast tumors were included in this study.

Group 1 contained 20 premenopausal patients with breast cancer.

Group 2 consisted 20 postmenopausal patients with breast cancer.

Group 3 comprised 20 cases of benign breast tumors.

All patients were admitted for treatment to AL-Karama Teaching Hospital, patients suffered from any disease that may interfere with our study were excluded.

The blood samples were collected from these patients, left for 30 min. at room temperature; blood clots were separated at 3,000 r.p.m. for 10 min. by using centrifuge. Sera were aspirated and stored in sterilized tubes at $-20°C$ until time of analysis.

ESTIMATION OF SERUM SUPEROXIDE DISMUTASE ACTIVITY:

Superoxide dismuatse (SOD) activity was estimated spectrophotometery by the method of Winterbourn et al , with our modification for serum. The method is based on the ability of the enzyme to inhibit the reduction of nitroblue tetra-zolium (NBT) by superoxide generated during the reaction of photoreduction riboflavin and oxygen.

Procedure:

1. Addition of 0.2 ml of EDTA/NaCN solution to 0.1 ml of serum sample was carried out, then 0.1 ml of NBT solution were added.

2. The assay tubes were brought to a standard temperature (20-22°C), after that 0.05 ml of riboflavin solution were added to each tube. The final assay volume of 3 ml was made up with phosphate buffer 0.067 M, pH = 7.8.

3. Subsequent exposure to bright lighting was controlled by placing the assay tubes in a white-light box where they received uniform illumination for 20 minutes with 18W fluorescent tube attached to the lid, then the absorbance was red at 560 nm against distilled water.

4. To determine the control value, the absorbance for another set of tubes containing the same mixture was read at 560 nm against distilled water immediately after the addition of riboflavin. (riboflavin was added after the addition of buffer).

5. To determine SOD unit, ten tubes containing (10, 20, 40, 60, 80, 100, 200, 300, 400 and 500 µL) of normal serum samples, and another tube containing no serum were treated as described in the steps 1, 2, and 3.

Calculations:

Percentage inhibition was calculated from each absorbance in the presence and absence of the enzyme:

Inhibition % = $(A_E - A_{NE}) \times 100$

Where;

A_E: The absorbance at 560 nm of the tubes containing different amounts of the enzyme.

A_{NE}: The absorbance at 560 nm in the absence of the enzyme.

The percentages of inhibition were plotted against the corresponding amounts of serum (Figure 95).

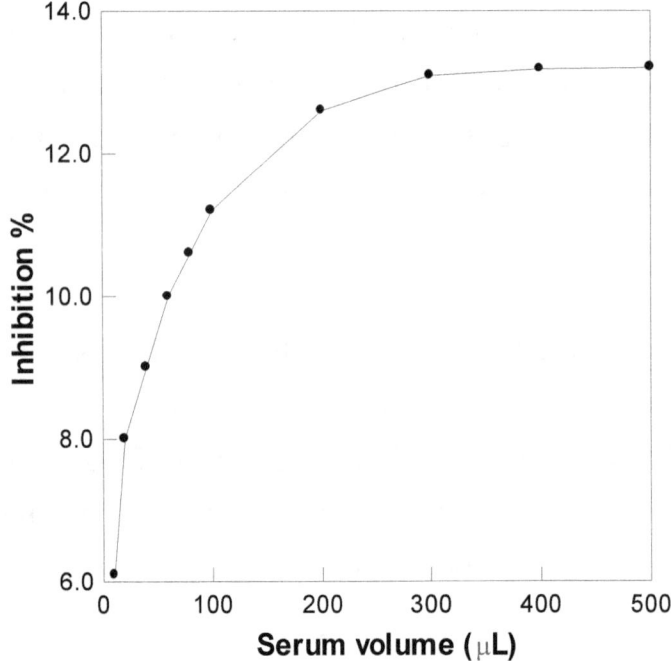

Figure (95): The Standard curve for the determination of SOD unit.

SOD unit was calculated from Figure (100) according to the following: the amount of serum (VµL) which gives half the maximum inhibition of NBT reduction (1 unit = 10.1 µL).

To calculate the SOD activity in sera of patients, the differences between absorbances before and after the light irradiation were, multiplied by the SOD unit.

DETERMINATION OF SERUM LIPID-BOUND SIALIC ACID:
Serum LBSA was determined by the method of Katopodis et al .

STATISTICAL METHODS:
Statistical analysis was performed by student's t-test .

RESULTS AND DISCUSSION:

The individual values of SOD activity and individual levels of LBSA for the normal healthy controls and patients with breast cancer are summarized in Table (57).

Figure (96) shows the distribution of the individual values of SOD activity in different types of breast cancer patients.

SOD levels in sera of patients with breast cancer differ significantly ($P < 0.001$) from those of healthy and patients with benign tumor, but the SOD levels in sera of benign patients shows no significant differences from those of healthy individuals.

It is clear from table (57), patients with breast cancer had lower values of SOD activity compared with that of healthy individuals, (premenopausal, 1.29 ± 0.14, and postmenopausal, 1.05 ± 0.23). Very low SOD activity in comparison with healthy was observed in sera from patients with postmenopausal. However, a reliable decrease in the SOD activity was found in patients with breast cancer (premenopausal and postmenopausal) as compared to that of benign patients ($P < 0.001$), on the other hand, patients with breast cancer have a higher levels of LBSA (premenopausal, 29.32 ± 3.55 and postmenopausal, 27.35 ± 5.29), compared with healthy individuals and benign patients. No statistically significant difference was found between LBSA levels in patients with benign and healthy individuals, also there was no statistically significant difference among LBSA levels between the different subgroups of breast cancer (premenopausal and postmenopausal).

Furthermore, the ratio of LBSA/SOD was higher in patients with breast cancer compared with those with benign and normal healthy controls, the mean values of this index are shown in table (57)

Table (57): Comparison of mean values for the superoxide dismutase (SOD) activity and Lipid-Bound sialic acid (LBSA) levels in sera from healthy and patients with breast cancer. All details are explained in the text.

	No.	SOD activity (mean±SD)	LBSA mg/dL (mean±SD)	LBSA/SOD (mean±SD)
Breast cancer: Premenopausal	20	1.29±0.14	29.32±3.55	22.74±2.25
Postmenopausal	20	1.05±0.23	27.35±5.29	30.05±5.87
Benign	20	1.53±0.17	18.73±2.29	12.34±3.64
Healthy (normal)	10	1.65±0.21	16.9±2.13	10.15±2.38

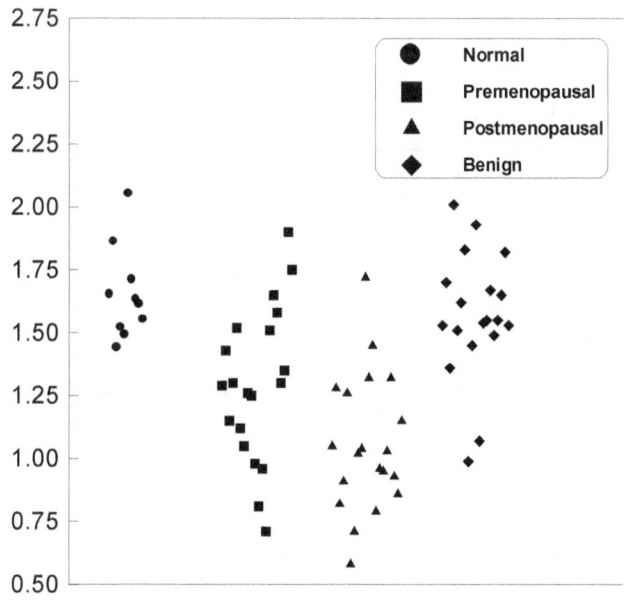

Figure (96): Distribution of the individual values of SOD activity in sera of patients with different types of breast cancer and normal healthy.

Knee et al and Bolzan et al reported that lower levels of SOD (especially the manganese dependent enzyme) are found in various malignant tumors compared with normal control cells.

High metastatic cell lines contain less SOD than low metastatic cell ones. Prognosis for patients with low SOD levels in the leukaemic blasts is unfavourable, and a negative correlation has been established between SOD activity in the whole blood and the chromosomal sensitivity after x-ray irradiation of lymphocytes from patients with cancer of the mammary gland. The results show that lower SOD activity in patients with tumors compared to the healthy controls (table 102), and the mean values for SOD activity in benign patients are near to those in healthy controls, these results are in agreement with the same results have been observed by Abella et al.

Table (58) shows the sensitivity and specificity of LBSA and SOD in normal controls and in patients with different types of breast cancer. Sensitivity and specificity of the determination of serum SOD activity in distinguishing patients with benign from premenopausal and postmenopausal patients are (85% and 95% respectively) for sensitivity, and it is possible that the lower specificity of this determination (15% and 5% respectively), while the sensitivity of LBSA for determination of benign from malignant is (95% for premenopausal and 100% for postmenopausal). Also, as with SOD activity measurement, the specificity of the determination (premenopausal 5%, and postmenopausal 0 %) is lower than its sensitivity.

Table (58): Specificity and sensitivity of LBSA and SOD activity in normal controls and in patients with breast cancer.

Groups	No.	Parameter	Negative results	Specificity % true negative	Positive results	Sensitivity % true positive
Healthy (normal)	10	LBSA	9	90	1	10
		SOD	8	80	2	20
Premenopausal	20	LBSA	1	5	19	95
		SOD	3	15	17	85

Postmenopausal	20	LBSA SOD	0 1	0 5	20 19	100 95
Benign	20	LBSA SOD	12 17	60 85	8 3	40 15

Note: Sensitivity and specificity of the SOD measuring as follows: sensitivity was calculated as % of patients with breast cancers who had positive results (lower SOD activity than healthy). Specificity was calculated as % of patients with breast cancers who had negative results (normal or higher SOD activity than healthy).

These results are due to the fact that patients with breast cancer have different diagnosis, also the differences in the serum activity of SOD in patients with malignant tumors and benign are probably due to the decreased enzyme level in the tumor cell. Such deficiency appears, for example, when enhanced serum levels of gangliosides (e.g. LBSA) are released by the tumor cells. LBSA binds to the plasma membranes of the mononuclear cells and inhibits their functions, which may be an important mechanism for immunosuppression in malignant diseases.

However, from these results, it possible to conclude that the serum levels of SOD and LBSA reflect the changes in the content of membrane glycolipids and cellular activity of SOD.

Some authors admit that these changes in the tumor cell are in closed connection; the membrane of the malignant cell has an altered lipid content and structure organization, which leads to decreased antioxidant protection.

The loss of cell differentiation leads to an increase of the cell glycolipids, on the other hand, to decrease of the intracellular SOD. Figure (97) shows a significant negative correlation between the two parameters in patients with breast cancer, and it is also indicates that patients with high levels of LBSA have low SOD content in the serum. Such negative correlation does not appear in benign. This is in accordance with the observations that changes in membrane glycolipids and cellular antioxidants occur in malignant, but not benign, tumors. These findings suggest that SOD and LBSA are good tumor markers.

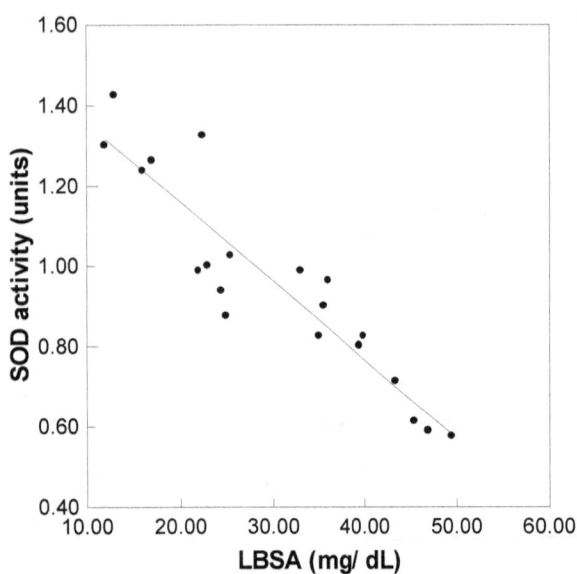

Figure (97): Correlation between superoxide dismutase (SOD) activity and the levels of lipid-bound sialic

LECTIN

INTRODUCTION

Lectin, a general term applied to haemo-agglutinating substances presents in extracts of seeds of certain plants, which specifically agglutinate RBC of certain blood group, are also glycoproteins and glycolipids.

The ability of lectins to interact with soluble glycoproteins and glycolipids have been used to isolate and fractionate glycoproteins and glycolipids. This ability is due to binding of lectin to the oilgosaccharides moieties of glycoproteins and glycolipids. Various lectins are known to be present in mammalian tissues and organs, the β-D-galactose and mannose-6-phosphate as a specific lectins from various tissues and organs have been studied extensively. However, a few lectins with high specificity for sialic acid have been identified, such as limulin and carcinoscropine. One major role of lectins, typified by the bacteria-legume symbiosis, appears to be to bind together cells of two different species. There is evidence that lectins acting in this way participate in both the prevention of plant infection, by binding to saccharides on bacteria or fungi, and in the promotion of bacterial infection of vertebrate cells. Bacterial lectins apparently mediate the adhesion of these microorganisms to oligosaccharides on animal cells, which could be a prelude to infection.

Thus, Escherichia coli contains a lectin that binds D-mannose and its α-glycosides, and that presumably mediates bacterial attachment to cells. Another evidence for binding of Vibrio cholera to intestinal cell surfaces by a reaction inhibited by L-fucose has been presented. Although a specific cellular function cannot yet be unequivocably assigned to any lectin, a large body of evidence indicates that lectins have been adapted for a variety of cell surface and intercellular functions in which the specific carbohydrate-binding site of the lectin binds a complementary saccharisde-containing substance as a prelude to one of a number of biological actions. Some lectins, such as those in root hairs, apparently bind complementary saccharides in a fairly discriminating way, which leads to highly specific symbiosis. In this case the lectin is apparently playing a highly refined recognition function, although its binding site is not so exclusive as to reject receptors on test erythrocytes. In contrast, evidence for different localizations and functions of the same lectin in different animal cells suggests that the specificity of lectin function is not dictated solely by the precise nature of its binding sites, but also by opportunistic factors that determine which of many receptors of adequate complementarity are available.

Furthermore, lectin which binds sialic acid residue of glycoproteins has also been isolated from wheat germ . Sialic acid occupies an outstanding position both sterically and with respect to biological function in various glycoproteins and glycolipids.

Also, a lectin which specificity recognizes terminal sialic acids residue is likely to be a useful in studying the biological functions of sialoglycoproteins.

MATERIALS AND METHODS

TISSUE COLLECTION AND PROCESSING:

Human tissues of breast cancer was removed by surgery from females patients, admitted for treatment at AL-Saddon Hospital, and diagnosed by specialist. Breast tissues were immediately immersed in ice-cold saline and then washed with phosphate-saline buffer pH 7.2 and kept at -20°C until time of homogenization process.

Preparation of Tissue Homogenate:

Three grams of frozen breast tissues were washed with 5 ml volume of 0.9% NaCl to remove surface mucus materials and contamination, then homogenized in 15 ml of 0.02 M phosphate buffered saline (0.075 M Na_2HPO_4/ 0.075 M KH_2PO_4 pH 7.2 containing 0.004 M of β-mercapto ethanol, 0.002 M EDTA and 0.075 M NaCl), using Tenbroeck ground-glass homogenizer to prepare the homogenate. The homogenate was centrifugated at 4000 r.p.m. for one hour. The supernatant was used as a source of lectin (crude or homogenate lectin) for binding (expressed as hemagglutination) studies. The supernatant was used through out study and stored at -20°C till using.

DETERMINATION OF BIOCHEMICAL CONSTITUENTS IN CANCEROUS BREAST HOMOGENATE:

Protein Determination:

Protein was determined by the method of Lowry et al , using bovine serum albumin as standard.

PRELIMINARY TEST FOR THE BINDING OF CANCEROUS BREAST LECTIN TO ERYTHROCYTE SUSPENSION:

Preparation of Standard erythrocytes Suspension for Hemagglutination:

Type A of human blood was obtained from physically normal volunteer. The erythrocytes were washed four times in 0.9% (w/v) NaCl and diluted with 0.9% NaCl to an absorbancy of about 2 at 620 nm.

The Hemagglutination Assay:

This assay is a modification of the procedure reported by Liener. The binding of cancerous breast lectin to erythrocyte suspension was preliminary checked by hemagglutination according to the following modification.

1. Half ml of diluted cancerous breast lectin by ((Tris-Saline buffer, 0.02 M Tris-HCl buffer pH 8 containing 0.15 M NaCl and 0.02 M $CaCl_2$) was incubated with 0.5 ml of washed erythrocytes at room temperature for 30 min. Then the cells were pelleted by centrifugation for 3 min., the supernatant was decanted.

2. The cells were resuspended in the same buffer and aggregated cells allowed to settle for 5 min., then the absorbance at 620 nm of the upper layer (free lectin and cells) of the assay solution was measured.

Calculations:

Total binding (expressed as % hemagglutination) represents the amount of lectin, which binds the erythrocytes and causing hemagglutination.

$$\text{Hemagglutination activity \%} = \frac{A - A^*}{A} \times 100$$

(total binding %)

Where:

A = The absorbance of standard erythrocyte suspension at 620 nm.

A* = The absorbance of free (unbound) erythrocytes at 620 nm.

Determination of non-Specific Binding of Cancerous Breast Lectin
Homogenate to Erythrocyte Surface Glycoconjugates:

The same steps mentioned in the text were followed to determine the percent of the non-specific binding, except, the whole blood were washed (3-4) times with normal saline (0.9% w/v NaCl), then two times with assay buffer (Tris-HCl buffer pH 8) and the suspension was prepared. Ten ml of this suspension placed in a test tube and 100 μL of neuraminidase was added (500 unit/ml) and the mixture was shaken for four hours at 25°C. Then centrifugation at 3000 r.p.m. for 3 minutes, was carried out. The supernatant was removed and the remaining blood cells were diluted with an appropriate amount of normal saline to an absorbancy of about 2 at 620 nm and then used for hemagglutination activity directly.

Calculations:

The percent of non-specific binding was calculated using the following equation:

$$\text{NSB \%} = \frac{A° - A^*}{A°} \times 100$$

Where:

NSB % = The percent of non-specific binding.

A° = The absorbance of neuraminidase-treated erythrocyte suspension at 620nm.

A* = The absorbance of free (unbound) erythrocytes at 620 nm.

Determination of The Specific Binding of Lectin to Erythrocyte Surface Glycoconjugates:

The percent of specific binding of lectin to glycoconjugates was calculated by subtracting the percent of non-specific binding from the percent of the total binding.

SB% = TB% - NSB%

Where:

SB% = The percent of specific binding of lectin to erythrocyte surface glycoconjugates.

TB% = The percent of total binding of lectin to erythrocyte surface glycoconjugates.

Determination of Hemagglutination Activity Unit:

The hemagglutination activity unit is the level of test solution (hence breast tumor homogenate concentration) which cause 50% of the standard cell suspension to sediment in (0.5-2 hours) as determined by Lis and Sharon method, or by plotting the hemagglutination activity data against the concentration of lectin, (Fig. 98).

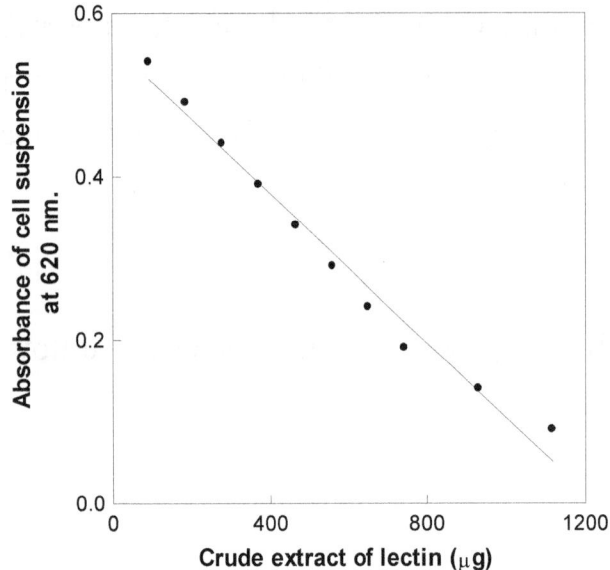

Figure (98): Hemagglutination assay of breast tumor homogenate lectin. All details are explained in text.

Factors Effecting on Lectin Binding to Erythrocyte Surface Glycoconjugates in Cancerous Breast Homogenate:

The Effect of Different Lectin Amounts on Its Binding To Erythrocyte Surface Glycoconjugates:

Half milliliter of erythrocyte suspension was incubated with different amounts of crude lectin (93, 186, 279, 372, 465, 558, 651, 744, 930 and 1116 µg) dissolved in assay buffer for 30 minutes at 22°C. The final reaction volume was 1 ml, then the cells were pelleted by centrifugation for 3 minutes.

Calculations:

1. The percent of specific binding (SB%) was calculated The percent of specific binding (SB%) was plotted against their corresponding lectin amount, as shown in Figure (98).

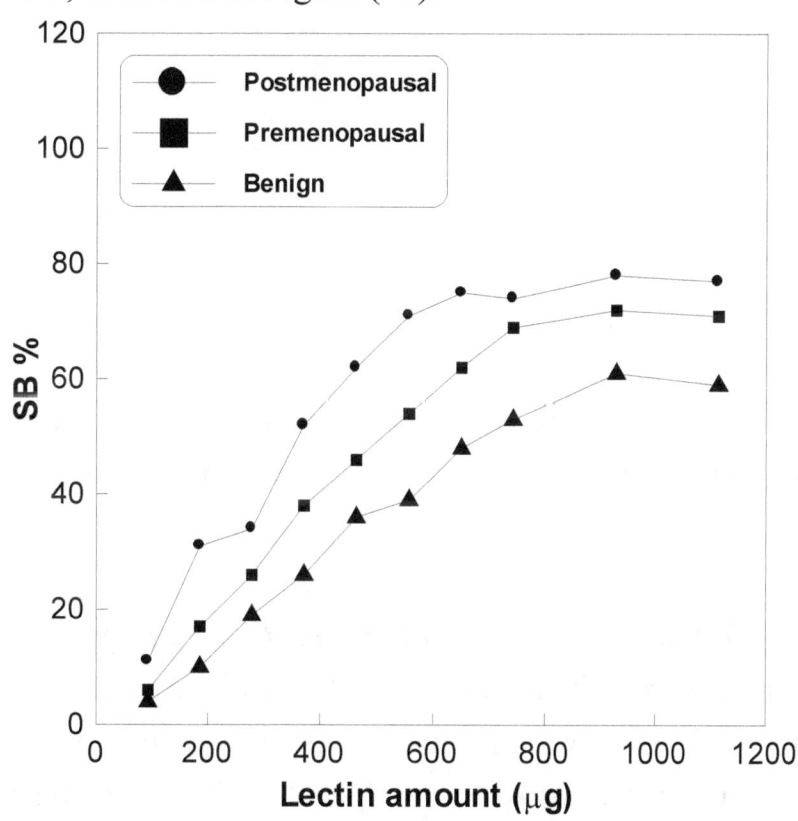

Figure (98): Effect of the amount of human breast tumor homogenate lectin on the hemagglutination activity. All details are explained in text.

Effect of pH on Hemagglutination Activity:

Fifty µL of human cancerous breast homogenate was incubated with 0.5 ml of erythrocyte suspension at 22°C for 30 minutes, using Tris-saline buffer of different pH from 7 to 9. The final reaction volume was 1 ml, then the cells were pelleted by centrifugation for 3 minutes. The percent of specific binding (SB%) was calculated.

1. The percentages of specific binding (SB%) were plotted against their corresponding pH values, as shown in Figure (99).

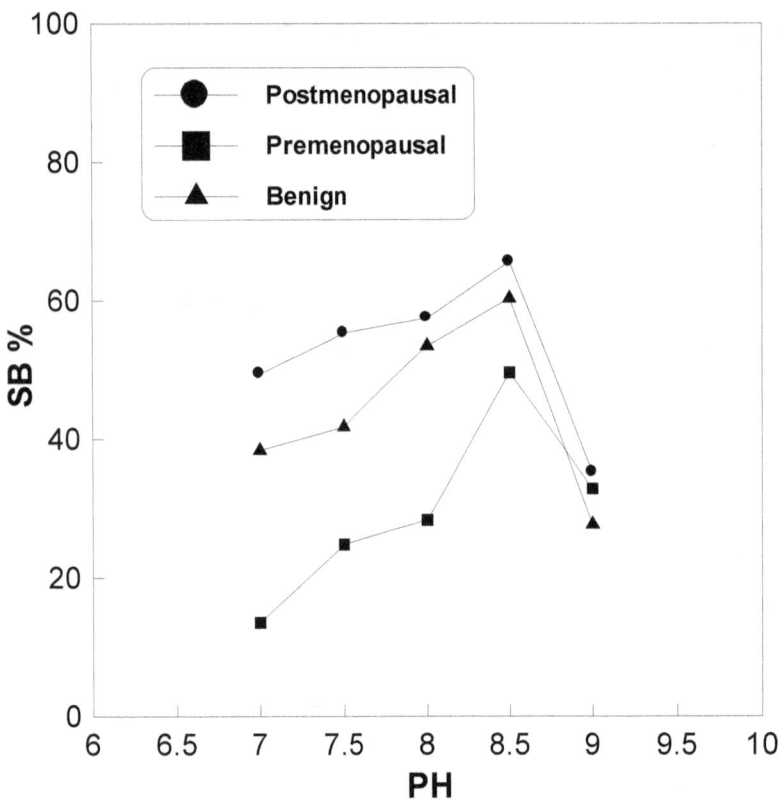

Figure (99): Effect of pH on the hemagglutination activity of human breast tumor homogenate lectin. All details are explained in text.

Effect of Temperature on Hemagglutination Activity:

Fifty μL of human cancerous breast homogenate was incubated with 0.5 ml of erythrocyte suspension for 30 minutes at different temperatures (5, 15, 20, 25, 30, and 35°C) using the assay buffer (pH 8.5). The final reaction volume was 1 ml, then the red cells were pelleted by centrifugation for 3 minutes.

Calculations:

1. The percent of specific binding (SB%) was calculated The percentages of specific binding (SB%) were plotted against their corresponding temperatures, as shown in Figure (100).

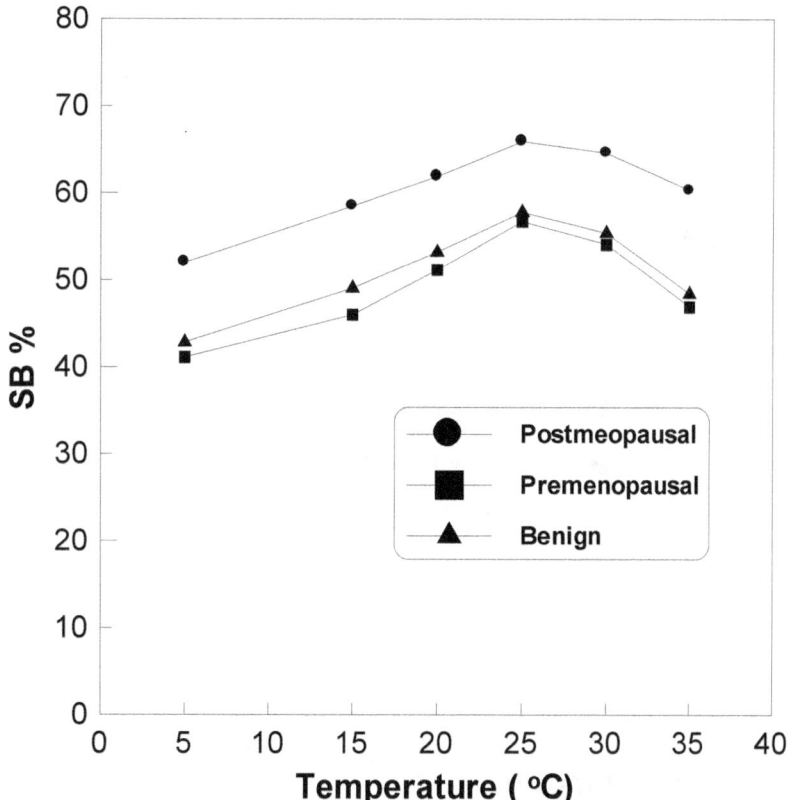

Figure (100): Effect of temperature on the hemagglutination activity of human breast tumor homogenate lectin. All details are explained in text.

Effect of Incubation Time on Hemagglutination Activity:

Fifty μL of human cancerous breast homogenate was incubated with 0.5 ml of erythrocyte suspension at 25°C for several time intervals (15, 30, 60, 90, 120, and 135 minutes) using the assay buffer (pH 8.5). The final reaction volume was 1 ml, then the red cells were pelleted by centrifugation for 3 minutes.

Calculations:

1. The percent of specific binding (SB%) was calculated The percentages of specific binding (SB%) were plotted against their corresponding times, as shown in Figure (101).

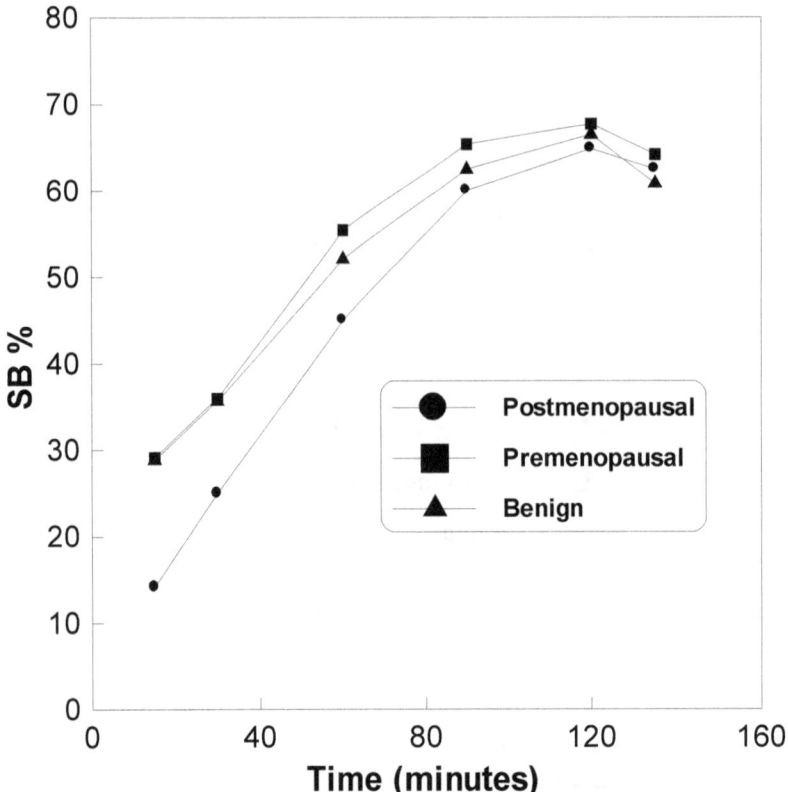

Figure (101): Dependence of hemagglutination activity of human breast tumor homogenate lectin on reaction time. All details are explained in text.

Effect of Exogenous Ca++ ions on Hemagglutination Activity:

Fifty μL of human cancerous breast homogenate was incubated with 0.5 ml of erythrocyte suspension at 25°C for 120 minutes using the assay buffer (pH 8.5), containing different Ca^{++} ions (2.5, 5, 10, 15, 20, 25,and 30 mM). The final reaction volume was 1 ml, then the red cells were pelleted by centrifugation for 3 minutes. The percent of specific binding (SB%) was calculated. The percentages of specific binding (SB%) were plotted against their corresponding Ca^{++} ions concentrations, as shown in Figure (102).

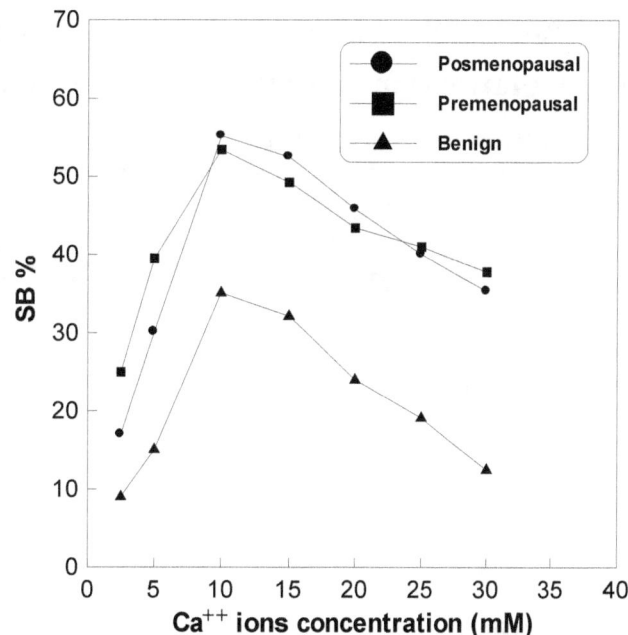

Figure (102): Effect of Ca^{++} ions concentration on the hemagglutination activity of human breast tumor homogenate lectin. All details are explained in text.

Effect of Denaturating Agents on Hemagglutination Activity:

Fifty µL of human cancerous breast homogenate was incubated at 25°C for 120 minutes with different concentration of denaturating agents (Urea, PEG, NaOH, and HCl), dissolved in the assay buffer (pH = 8.5), then 0.5 ml of erythrocyte suspension was added. The final reaction volume was 1 ml, then the red cells were pelleted by centrifugation for 3 minutes. The percent of specific binding (SB%) was calculated . The data of percent specific binding (SB%) were summarized in Table (59):

Table (59): Effect of denaturating agents on the hemagglutination activity of breast homogenate lectin. All details are explained in the text.

Groups	Type of test	Reagent added (M)	Specific binding %
Premenopausal	Control	-	100
	urea	3	74.59
		4	66.77
		5	57.13
		6	56.54
Postmenopausal	=	3	75.46
		4	71.65
		5	58.48
		6	47.58
Benign	=	3	73.65
		4	58.44
		5	29.17
		6	20.19
Premenopausal	Polyethlene glycol	0.5%	76.82
		1%	75.43
		2%	73.82
		4%	67.1
Postmenopausal	=	0.5%	76.82
		1%	75.65
		2%	74.27
		4%	71.59
Benign	=	0.5%	75.35
		1%	71.67
		2%	70.85
		4%	68.01
Premenopausal	NaOH	0.15	74.25
	HCl	0.15	71.38
Postmenopausal	NaOH	0.15	74.62
	HCl	0.15	71.59
Benign	NaOH	0.15	73.03

	HCl	0.15	70.98

Inhibition Studies of Hemagglutination Activity:

A number of carbohydrate were used as inhibitors (sialic acid, glucuronic acid, fructose, mannose and xylose) for the binding of lectin to glycoconjugates of erythrocyte surface.

The first step in this type of assay was carried out by addition of high concentration of lectin, which gives more of hemagglutination. Before addition the sugar, which used as inhibitor, 50 µL of human cancerous breast homogenate was incubated with 0.5 ml of erythrocyte suspension at 25°C for 120 minutes. The final reaction volume was completed to 1 ml by adding Tris-saline buffer (pH 8.5), then the cells were pelleted by centrifugation for 3 minutes. The second step in this type of assay was carried out according to the following:

Fifty µL of human cancerous breast homogenate was incubated with 0.5 ml of erythrocyte suspension at 25°C for 120 minutes The final reaction volume was completed to 1 ml by adding Tris-saline buffer (pH 8.5) which contain the desired concentration of sugar used (inhibitor). Then the cells were pelleted by centrifugation for 3 minutes.

Calculations:

1. The same mathematical formula was used to calculate the percent of total binding before and after addition of the inhibitor.

2. The percent of specific binding (SB%) was calculated before and after addition of inhibitor, The percent of inhibition of hemagglutination represents the difference between the percent of specific binding (SB%) with lectin alone and that obtained with lectin plus the inhibitor. The data of % inhibition was summarized in table (60) and these data were plotted against their corresponding sugar concentration, as shown in Fig. (103).

Table (60): Inhibition of hemagglutination activity of human breast homogenate lectin.
All details are explained in the text.

Groups	Type of carbohy. test	Carbohy. conc. (mM)	Inhibition %
Premenopausal	Sialic acid	1	6.6
		1.5	10.4
		2	15.3
		2.5	19.7
		3	24.4
		3.5	29.1
Postmenopausal	=	1	15.45
		1.5	22.41
		2	30.6
		2.5	36.1
		3	44.85
		3.5	52.1
Benign	=	1	4.65
		1.5	6.61
		2	8.1
		2.5	10.55
		3	12.56
		3.5	14.65
Premenopausal	D-glucuronic acid	1	3.25
		10	6.1
		15	9.9
		20	12.2
Postmenopausal	=	1	4.1
		10	8.75
		15	11.1
		20	13.15
Benign	=	1	2.4
		10	5.9

		15	9.3
		20	11.9
Premenopausal	D-Fructose	30	59.15
	D-Mannose	30	55.8
	D-Xylose	30	51.05
Postmenopausal	D-Fructose	30	28.35
	D-Mannose	30	20.45
	D-Xylose	30	19.95
Benign	D-Fructose	30	-
	D-Mannose	30	-
	D-Xylose	30	11.6

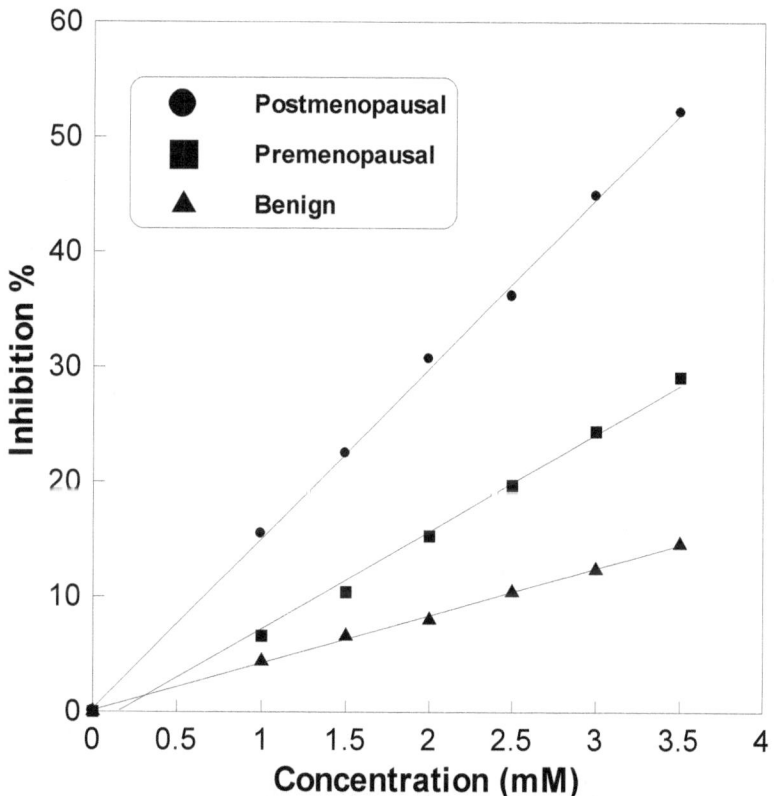

Figure (103): Effect of sialic acid on the hemagglutination activity of human breast tumor homogenate lectin. All details are explained in text.

Effect of Ionic Strength and Different Salts on Hemagglutination Activity:

Effect of monovalent salts on hemagglutination activity:

Fifty μL of human cancerous breast homogenate was incubated with 0.5 ml of erythrocyte suspension at 25°C for 120 minutes using the assay buffer (pH 8.5), which contains NaCl of various concentrations (0.05 M to 0.3 M). The total volume was 1 ml, then the cells were pelleted by centrifugation for 3 minutes.

Calculations:

1. The same equation mentioned was used to calculate the percent of total binding.

2. The percent of specific binding (SB%) was calculated.

3. The percentages of specific binding (SB%) were plotted against their corresponding NaCl concentrations, as shown in (Figure 104 and table 61):

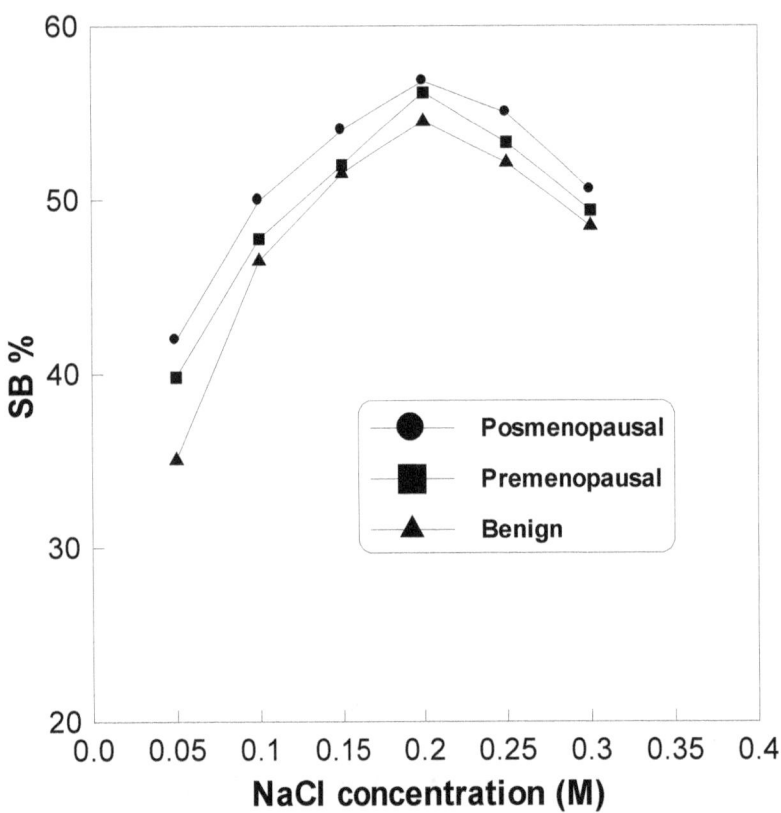

Figure (104): Effect of monovalent salt concentration on the hemagglutination activity of human breast tumor homogenate lectin. All details are explained in text.

Effect of divalent salts on hemagglutination activity:

Fifty μL of human cancerous breast homogenate was incubated with 0.5 ml of erythrocyte suspension at 25°C for 120 minutes in presence of 10 mM $CaCl_2$ and different concentrations from (0.005 M to 0.02 M) of $MgCl_2$ (dissolved in the assay buffer pH 8.5). The final reaction volume was 1 ml, then the cells were pelleted by centrifugation for 3 minutes

Calculations:

1. The same equation mentioned was used to calculate the percent of total binding.

2. The percent of specific binding (SB%) was calculated by using the equation in the text

3. The percentages of specific binding (SB%) were plotted against their corresponding $MgCl_2$ concentrations, as shown in (Figure 111 and table 64):

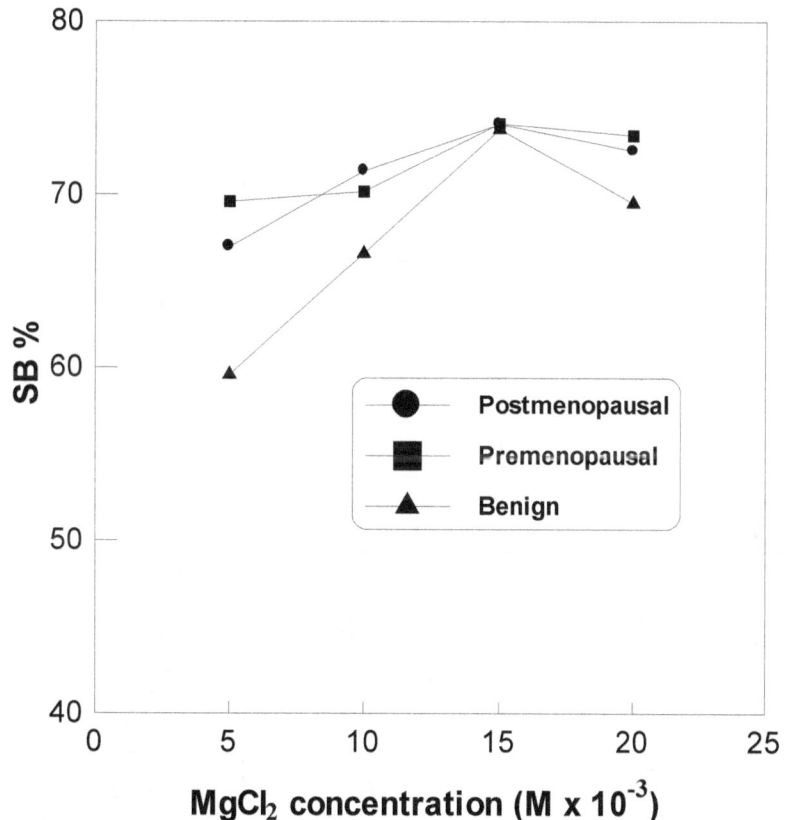

Figure (105): Effect of divalent salt concentration on the hemagglutination activity of human breast tumor homogenate lectin. All details are explained in text.

Table (61): Effect of mono and divalent salts on the hemagglutination activity of breast homogenate lectin. All details are explained in the text.

Groups	Type of salt test	Salt conc. (M)	Specific binding %
Premenopausal	NaCl	0.05	36.82
		0.1	47.78
		0.15	52.05
		0.2	56.17
		0.25	53.3
		0.3	49.39
Postmenopausal	=	0.05	41.93
		0.1	49.95
		0.15	54.01
		0.2	56.81
		0.25	55.03
		0.3	50.56
Benign	=	0.05	35.05
		0.1	46.5
		0.15	51.52
		0.2	54.96
		0.25	52.13
		0.3	46.19
Premenopausal	$MgCl_2$	5×10^{-3}	69.58
		10×10^{-3}	70.16
		15×10^{-3}	74.09
		20×10^{-3}	73.4
Postmenopausal	=	5×10^{-3}	66.98
		10×10^{-3}	71.35
		15×10^{-3}	74.06
		20×10^{-3}	72.52
Benign	=	5×10^{-3}	59.6
		10×10^{-3}	66.58
		15×10^{-3}	73.76
		20×10^{-3}	69.49

RESULTS AND DISCUSSION:

The homogenate of human tissue of breast cancer was characterized by significant hemagglutination activity toward human red blood cells (group A). Homogenization of breast tumor tissues was carried out in cold medium to protect of protein from denaturation due to proteolytic enzyme activity. However, addition of β-mercaptoethanol and EDTA to the extracting solution were necessary to prevent the inactivation of lectin and to achieve the maximal extraction of the lectin.

The total binding of lectin to glycoconjugates was estimated according to the hemagglutination assay. Hemagglutination assay is a semiquantitaive procedure and has been widely used as a laboratory test because of its ease and versatility. It depends on aggregating and sedimentation of the erythrocytes after reaction with the bivalent or multivalent lectin.

However, the non-specific binding was determined by using neuraminidase to be incubated with the erythrocyte suspension before the assay, this enzyme is responsible for the release of terminal sialic acid residue from the erythrocyte surface glycoconjugates, and hence, the penultimate N-acetylgalactosamine will be exposed for the lectin binding.

Until now relatively few sialic acid specific lectins have been identified, among which, only one is commercially a available. Most of them show a broad specificity. Basu et al have reported a novel sialic acid binding lectin Achatinin$_H$ from the hemolymph of <u>Achatina Fulica</u> snail. The types of sialic acid found on the erythrocytes, a striking correlation has been found between the ability of the agglutinin to agglutinate cells and the presence of O-acetylneuraminic acid residue on the mammalian cell surface. Indeed the lectin agglutinated only those erythrocytes which contain 9-O-acetylneuraminic acid residues. Furthermore, human erythrocytes which contain only nueraminic acid [211], but no 9-O-acetylneuraminic acid were not agglutinable even by enzyme treatment. Similarly horse erythrocyte known to contain high amount of 4-O-acetylneuraminic acid were not affected.

From these results it was believed that this active agglutination may be due to the O or N-acetyl group which is present on the structure of sialoglycoproteins present on outer surface of the red cellsit shows a rapid steep decline in absorbance with an inflection point appeared in the curve was due to the lectin concentration used and in 93 µg, which has a good agglutination activity.

Optimum Conditions of Hemagglutination:

Through figures the results show that the maximum hemagglutination activity of lectin required pH 8.5, the temperature at 25°C and the time for 120 minutes and it is clear that the lectin binding is dependent on pH and temperature. It should also be noted that the percent of specific lectin binding decreases with the change of pH (below 7 and upper 9.5), this suggests that the abundance of H^+ ions in the acidic medium may inhibit the binding sites on both glycoconjugate and lectin molecules, OH^- ions in more basic medium may influence in the same manner, and that the sialic acid which involved in binding is unstable to both acid and alkaline pH.

In general the results has led to the conclusion that the high binding process of lectin with glycoprotein is pH dependent and any shift in the pH of environment my affect the stability of the macromolecules involved in the binding. This effect includes the induction of protonation-deprotonation process occurring within the ionizable groups of the amino acids present in the binding groups of these macromolecules.

Effect of Ca^{++} Ions Concentration on Hemagglutination Activity:

It represents the effect of Ca^{++} ions concentration on the binding activity of human breast homogenate lectin. The results in this figure show that the highest binding of lectin was found in the presence of 10 mM Ca^{+2} ions. However, the Ca^{+2} ions plays an important role in stabilization the complex formed between the lectin and the glycoprotein present on the red cell surface, also the stabilization is due to the conformational changes in the protein due to the binding of Ca^{+2} ions. Dolichos, observed that the Ca^{+2} ion will stabilize the native structure of the protein itself.

Furthermore, there are many isolated lectins requires Ca^{+2} ions in their binding or their physiological roles, also, different Ca^{+2} ions dependent lectins have been purified from various sources and most of these possess multimeric structures and are capable of forming cross-linked complexes .

Effect of Denaturating Agents on the Hemagglutination Activity:

The data presented shows the effect of different denaturating agents on the binding activity of breast lectin (Urea, PEG, NaOH and HCl). The results show that all of denaturating agents were effected on specific binding for lectin as compared to the control, but this effect was different between types of denaturating agents. An analysis of the data shows that the percent of specific binding for lectin to glycoconjugates decreased with increasing urea concentrations, this effect can be attributed to the effect of urea on the hydrophobic forces between protein molecules.

Also, increasing concentrations of polyethylene glycol may results in precipitation of protein molecules which leads to decrease the interaction between lectin and glycoconjugates, and hence decrease the percent of specific binding.

Furthermore, the effect of 0.15 M of NaOH and HCl on the binding activity of breast lectin was investigated in this work, the results indicated that NaOH and HCl considerably reduce the percent of specific binding, their denaturating effect is due to great changes in pH of icubation medium.

Inhibition Studies of Hemagglutination Activity:

The inhibition percent of hemagglutination activity of human breast lectin by various carbohydrates (sialic acid, D-glucuronic acid, fructose, mannose and xylose).

The results in this table are classified according to their respective of group patients.

It is clear from this table, sialic acid and D-glucuronic acid were found to be the most potent inhibitors and gives high inhibition percent of hemagglutination at 3.5 mM and 20 mM respectively. However, the results obtained from this assay demonstrated that D-fructose, D-mannose and D-xylose tested at 30 mM in patients with breast cancer (premenopausal and postmenopausal) have the high activities to inhibit the binding of cancerous lectin, as compared to the benign patients. On the other hand, the inhibition of D-glucuronic acid suggests that it might be used as the eluting sugar in the purification of lectin.

Furthermore, 0.953 mM, 9.64 mM and 3.12 mM of sialic acid, sodium glucuronate and EDTA respectively were enough to produce 50% inhibition for lectin isolated from rat uterus.

Goebal et al), have suggested that the inhibition studies could be used as a technique for demonstrating and measuring the reactions occurred between lectin and its binding protein, for example, coupled disaccharides binds to protein via their p-aminophenyl glycosides and the terminal non-reducing hexose to play the predominate role in the specificity.

Lectin Purification

Lectin which binds sialic acid residue of glycoproteins nature has been isolated and purified from human cancerous breast homogenate by a combination of adsorption to formalinized human red cells and gel filtration, elution with D-glucuroinc acid and subsequent fractionation on sephadex G_{150}. The recovery percent of lectin was determined and found in the range of 80%. The purification and identification of lectin was performed in the presence of Ca^{+2} ions, the elution volume (V_e) and then K_{av} value, for elution of breast homogenate which contain Ca^{+2} ions from sephadex G_{150}, was calculated and found to be 54 ml and 0.265 respectively.

The purified lectin appeared to be homogenous by electrophoresis in the presence of sodium dodecyl sulphate (SDS), was characterized by a single band whose molecular weight was (51,500) KDa.

However, the determination of physiochemical properties of human malignant breast lectin, such as stoke radius, purification folds and effect of Ca^{+2} ions were determined. On binding of glycoprotein to purified human cancerous breast lectin, heat stability and effect of EDTA were studied.

There are several procedures available for the isolation and purification of lectin from human tissue , but most of these are conventional methods. However, other methods such as affinity chromatography are used for purification of lectin from mammalian tissues and organs . Of known lectins which have been purified and characterized, few bind sialic acid. Wheat germ agglutinin is the only plant lectin that binds to sialic acids, also it binds to oligosaccharides containing terminal N-acetylglucosamine and N-acetylgalactosamine .

Two lectins specific for sialic acids have been purified from animal sources, namely from the hemolymph of the horseshoe carbs of the class arachinda and slug, Liman Flavus.

Recently two sialic acid-specific lectins have been purified from tritrichomonas mobilensis and from the mushroom hericium exinaceum .

In general, this chapter was planned to develop a method for purification of lectin from breast cancer by using formalinized erythrocytes as affinity adsorbant accompanied by gel filtration.

MATERIALS AND METHODS

GEL PREPARATION AND COLUMN PACKING:

A. Preparation of the column: The dimension of the column were chosen according to the following equation :

diameter = $\sqrt[3]{m/10}(cm)$, where:

m = amount of protein in mg.

Length = $30 \times$ diameter

B. Preparation of the gel: The gel (Sephadex G_{150}) was allowed to swell in excess of 0.02 M Tris-saline buffer pH 7.2 containing 0.01 M $CaCl_2$ and left to the stand for three days at room temperature without stirring, then the slurry was poured carefully into a vertical glass-column down the wall using a glass-rod. After the gel has settled, the column was equilibrated with Tris-saline buffer pH 7.2 for 24 hours with the dimension of (1.5 x 50 cm).

Void Volume (V_o) Determination:

The volume of the gel column was determined by using blue dextrin 2000 at concentration of 1mg/ml in Tris-saline buffer pH 7.2. One ml of blue dextrin solution was applied to the column surface carefully, then the elution was carried out with the same buffer using a flow rate of 8 ml/hour, fractions of three ml were then collected and their absorbance was measured at 600 nm to determine the void volume (V_o).

Column-Calibration:

The column was calibrated by the calibration Kits, purchased from Pharmacia Fine Chemicals using six standard proteins. Standard protein solutions were prepared according to manufactures instructions, then applied through two separate runs (2 ml portion). First run include the following proteins, ribonuclease A, ova albumin and aldolase, while the second run include the following proteins, ferritin, albumin and catalase. Elution was carried out by Tris-saline buffer pH 7.2 with flow rate of 8 ml/hour. The absorbance of the fractions collected were measured at 280 nm to determine the elution volume (V_e) of the standard proteins, .

Calculations:

1. The K_{av} values of the proteins eluted were determined by the following formula:

$$K_{av} = \frac{V_e - V_o}{V_t - V_o}$$

Where:

V_o = Void volume

V_e = Elution volume of each protein

V_t = Total gel-bed vloume: determined from the following equation:

$$V_t = \left(\frac{\text{column diametr}}{2}\right)^2 \times \frac{22}{7} \times \text{column length}$$

2. Tow calibrations curves were plotted:

a. K_{av} values against log. M.Wt. of the standard proteins, as shown in Fig. (6.1).

b. $(-\log K_{av})^{1/2}$ values against the stokes radius (R_s) of the standard proteins, as shown in Figure (107).

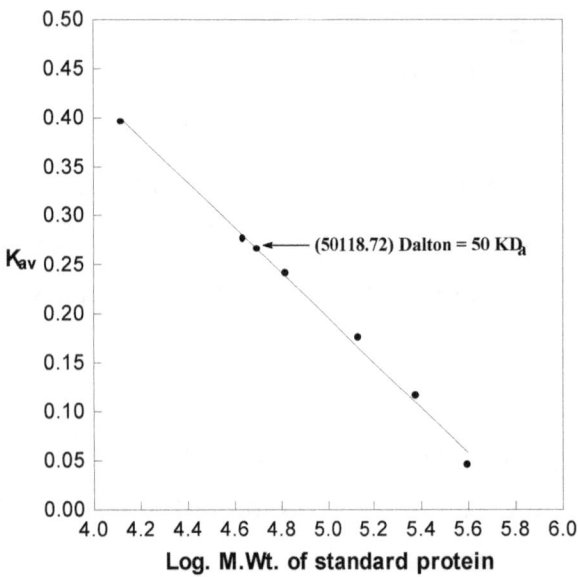

Figure (106): Calibration curve for estimation molecular weight of lectin by gel-filtration using standard proteins. All details are explained in the text.

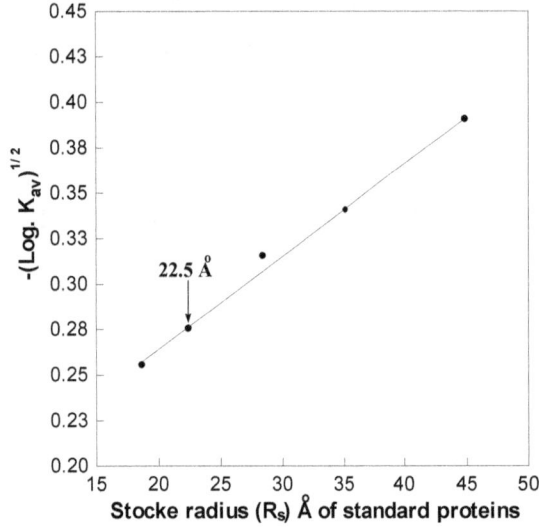

Figure (107): Calibration curve for estimation of Mol. Wt. and stocke radius (R_s) of cancerous breast lectin by gel-filtration technique using sephadex G-150 column (1.5x50cm). All details are explained in the text.

PURIFICATION AND IDENTIFICATION OF HUMAN CANCEROUS BREAST LECTIN USING FORMALINIZED ERYTHROCYTES AND GEL FILTRATION:

In each step of purification, the assay of binding was carried out and accompanied by protein determination, the other calculations were also performed, as shown in table (62)

Table (62): Scheme of lectin purification form breast cancer. The hemagglutination activity of the lectin obtained by each step was determined as described under method.

Step	Volume of fraction (ml)	Protein (mg/ml)	Total amount of protein (mg)	Activity* Per mg protein	H.U** Total	purification	Recovery %
Crude cell free extract of breast cancer	10	18.6	186	11	2000	1	100
Elute from red cells	60	0.113	6.78	251	1700	23	85
Concentration elute	4	1.62	6.48	262	1700	24	85
Pooled peak from sephadex column	9	0.38	3.42	468	1600	43	80

Note: * Hemagglutination activity was determined
** Hemagglutination activity unit was determined as Lis and Sharon method (.

Preparation of Formalinized Erythrocytes as Affinity Matrix:

The method used for preparation of formalinized erythrocytes was essentially as described by Csizman and Nowak et al .

Human erythrocytes (group A) were obtained from blood bank, washed four times in twenty volume of 75 mM NaCl, 75 mM Na_2HPO_4/75 mM K_2HPO_4 (phsphate-saline buffer, pH 7.2), per packed cell volume by centrifugation at 3000 r.p.m. for ten minutes. Twenty five ml of the packed red cells were resuspended to 200 ml in phosphate-saline buffer, pH 7.2 and placed in a 500 ml conical flask. Fifty ml of commercial formaline (40% formaldehyds) was introduced and the mixture was incubated at 37°C for 20 hours. The cells where then washed five times in five volumes phosphate-saline buffer, pH 7.2 per packed cell volume and kept at 4°C as 10% suspension in this buffer.

Step 1:

The stored formalinized cells prepared were used as an affinity reagent by washing the cells six times in ten volumes of 50 mM Tris-HCl, 100 mM NaCl (Tris-saline buffer). PH 7.2 containing 0.01 M $CaCl_2$. Ten ml of cancerous breast homogenate was added to thirty ml of cell suspension (10% suspension in buffer) for three hours at room temperature then washed three times with twenty volumes of Tris-saline buffer, pH 7.2 containing 0.01 M $CaCl_2$. Elution of the adsorbed lectin was accompanied by incubation the cells with 50 ml of 0.15 M D-glucuronic acid in Tris-saline buffer, containing 0.01 M $CaCl_2$ which has been brought to pH 7.2 with NaOH before the addition to the cells and placed in refrigerator for overnight. The elution mixture was then centrifuged for ten minutes at 3000 r.p.m., and the resultant supernatant was referred as fraction 2 whereas the breast homogenate being fraction 1.

Step 2:

Dialysis for concentration:

Dialysis tube of suitable length was prepared and used to concentrate fraction 2 to a volume of about 4 ml against solid sugar at 10°C for 24 hours. The resulted concentrated lectin was considered as fraction 3.

Step 3:

Gel Filtration:

Before applying the sample to a sephadex G_{150} column (1.5 x 50cm) the column had been equilibrated with Tris-saline buffer, pH 7.2, containing 0.01 M $CaCl_2$, after that the sample was transferred at the top surface of the column and then eluted with this buffer at an elution rate of 8 ml/hour. Fractions of three ml volume were collected then identified by the assay method as well as the absorbance at 280 nm and protein determination were carried out. The elution volume (V_e) of the lectin in each fraction was determined by the following formula:

V_e = Fraction volume (3 ml) x Fraction number containing the highest level of the lectin.

The Assay Methods:

In order to identify the fractions which contain lectin, the % of specific binding (S.B %) for each fraction was determined as follows:

a. Half ml of each fraction isolated by gel filtration was incubated with 0.5 ml of erythrocytes in a final volume of one ml at 25°C for 120 minutes, then determining absorbance at 620 nm of the upper layer (free lectin and cells) of the assay solution.

b. Parallel experiments was performed to determine the amount of non-specific binding for each fraction.

Calculations:

The same equations mentioned were used to calculate the specific and non-specific binding of lectin.

Determination of Molecular Weight of Lectin Using K_{av} Values:

The same formula mentioned were used to calculate the K_{av} value of lectin and then using calibration curve to determine molecular weight of lectin.

Determination of Lectin Stoke Radius:

$(-\log . K_{av})^{1/2}$ values for standard proteins were determined and applied to stokes radius (R_s) in the standard curve to determine lectin stoke radius.

Determination of Purification Folds:

The purification fold for the lectin was determined using the following formula:

$$\text{Purification fold of lectin} = \frac{\text{Specific binding of purified lectin}}{\text{Specific binding of homogenate lectin}}$$

Determination of % Recovery of Lectin:

The % recovery was determined as follows:

Total unit of purified lectin

$$\text{Recovery \%} = \frac{\text{Total unit of homogenate lectin}}{} \times 100$$

ANALYSIS OF THE PURIFIED FRACTION BY SDS POLYACRYLAMIDE GEL-ELECTROPHORESIS (SDS-PAGE):

Gel electrophoresis in the presence of sodium dodecyl sulphate (SDS) was performed according to the method of Laemmli. Using 7.5% polyacrylamide separating gel. To determine the lectin molecular weight after comparison with molecular weights of standards proteins, the standard proteins and their molecular weights are detailed below:

Protein	M.Wt. (KD$_a$)
Phosphorylase b	94
BSA	67
Ova albumin	46
Trypsin inhibitor	20
α-lactoalbumin	14

Procedure:

a. Polyacrylamide gel (concentration 7.5%) was prepared. Standard proteins solutions: Pharmacia electrophoresis calibration kit for the determination of native molecular weight of protein by polyacrylamide-gel electrophoresis was used. The content of each vial was redissolved in 0.3 ml of sample buffer (0.0625 M Tris-HCl, pH 6.8, containing 2% SDS, 10% glycerol, 5% 2-mercaptoethanol and 0.001% bromophenol blue as the dye).

b. Sample preparation: Purified lectin was concentrated by dialysis against sucrose, then diluted with sample buffer to protein concentration range (0.2-2 mg/ml).

Calculations:

1. The relative mobility (R_m) of each protein was measured as follows:

$$R_m = \frac{\text{Distance of protein migration}}{\text{Length after dying}} \times \frac{\text{Length before fixation}}{\text{Distance of dye migration}}$$

2. Molecular weights of the marker standard proteins was plotted against relative mobilities (R_m), typically gives a straight line (Fig.108) from which the lectin M.Wt. was estimated.

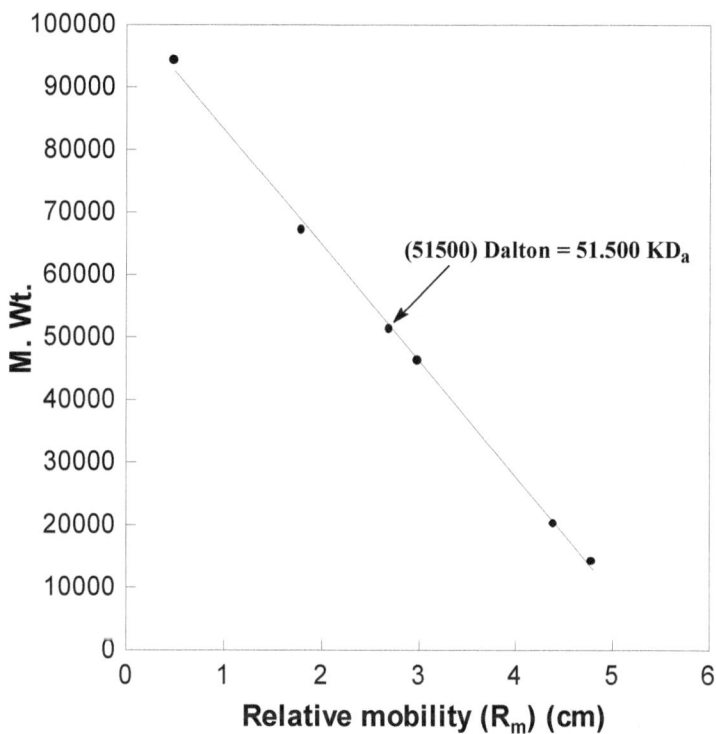

Figure (108): Molecular weight determination of purified breast lectin by sodium dodecyl sulphate-7.5% acrylamide gel analysis. All details are explained in the text.

THE EFFECT OF C^{a++} IONS ON BINDING OF GLYCOPROTEIN TO PURIFIED HUMAN CANCEROUS BREAST LECTIN:

a. Fifty microliters of purified cancerous breast lectin was incubated with 0.5 ml of red cell suspension at 25°C for 120 minutes. Different concentration of Ca^{++} ions (5, 10, 15, 20 and 25) $\times 10^{-3}$ M were dissolved in the assay buffer (pH = 8.5) and were added to the sample. The final reaction volume was one ml. Then the cells were pelleted by centrifugation for 3 min. Parallel incubations were performed to determine the non-specific binding of the lectin, .

Calculations:

The same mathematical formula mentioned in experiment was used to calculate the % of specific binding of purified lectin. The % specific binding was plotted against their corresponding Ca^{++} ions concentration, as shown in Fig. (109).

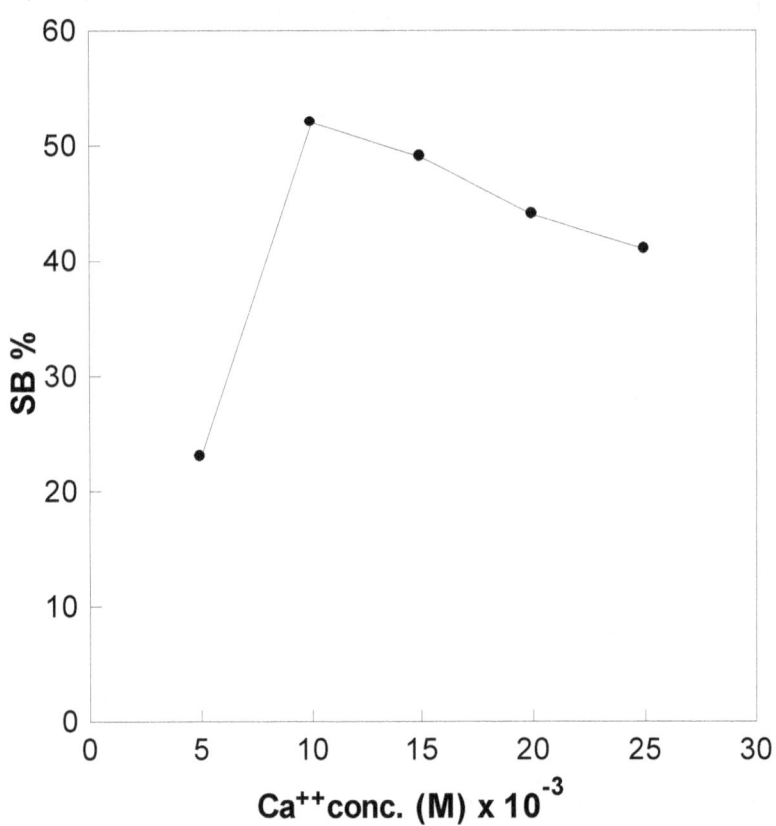

Figure (109): Effect of Ca^{++} ions concentration on the hemagglutination activity of human purified cancerous breast lectin. All details are explained in the text.

THE EFFECT OF TEMPERATURE ON THE STABILITY OF THE PURIFIED CANCEROUS BREAST LECTIN:

The heat stability of lectin were carried out according to the following:

a. Fifty microliters of purified cancerous breast lectin was incubated with 0.5 ml of red cell suspension for 10 min. at different temperatures (15, 25, 37, 50, and 60°C), using the assay buffer (pH = 8.5). The final reaction volume was one ml, then cooled on ice-bath for 5 min. After that the cells were pelleted by centrifugation for 3 min. Parallel incubations were performed to determine the non-specific binding of the lectin,.

Calculations:

The same mathematical formula mentioned in the text was used to calculate the % of specific binding of purified lectin. The % specific binding was plotted against their corresponding temperatures, as shown in Fig. (110).

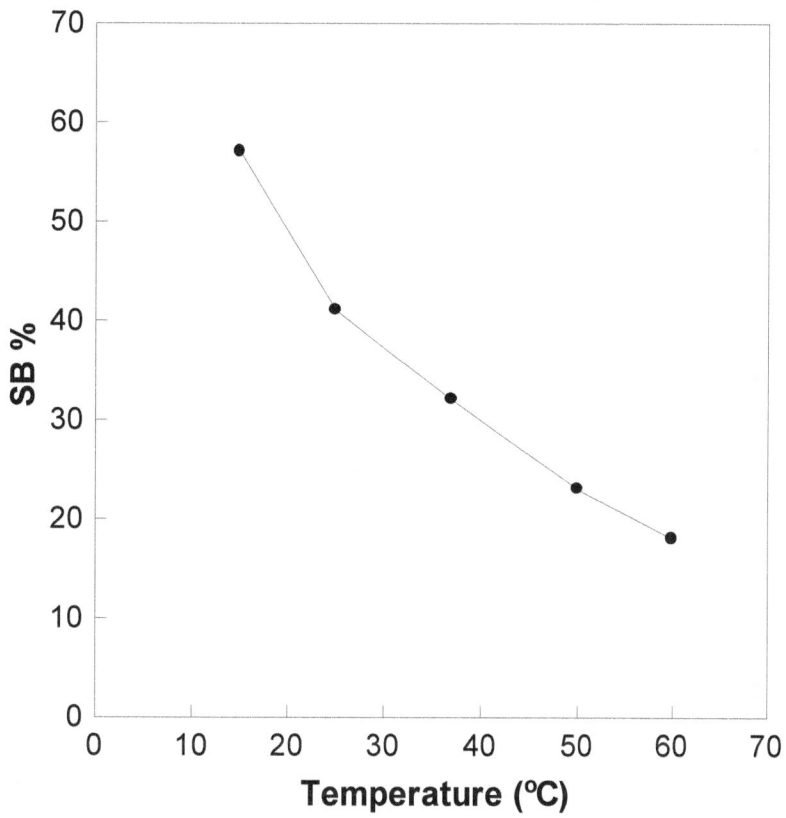

Figure (110): The effect of temperature on the stability of purified cancerous breast lectin. All details are explained in the text.

EFFECT OF EDTA ON HEMAGGLUTINATION ACTIVITY OF PURIFIED LECTIN:

a. Fifty microliters of purified human cancerous breast lectin which gave a high percent hemagglutination was incubated with 0.5 ml of red cell suspension at 25°C for 120 min. The final reaction volume was completed to one ml, by adding Tris-saline buffer pH 8.5 (Assay buffer) which contains the desired concentration of EDTA (0.1, 0.15, 0.2, 0.25, 0.3 and 0.35 mM). Then the red cells were pelleted by centrifugation for 3 min.

b. The same experiment was carried out in the absence of the EDTA.

Calculations:

The same mathematical formula was used to calculate the % hemagglutination before and after addition of EDTA. The % inhibition of hemagglutination represents the difference between the % hemagglutination with lectin alone and that obtained with lectin plus EDTA.

RESULTS AND DISCUSSION

Purification and identification of lectin from human breast cancer tissue depends on the existence formalinized erythrocytes as affinity adsorbant accompanied by gel filtration, elution with D-glucuronic acid and subsequent fractionation on sephadex G_{150}.

The purification of lectin was performed in the presence of Ca^{+2} ions. Typical yields and purity data from the different steps in the purification of lectin from human breast cancer tissue are summarized in table (63). On the other hand, it has estimated the elution volume (V_e) and K_{av} value for elution of breast homogenate from sephadex G_{150} and found to be 54 ml and 0.265 respectively, whereas this K_{av} value of the lectin was obtained from the calibration curves in the Figures, to determine molecular weight and stoke radius for lectin. These parameters were estimated and found to be 50118.72 Dalton and 22.5 Å respectively. Furthermore, other investigators have purified lectin form different human tissues using the same method and their results revealed that the M.wt. and stoke radius (R_s) of the purified lectin were (60000 dalton, 34 Å) and (40180 Dalton, 28.8 Å) respectively.

Analysis of Purified Fractions by SDS-PAGE:

Molecular weight of subunit for the purified human breast lectin was determined by sodium dodecyl sulphate-7.5% acrylamide gel electrophoresis, calibration curve for SDS-PAGE (7.5%), using low molecular weight groups of known subuints proteins as markers (14-94 KDa). From this curve, application of the relative mobility (R_m) values for each standard protein, molecular weight of purified lectin was estimated and found to be 51500 Dalton. However, and from the results obtained, there is only one lectin in breast homogenate specific for sialic acid.

On the other hand, the results were confirmed by Nassir and Huda where they found one lectin in human kidney and thyroid gland respectively, using the same method. Also, their results revealed that the molecular weight of purified lectin was 64000 and 44157.05 Dalton respectively.

Moreover, other investigators, were able to purify human kidney lectin with molecular weight of 63000 Dalton and whom using gel chromatogarphy and then adsorption with immobilized glycoconjugates and ion-exchange gel.

Stability of Cancerous Breast Lectin:

The effect of temperature on the stability of purified cancerous breast lectin, is clear form that there was only slight decrease in stability on increasing the temperature up to 60°C, whereas, above this temperature the stability of purified lectin may decrease due to the denaturation process of lectin.

Effect of Ca^{+2} Ions and EDTA on Hemagglutination Activity of Purified Lectin:

The effect of Ca+2 ions concentration on the hemagglutination activity of human purified cancerous breast lectin.

The results indicated that the breast lectin requires Ca^{+2} ions for binding activity. It was found that Ca^{+2} ions had a marked effect on the hemagglutination activity of the lectin [218], and this effect was proved by loosing the hemagglutination activity, when the chelating agent EDTA was added which may remove the available Ca^{+2} ions, this effect is illustrated in Table (63).

Table (63): Inhibition of hemagglutination activity of purified human cancerous breast lectin by EDTA.

EDTA conc. (mM)	Inhibition %	Conc. Of EDTA that gives 50 % inhibition (mM)
0.10	22.18	
0.15	27.91	
0.20	44.37	0.3
0.25	46.08	
0.30	53.43	
0.35	62.87	

froer for diagnosis of the d

The present investigation is carried out to describe the kinetics and thermodynamic properties associated with the binding of lectin to glycoprotein, and then determination of their parameters.

The results show that the association constant (K_a) and the specific binding (B) were increased as temperature increased. However, the time course data for the binding follows, the pseudo-first order kinetics accompanied with increase in K_{obs}. Whereas, the Hill plot data revealed that there was a weak positive cooperativity between lectin binding sites.

Also, the Van't Hoff plot demonstrated a linear relationship between $\ln K_a$ and $\frac{1}{T}$.

On the other hand, the parameters for the equilibrium reaction described by $\Delta G°$, $\Delta H°$ and $\Delta S°$ were determined. Furthermore, the Arrhenius plot indicates that there was a linear relationship between $\ln k_{+1}$ and $\frac{1}{T}$ from which the transition state thermodynamic parameters for the formation of the lectin-glycoprotein complex represented by; E_a, ΔG^*, ΔH^* and ΔS^* were determined.

KINETIC STUDIES OF LECTIN

Lectins are carbohydrate-binding proteins that are grouped together. They agglutinate cells or other materials that display more than one saccharide of sufficient complementarity. They are found in many categories of living things. The major role of lectins are the ability to bind saccharides specifically, and interact with RBC of certain blood group, also glycoproteins and glycolipids . However, binding of lectins at the cell surface may cause other changes in cell function than mitogenic stimulation or the release of stimulating factors .

In this investigation it has attempted to explain the mechanism of binding lectin to glycoprotein to form lectin-glycoprotein complex, and then to determine the kinetics and thermodynamic parameters, and describe the molecular basis of lectin interaction, through the effect of time-course, temperature and the other factors.

MATERIALS AND METHODS

KINETIC STUDIES:

The Time-Course of cancerous breast lectin binding to glycoprotein present on red cell surface:

Reagents:

The standard erythrocyte suspension and the assay buffer (Tris-HCl pH 8.5) containing 10 mM $CaCl_2$ and 0.2 M of NaCl were prepared as described in sections (5.6.1 and 5.3).

Procedure:

1. At zero time, 0.5 ml of erythrocyte suspension was added to 50 µL of purified lectin (930 µg protein), the final volume of the assay mixture was made up to 1 ml with Tris-HCl buffer pH 8.5. The reaction mixture was incubated at 25°C for several time intervals (15, 30, 60, 90, 120 and 150 minutes).

2. After each time interval the assay tubes were treated .

3. Parallel experiments were carried out to determine the amount of non-specific binding.

4. To determine the time-course of lectin binding to glycoprotein present on red cell surface at different temperatures, the above steps were performed at (4, 11, 18 and 25°C).

Calculations:

1. The molar concentration of lectin involved in total binding to erythrocyte surface glycoconjugates, was calculated according to the following formula:

$$\text{The concentration of lectin (M) involved in total binding} = \frac{A - A^*}{A} \times \text{The total concentration of lectin (M) used in the assay}$$

Where:

A: The absorbance of standard erythrocyte suspension at 620 nm.

A*: The absorbance of unbound (free) erythrocytes at 620 nm.

2. The concentration of lectin in molar involved in non-specific binding to erythrocyte surface glycoconjugates, was calculated according to the following formula:

$$\text{The concentration of lectin (M) involved in non-specific binding} = \frac{A' - A^*}{A'} \times \text{Total lectin concentration (M) used in the assay}$$

Where:

A': The absorbance of neuraminidase-treated erythrocytes suspension at 620 nm.

A*: The absorbance of unbound (free) erythrocytes at 620 nm.

3. The concentration of specifically bound lectin in molar was calculated by subtracting the concentration of lectin involved in non-specific binding from the concentration of lectin involved in total binding:

Concentration of specifically bound lectin (M)	Concentration of lectin (M) involved in total biding	Concentration of lectin (M) involved in non-specific binding

4. The concentrations of specifically bound lectin (lectin-glycoconjugate) complex in molar were plotted against their corresponding incubation times ..

Scatchard Analysis:

Procedure:
1. Half ml of erythrocyte suspension was added to increasing amounts of lectin (0.37-3.7×10^{-7} M), the final volumes were made up to 1 ml with tris-HCl buffer (pH 8.5, contains 0.2 M NaCl and 10 mM $CaCl_2$).

2. The assay tubes were incubated for 120 minutes at 25°C, then they treated as mentioned in the text

3. The previous steps were repeated at different temperatures (4, 11 and 18°C).

Calculations:
1. The concentration of specifically bound lectin (molar) was calculated for each tube according to the calculations .

2. The concentration of free (unbound or unreacted) lectin was calculated by subtracting the concentration of lectin (M) involved in total binding from the total concentration of lectin (M) used in each experiments:

The concentration of free lectin (M)	Total concentration of lectin (M)	The concentration of lectin (M) gives total binding

3. The concentration of lectin binding sites (B_{max}) and the affinity constant (K_a) were determined according to Scatchard equation .

$$\frac{B}{F} = \frac{1}{K_d} \times (B_{max} - B)$$

$$K_a = \frac{1}{K_d}$$

Where:
B: The concentration of specifically bound lectin.
F: The concentration of free lectin.
K_a: The affinity constant.

B_{max}: the maximal binding capacity.

K_d: The dissociation constant.

4. The plot of B/F values against the values of B, gives liner relationship. The total concentration of lectin binding sites (B_{max}) was calculated from the intercept on the x-axis, while the value of affinity constant was calculated from the slope of the straight line.

5. The K_a and B_{max} values were also determined from the Eadie-Hofstee plot of data getting from Scatchard plots, using the following equation:

$$B = -K_d \frac{B}{F} + B_{max}$$

The values of K_a and B_{max} were calculated from the slope of the straight line and the intercept on Y-axis respectively.

Determination of Hill-Coefficient (n) of Lectin binding to Glycoconjugates:

Calculations:

1. The value of Hill-coefficient (n) was calculated according to Hill equation;

$$\log\left(\frac{B}{B_{max} - B}\right) = n \log F - \log K_d$$

2. The values of log (B/B_{max}-B) were plotted against the values of logF, the Hill coefficient (n) was calculated from the slope of the straight line.

THE THERMODYNAMIC STUDIES:

Calculations:

1. The thermodynamic parameters of standard state were obtained from Van't Hoff plot, the values of the natural logarithm of affinity constant (K_a) obtained at different temperatures were plotted against the reciprocal values of the absolute temperature in Kelvin (1/T), according to the following equation:

$$\ln K_a = \frac{\Delta S°}{R} - \frac{\Delta H°}{RT}$$

Where:

$\Delta H°$: The enthalpy change of the standard state.

$\Delta S°$: The entropy change of the standard state.

R: The gas constant (8.31441 J.K^{-1}.mol^{-1}).

$\Delta H°$ value was obtained from the slope of the linear relationship of the plot.

The change in Gibbs free energy of the standard state ($\Delta G°$) was obtained from the following equation:

$$\Delta G° = -RT \ln K_a$$

While the entropy change of the standard state $\Delta S°$ was obtained from:

$$\Delta S° = \frac{\Delta H° - \Delta G°}{T}$$

2. The thermodynamic parameters of the transition state were obtained from Arrhenius plot of $\ln k_{+1}$ values against 1/T values, that gives a linear relationship according to the following equation:

$$\ln K_{+1} = \ln A - \left(\frac{E_a}{RT}\right)$$

Where:

A: Arrhenius constant.

The value of apparent energy of activation (E_a) of the binding reaction can be determined from the slope of the straight line. The enthalpy of the transition state ΔH^* was obtained from:

$$\Delta H^* = E_a - RT$$

The free energy change of the transition state ΔG^* is calculated from the following equation:

$$\Delta G^* = -RT \ln K_{+1} + RT \ln\left(\frac{KT}{h}\right)$$

Where:

K: is Boltzmann constant (1.38x10^{-23} Jdeg^{-1})

h: is Plank constant (0.622x10^{-33} Js^{-1}).

The change in entropy of the transition state ΔS^* is calculated from the following formula:

$$\Delta S^* = \frac{\Delta H^* - \Delta G^*}{T}$$

Results and discussion

Kinetics of lectin binding to erythrocyte surface glycoconjugate:

The Time-Course of Cancerous Breast Lectin Binding to Erythrocyte Surface Glycoconjugates:

Figure (111) shows the time course of the formation of lectin binding to erythrocyte surface to glycoconjugate complex at four different temperatures (4, 11, 18 and 25°C), in breast homogenate sample.

The concentration of lectin-glycoconjugate complex that formed after time (t) was calculated from the following equation:

The concentration of lectin-glycoconjugate complex formed after time (t) in molar	The concentration of lectin involved in total binding (M)	The concentration of lectin involved in non-specific binding (M)

The results of time-course pattern at different temperatures indicated that the lectin binding to erythrocyte surface glycoconjugate is a temperature and time dependent process, since a maximum binding can be obtained at 25°C after incubation for 120 minutes, there is no analogous studies are available to compare our results.

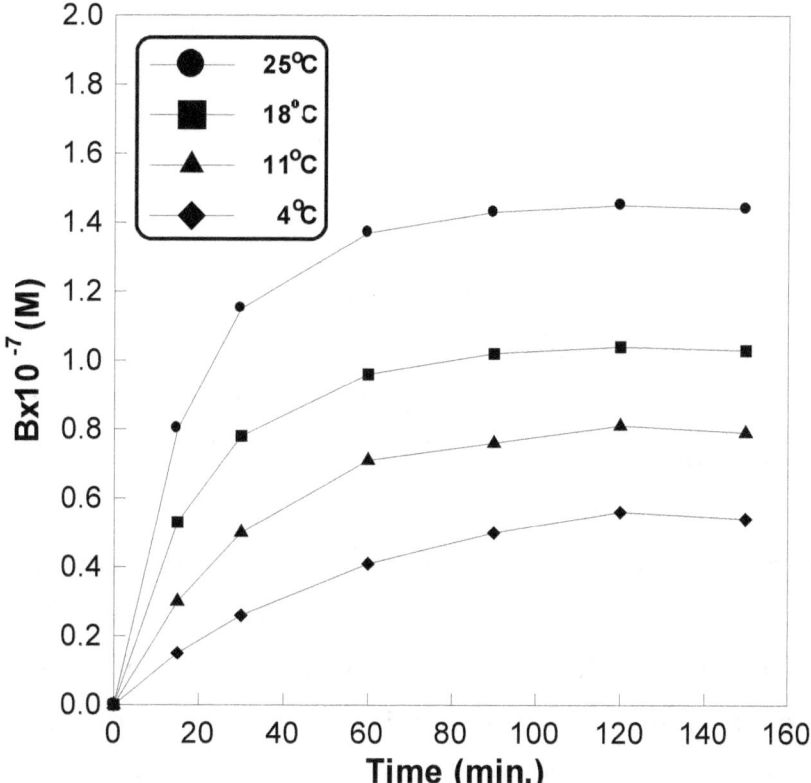

Figure (111): Time-course of lectin binding to erythrocyte surface to glycoconjugates at different four temperatures. All details are explained in the text.

Determination of Kinetic Parameters of Lectin-Glycoconjugate Complex Formation:

The time-course of lectin-glycoconjugate complex formation was carried out to describe the kinetic parameters of the binding (expressed as specific binding). The simplest proposed model representing this interaction was:

$$\text{Lectin} + \text{glycoconjugate} \underset{K_{-1}}{\overset{K_{+1}}{\rightleftharpoons}} [\text{Lectin-glucoconjugate}] \tag{1}$$

Where:

K_{+1}: is the association rate of lectin to glycoconjugate.

K_{-1}: is the dissociation rate of lectin-glycoconjugate complex.

At equilibrium:

$$K_a = [\text{Lectin-glycoconjugate}]/[\text{Lectin}][\text{glycoconjugate}] \tag{2}$$

$$K_d = [\text{Lectin}][\text{glycoconjugate}]/[\text{Lectin-glycoconjugate}] \tag{3}$$

Thus;

$$K_a = \frac{1}{K_d} = \frac{K_{+1}}{K_{-1}}$$

(4)

Where:

K_a: is the equilibrium constant (affinity constant).

K_d: is the equilibrium constant of dissociation of the complex.

The values of K_a and total concentration of lectin binding sites (B_{max}) were calculated from Scatchard and Eadie-Hofstee plots (112,113) at four different temperatures (4, 11, 18 and 25°).

It is clear from table (64), the results show that the affinity constant (K_a) is depended on temperature (K_a increased from 0.847×10^7 M^{-1} at 4°C to 0.926×10^7 M^{-1} at 25°C). Whereas the value of dissociation constant (K_d) was calculated by using equation (4), and show that the lowest K_d value of lectin-glycoconjugate complex occurs at 25°C after incubation for 120 minutes.

Table (64): The kinetic parameters of lectin binding to erythrocyte surface glycoconjugates. All details are explained in the text.

T°C	$K_d \times 10^{-7}$ (M)	$K_a \times 10^7$ (M^{-1})	$B_{max} \times 10^{-7}$ (M)
4	1.18	0.847	0.97
11	1.14	0.877	1.05
18	1.11	0.901	1.32
25	1.08	0.926	1.47

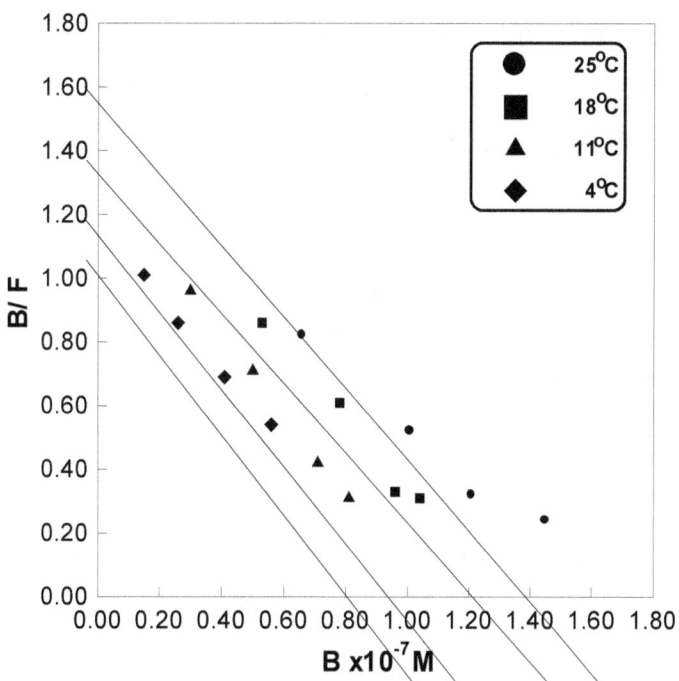

Figure (112): Scatchard plot of lectin binding to erythrocyte surface glycoconjugate at different four temperatures. All details are explained in the text.

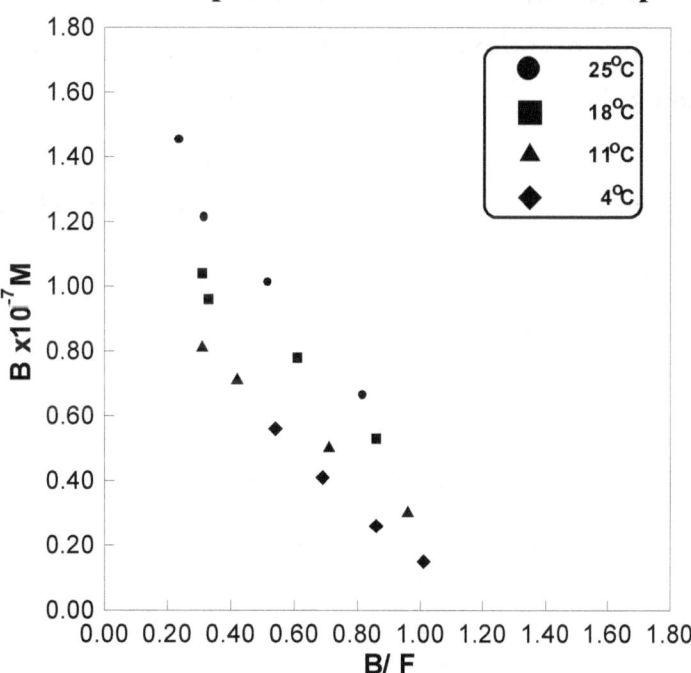

Figure (113): Eadie-Hofstee plot of data from Scatchard plot of lectin binding to erythrocyte surface glycoconjugate at different four temperatures. All details are explained in the text.

However, the time-course data shown in Figure (114) could be used to determine the reaction order of lectin binding to erythrocyte surface glycoconjugates using the following equation:

$$Ln[lectin\text{-}G]_e \left[\frac{(lectin)_t - (lectin\text{-}G)_t (lectin\text{-}G)_e / (G)_t}{(lectin)_t [(lectin\text{-}G)_e - (lectin\text{-}G)_t]} \right]$$

$$K_{+1}t \left[\frac{(lectin)_t (G)_t}{(lectin\text{-}G)_e} \quad (lectin\text{-}G)_e \right] \quad (5)$$

Where:

K_{+1}: is the kinetic association constant in M^{-1} min.$^{-1}$.

$(Lectin\text{-}G)_e$: is the concentration of the complex formed at equilibrium.

$(Lectin\text{-}G)_t$: is the concentration of the complex formed after time (t).

The equation (5) represents the second order kinetics, but in this work the percent of specific binding was in some cases, small and most of the lectin remains free and only small fraction binds even at equilibrium, i.e, $(lectin)_t \gg (lectin\text{-}G)_e$ thus, $(lectin)_t (lectin\text{-}G)_t (lectin\text{-}G)_e$

and $\dfrac{(lectin)_t (G)_t}{(lectin\text{-}G)_e} \quad \dfrac{(G)_t}{\gg (lectin\text{-}G)_e}$

So that the following equation could be used in order to fit the pseudo-first order kinetics:

$$Ln \frac{(lectin\text{-}G)_e}{(lectin\text{-}G)_e - (lectin\text{-}G)_t} \quad \frac{(lectin)_t (G)_t}{(lectin\text{-}G)_e} = K_{+1}t$$
(6)

On the other hand, Figure (114) shows the plot of $\ln \dfrac{(lectin - G)_e}{(lectin - G)_e - (lectin - G)_t}$ against time (t) gives a straight line with a slope equal to the observed value of first rate constant (K_{obs}) in min^{-1}. The rate constant (K_{+1}) in M^{-1} min^{-1} was calculated at four different temperatures by using the following formula:

$$K_{obs} = K_{+1} \frac{[\text{lectin}]_t [G]_t}{[\text{lectin} - G]_e}$$

$$\therefore K_{obs} = K_{+1} [\text{lectin}]_t$$

(7)

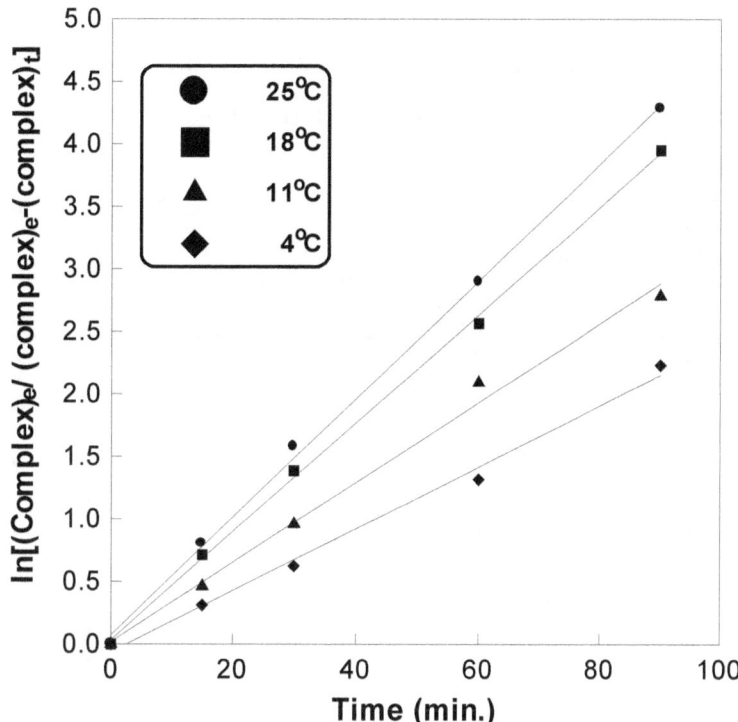

Figure (114): Kinetics of lectin binding to erythrocyte surface to glycoconjugate at different four temperatures. All details are explained in the text.

Also, the value of K_{-1} at four temperatures were calculated by using equation (4). Whereas, the half life time of association $(t\frac{1}{2})_{ass.}$, which represents the time needed for the formation of half amount of the complex at equilibrium, was determined from the concentration of the complex at equilibrium and the time-course curve. Also, the half life time of dissociation $(t\frac{1}{2})_{diss.}$, was calculated from the following relation:

$$(t\frac{1}{2})_{diss.} = \ln \frac{2}{k_{-1}} = \frac{0.693}{k_{-1}}$$

(8)

The values of $K_{obs.}$, K_{+1}, K_{-1}, $(t\frac{1}{2})_{ass.}$, and $(t\frac{1}{2})_{diss.}$ at four different temperatures are summarized in table (65). Data analysis of this table show that the highest rate for the association reaction occurs at 25°C, while the lowest rate occurs at 4°C, where the reaction temperature was increased from 4°C to 25°C, the value of K_{+1} increased from $(0.676 \times 10^5$ $M^{-1}.min^{-1})$ to $(0.973 \times 10^5$ $M^{-1}.min^{-1})$ which means the dependence of reaction rate on temperature. Also, the rate of dissociation of lectin-glycoconjugate complex (K_{-1}) is temperature dependent.

Table (65): The effect of temperature on the kinetic parameters of lectin binding to erythrocyte surface glycoconjugates. All details are explained in the text.

T °C	$K_{obs.}$ (min^{-1})	K_{+1} $(M^{-1}.min^{-1}) \times 10^5$	K_{-1} $(min^{-1}) \times 10^{-3}$	$(t_{1/2})_{ass.}$ (min)	$(t_{1/2})_{diss.}$ (min)
4	0.025	0.676	7.98	30	86.84
11	0.031	0.838	9.56	26	72.49
18	0.033	0.892	9.90	15	70
25	0.036	0.973	10.51	12	65.94

Scatchard Analysis:

Figure (115) shows Scatchard plot of lectin binding to erythrocyte surface to glycoconjugate in the presence of 10 mM Ca^{++} ions at different temperatures (4, 11, 18 and 25°C) after incubation for 120 minutes. This figure could be used to determine the kinetic parameters of lectin binding such as, the equilibrium constant of dissociation of the complex (K_d) and total concentration of lectin binding sites (B_{max}) of human cancerous breast lectin by using the following equation:

$$\frac{B}{F} = \frac{1}{K_d} \times (B_{max} - B)$$

The values of these parameters at different temperatures in presence of Ca^{++} ions are summarized in table (66). Through analysis the results in this table show that the total concentration of lectin binding sites (B_{max}) is temperature dependent, when the temperature was increased from 4°C to 25°C, B_{max} was increased from $(0.97 \times 10^{-7}$ M to 1.47×10^{-7} M), this fact could be explained according to the number of molecules possessing the activation-energy for interaction, increase with increase the temperature. On the other hand, the affinity constant (K_a) is also depended on temperature, this indicates that the reaction is slightly endothermic and explained by the fact that affinities of endothermic reactions enhanced by increasing temperatures. However, the values of B_{max} and K_d for human cancerous breast lectin at different temperatures obtained from Scatchard analysis were similar to those obtained from the Eadie-Hofstee plot (Fig. 115).

Determination of Hill-Coefficient (n) of Lectin Binding to Glycoconjugates:

Figure (115) represents the Hill plot of lectin binding to erythrocyte surface glycoconjugate in the presence of 10 mM Ca^{++} ions at four different temperatures (4, 11, 18 and 25°C), The value of Hill-coefficient (n) equals the slope of the resulting straight line. The values were (1.54, 1.63, 1.71, 1.8) respectively. However, on application of the Hill equation and using the results obtained from Scatchard analysis, it would be possible to evaluate the cooperativity of lectin binding sites through the determination of Hill coefficient (n). Furthermore, the results obtained in this work indicates that the cooperativity of lectin binding sites was low and affected by temperatures.

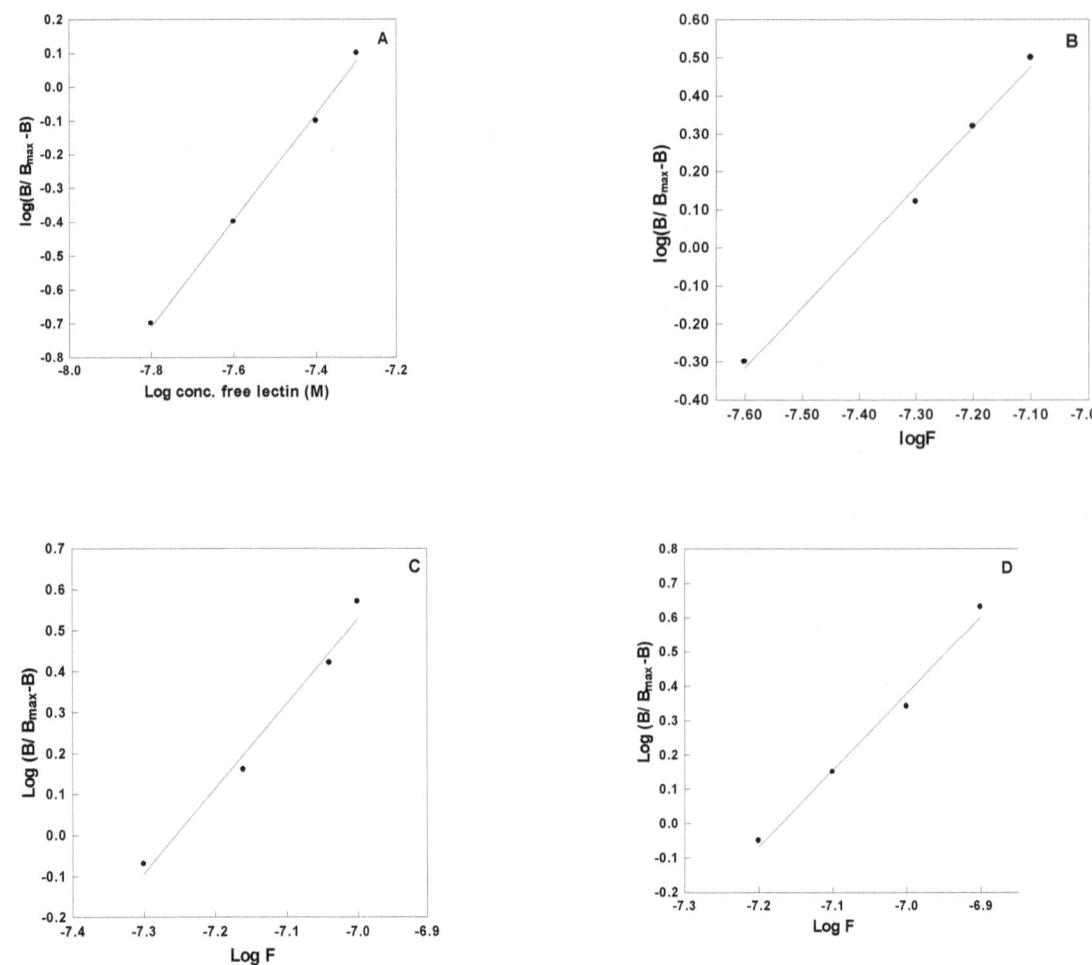

Figure (115): Hill plots of lectin binding to erythrocyte surface glycoconjugates at different temperatures, -A- 4°C, -B-11°C, -C-18°C, -D-25°C. All details are explained in the text.

THE THERMODYNAMICS OF THE LECTIN BINDING TO ERYTHROCYTE SURFACE GLYCOCONJUGATES:

A. Thermodynamic parameters of standard state:

Figure (116) shows Van't Hoff plot for the binding of lectin to erythrocyte surface glycoconjugates at different temperatures (4, 11, 18 and 25°). This figure revealed that the equilibrium binding constant (affinity constant) for lectin binding to glycoconjugates is temperature dependent. The results obtained from Van't Hoff plot indicated that $\Delta H°$ in general had a positive value of 9.15 KJ/mol, and that the reactions were nearly endothermic. However, the small positive value of $\Delta H°$ may indicate a favorable interaction between the lectin and glycoconjugate subgroups. These include the non-covalent interaction which are fundamentally electrostatic in nature such as charge-charge, charge-dipole, dipole-dipole, charge-induced dipole, dipole-induced dipole interactions and hydrogen bonds. The sum of these types of interactions can yield some stabilization to the folded structure of the complex.

The other values of thermodynamic parameters of standard state at four different temperatures (4, 11, 18 and 25°C), such as $\Delta G°$ values and $\Delta S°$ are summarized in table (66). From the analysis, the results in this table shows that the $\Delta G°$ values increases with decreasing temperatures, since the lectin binding to erythrocyte surface glycoconjugates needs higher energy at low temperatures. Whereas, the negative values of $\Delta G°$ indicates the stability of lectin glycoconjugate complex, subsequently the high affinity of the reactant.

Also, the high negative values of $\Delta G°$ indicates that the binding of lectin to glycoconjugates is a spontaneous reaction. Furthermore, these values are controlled by a high positive $\Delta S°$ values, table (66). The results show that the values of $\Delta S°$ decrease with increasing temperatures, this can be attributed to the more stable and more arranged status of lectin-glycoconjugate complex. On the other hand, the high positive value of $\Delta S°$ may be indicated that the binding spontaneity was enropically driven.

Entropy was the driven force for the occurrence of the binding, this shows that the hydrophobic interactions played an important role in the stability of complex formation .

Table (66): Thermodynamic parameters at standard state of lectin binding to erythrocyte surface glycoconjugates. All details are explained in the text.

T °C	ΔH° (KJ/mol.)	ΔG° (KJ/mol.)	ΔS° (J/mol.K)
4	9.15	-36.74	165.67
11	9.15	-37.75	165.14
18	9.15	-38.74	164.57
25	9.15	-39.74	164.06

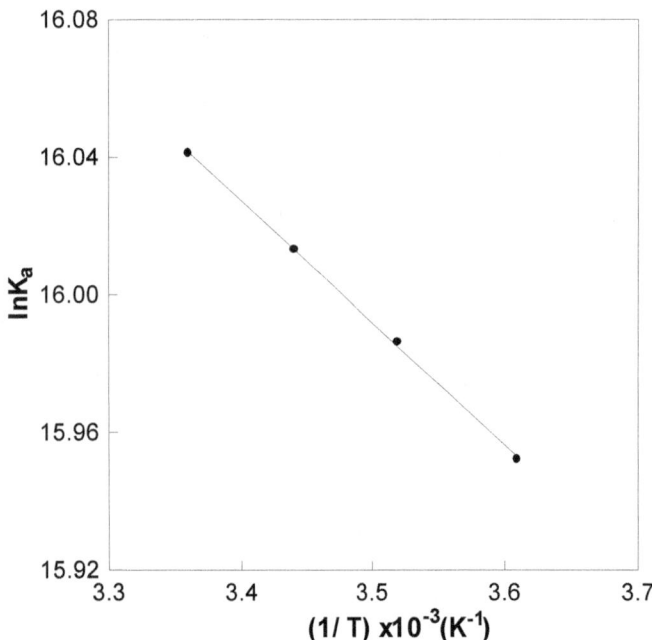

Figure (116): Van't Hoff plot for the binding of lectin to erythrocyte surface glycoconjugates. All details are explained in the text.

B. Thermodynamic parameters of transition state:

Through the transition state, the interaction of two substances leads to the formation of an activated complex (transition state), then the formation of the final product, i.e.: (the association of lectin with erythrocyte surface glycoconjugates) can be represented as follows:

Lectin + glycoconjugate ⟶ [Lectin-glycoconjugate]
an activated-complex
(transition state)

⟶ Lectin-glycoconjugate

(final product)

According to Arrhenius equation and kinetic constant, it could be calculated the thermodynamic parameters of the transition state ($\Delta H^*, \Delta G^*$ and ΔS^*) at four different temperatures (4, 11, 18 and 25°C). Figure (124) shows Arrhenius plot for the binding of lectin to erythrocyte surface glycoconjugates, the slope of the straight line represents the activation energy (E_a) of the binding reaction.

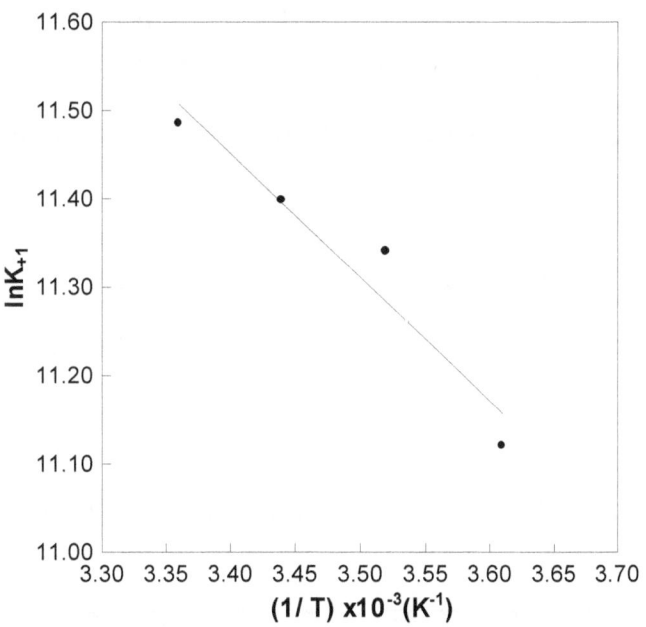

Figure (117): Arrhenius plot for the binding of lectin to erythrocyte syrface glycoconjugates. All details are explained in the text.

The values of thermodynamic parameters of the transition state (E_a, ΔH^*, ΔG^* and ΔS^*) are summarized in table (67). It is clear from this table, the high value of activation energy (13.05 KJ/mol) represents the required energy to overcome the transition state energy barrier and then giving the final product (lectin-glycoconjugate complex).

However, the value of activation energy is accordance with the high positive values of ΔG^* indicates that the formation of an activated complex is a non-spontaneous process.

Also, table (68) shows the values of ΔH^* at four different temperatures (4, 11, 18 and 25°C), the results revealed that the ΔH^* values decreased with increasing temperature. The slight changes in the values of ΔH^* at different temperatures could be attributed to the dependence on ΔH^* an activation energy (E_a) through the equation:

$$\Delta H^* = E_a - RT$$

Since the numerical value of RT is too small in comparison with the value of activation energy for the binding of lectin to glycoconjugates.

An analysis of the data in table (68), the results show that the ΔS^* values increases with decreasing temperature, was (-113.03 J/mol.K) at 4°C, -113.31 J/mol.K at 11°C, -113.44 J/mol.K at 18°C and -113.95 J/mol.K at 25°C). On the other hand, the high negative values of ΔS^* revealed that the activated complex had a more arranged structure than the reactants.

Finally, it could be concluded that the values of the thermodynamic parameters obtained from the study of lectin binding to erythrocyte surface glycoconjugates, give a distinct idea about the nature of forces that regulate the fromation of the complex.

Table (67): Thermodynamic parameters at transition state of lectin binding to erythrocyte surface glycoconjugates. All details are explained in the text.

T °C	E_a (KJ/mol.)	ΔH^* (KJ/mol.)	ΔG^* (KJ/mol.)	ΔS^* (KJ/mol.K)
4	13.05	10.75	42.06	-113.03
11	13.05	10.69	42.87	-113.31
18	13.05	10.63	43.64	-113.44
25	13.05	10.57	44.53	-113.95

5				

In order to compare the values of transition state with those of standard state, it is suggested to have the thermodynamic model to describe the formation of the complex.

This model is illustrated in Figure (117). The thermodynamic model proposes that the formation of lectin-glycoconjugate complex undergoes three thermodynamic states.

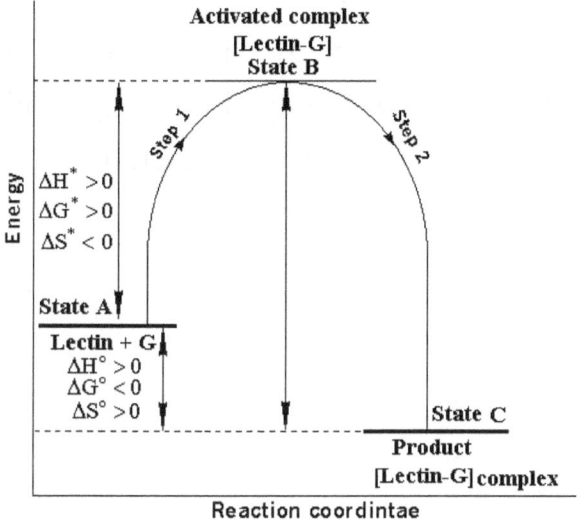

Figure (117): General energy diagram and thermodynamic model applied to the complex formation between lectin and erythrocyte surface glycoconjugates.

The thermodynamic state A, represents the initial energy level of lectin and glycoconjugate. The thermodynamic state B, represents the association of the two species to form the activated complex (lectin-glycoconjugate). The thermodynamic state C, represents the complete binding of lectin with glycoconjugate and formation of the complex.

(lectin-glycoconjugate complex). However, this model involves two steps, in step 1 of the reaction, the binding of lectin to glycoconjugates is associated with positive ΔG^* value and thus requires external energy. Also, in step 1, the lectin binding, shows negative value for entropy change (ΔS^*), this negativity indicates the alteration in the structure of lectin-glycoconjugate transition complex to a more arranged structure. At step 2, the contribution of more interactions, gives a fully interacting complex (lectin-glycoconjugate).

The formation of a protein-ligand complex is proposed to occur in two steps, the first is, the stabilization of the complex by hydrophobic interactions and the second is the stabilization by short range interactions, such as electrostatic interactions, protonation, hydrogen bonding and Van der Waals interactions .

Hydrophobic interactions contribute to the complex stability via high positive entropy ($\Delta S° > 0$), whereas, the electrostatic interactions, protonation, hydrogen bonding and Van der Waals interactions contribute to the complex stability via negative entropy change ($\Delta S° < 0$) .

It is clear from the thermodynamic data, the binding of lectin with glycoconjugate is entropy driven and it is in agreement with the concept that the hydrophobic interactions play an important role in such reactions.

Chapter Seven

Prostate specific antigen (PSA)

Introduction

There are several tumor markers correlate with the incidence of breast cancer, but the most important markers are:

. Carcinoembryonic antigen (CEA)

Carcinoembryonic antigen (CEA), one of the onco-fetal proteins, is a cell surface glycoprotein that belongs to the immunoglobulin gene superfamily, with a high molecular mass of 150-300 kDa . It is normally expressed in the early embryonic development and tends to disappear with the onset of the differentiation of fetal tissues into adult ones. Abnormal CEA increment is present in breast cancer and other kinds of cancers and is associated with tumor extent of breast cancer.

Carbohydrate Antigen 15.3 (CA 15.3) CA 15.3 is a mucin-like membrane glycoprotein recognized by a pair of monoclonal antibodies: the murine antibody DF3 and 115D8 . Distinct epitopes of this high molecular-weight mucin-like glycoprotein of 300-400 kDa, which carbohydrate side chain account for about 50% . It is associated with menopausal status and commonly used prognostic factors is also associated with cancer recurrence, disease-free survival and overall survival . Recent data show that pre-operational CA15.3 level is an independent prognostic factor for disease-free survival .

Tissue polypeptide antigen (TPA specific antigen (TPS) is a degradation product of cytokeratin (CK)18 . TPS level in serum of women without cancer is less than 80 U/l. It increases in pregnant women, in all conditions in which cell proliferation is involved and in some cases with benign breast tumor. TPS correlates with tumor cell activity since it is associated with the degree of tumor cell differentiation and is more valuable than CEA and CA15.3 in predicting the response of breast cancer to treatment. Therefore, TPS seems to be a promising tumor marker for breast cancer

. Cancer Antigen 27.29 (CA 27.29)

Cancer antigen (CA) 27.29 is elevated in breast carcinoma, ovarian and lung cancer, in normal pregnancy (1st trimester), benign breast disease, cirrhosis and hepatitis . CA 27.29 is associated with the early detection of recurrent breast carcinoma.

Human Epidermal growth factor Receptor 2/neu (HER-2/neu)

HER-2/neu is an oncogene-encoded growth factor receptor (homologue of epidermal growth factor (EGF) receptor), also known as c-erbB-2. It is overexpressed in breast cancers as a result of HER 2 proto-oncogene amplification. It is measured in the tissue from a biopsy either by immunological assays of the protein or polymerase chain reaction PCR.

P53

P53 is a tumor suppressor gene that is mutated or changed in more than 50 percent of tumors. Studying p53 as a tumor marker helped researchers understand how tumors form, but measuring p53 levels in cancer patients has not been shown to predict differences in survival or quality of life. p53 was indicated as responsible for tamoxifen resistance in breast cancer suggesting that it can interfere in treatment response .

Cathepsin-D

Cathepsin-D: High levels of this lysosomal enzyme may indicate breast cancer. There is not enough information to recommend using cathepsin-D levels to make treatment decisions for patients with primary or metastatic breast cancer and especially to diagnose the disease but studies have shown its association with reduced disease-free and overall survival of breast cancer patients .

Steroid hormone receptors·:

Steroid (estrogen and progesterone) receptors are specific proteins located in the nucleus of the cells which form the target organs for steroid action . Their function is to receive active steroid (estradiol-17B and progesterone) hormones entering into the cells from the circulation and subsequently mediated the intracellular response to such hom1ones .

One of the most important aspects is that, following ligand binding these receptors become activated with increased affinity for nuclear structure. On the other hand there is much interest in the clinical usefulness of steroid receptors for the diagnosis and treatment of breast cancer. However it is now generally accepted that tumors which do not contain significant amounts of these receptors are unlikely to response to endocrine therapy.

Subcellular localization of steroid hormone receptors:

In subcellular preparations from tissues exposed in vivo to relatively low concentrations of steroid hormones, it was found that steroid receptors behaved as soluble cytosolic proteins. Following steroid stimulation the majority of these receptors are recovered in the nuclear fraction, therefore a translocation hypothesis for steroid receptors was developed. Using immunohistochemical cell enucleation techniques which lead to minimal disruption of normal cellular architecture, it has been demonstrated that steroid receptors are nuclear associated proteins. Two forms of receptors with different affinities for nuclear structures have been proposed. The unliganded receptors with lower affinity for nuclear structures is thought to become soluble during tissue disruption while a higher affinity form induced by ligand binding, remains associated with the nucleus. Exposure to steroid hormones in vivo may change the relative proponions of these two forms of receptors resulting a variation in the partitioning during subcellular fractionation and hence an apparent decrease in receptors of the cytosolic fraction following steroid hormone stimulation. Autoradiographic studies of the subcellular distribution of labelled steroid hormones combined with immunohistochemistry provide evidences that steroid receptors are nuclear associated proteins. However the cellular mechanisms that determine the nucleocytoplasmic segregation of proteins remain to be established.

Several methods are also available to measure nuclear receptor concentrations, however the clinical significance of these receptors is somewhat less than that of the cytoplasmic receptors.

Other additional methods for receptor assay also being evolved. These methods may be more preferable, since specific poly-or monoclonal antibodies against steroid hormone receptors were used. Such procedures enable the detection of both empty and filled (by endogenous hormones) receptor sites in the cytoplasmic and nuclear fractions histochemically. Quantitation by these techniques demonstrate that breast cancers are heterogeneous with respect to receptor - positive cells.

Other studies revealed that cytosolic steroid hormone receptors prepared from tissues and cells not previously exposed to steroid hormones exist as oligomeric structures of molecular weight 250-300 kD). However exposure to steroid hormones activates receptors, a process which confers increased affinity for nuclei and DNA and which is accompanied by a decrease in molecular size and net charge. Oligomeric non activated forms of steroid receptors contain a non hormone binding phosphoprotein of moleuclar weight 90 kD.

Prostate specific antigen is a 32-34 kDa single chain glycoprotien [68-70] comprising 237 amino acid residues, with a molecular weight 26.079 for the peptide moiety, with five inter-chain disulfide bonds. Moreover, although the composition of carbohydrates was estimated at 7 to 12% of the total molecular weight and the presence of a single N-linked oligosaccharide side chain at asparagine 45.

Van Halbeek *et al*. suggested that PSA is composed of di-, tri- and tetra-antennary carbohydrate structure, a possible explanation for this conclusion may be the presence of contamination in the protein preparations. The N-terminal amino acid is isoleucine, while the C-terminal residue is proline.

Lundwall and Lilja proposed the primary sequence of amino acids and carbohydrate structure. In seminal plasma PSA can be shown to exist in five isoforms based on differences in isoelectric point, two biologically active and differing in the degree of glycosylation, ranging

from non-glycosylated to fully glycosylated , and three biologically inactive or "nicked" forms .

Prostate specific antigen exhibits serine protease activity (EC 3.4.21.77) similar to chymotrypsin.

Prostate specific antigen is a member of the human kallikrein family, with coded for by 6 kilobase length of DNA on human chromosome 19 that consists of five axons and two promoter regions, one of which contains the androgen response element (ARE). The androgen receptor binds to the ARE and up-regulates the transcription of the gen.

Prostate specific antigen has a high degree of homology with pancreatic/renal kallikrein (termed hK1) and a prostate specific glandular kallikrein (hK2). For this reason PSA is also given the alternative name of hK3. Genes for all three proteases are closely grouped together on chromosome 19. The coding region of PSA and hK2 genes has 85% identical with 91% of their promoter region are the same, and 63% with hK1.

. Synthesis

Prostate specific antigen is synthesized as a 261 amino acid preproform from which the 17-amino acid single peptide is cleaved in the secretion process then, the remaining zymogen form of PSA is activated to an active serine protease enzyme by cleavage of the 7 amino acids of propeptide

Prostate specific antigen is synthesized in the ductal and acinar epithelium of the prostate gland, hence it is secreted into the seminal plasma at a concentration of 0.5-2.0 g/L , secretion has been identified in the paraurethral and perianal glands as well as in apocrine sweat glands and the mammary glands. Synthesis of PSA has also been demonstrated in a number of tumour cell-lines, notably neuroblastoma.

Function

Prostate specific antigen in semen degrades seminal vesicle proteins and fibronectin, and proteolysis of these molecules by PSA , with release of the entrapped spermatozoa , and may have a bioactive role in fertilization. PSA in seminal vesicle fluid also stimulates the contraction of smooth muscle. It has been shown that PSA is able to proteolyse a major insulin-like growth factor binding protein-3 (IGFBP-3) releasing free insulin-like growth factor-1 (IGF-1) which have strong mitogenic and antiapoptotic effects on a variety of cells and are involved in the regulation of cell proliferation and differentiation PSA may also activate the transforming growth factor (TGF-ß) and regulate the bioactivity of parathyroid hormone-related protein

Recent experimental evidence suggests that PSA may also have antiangiogenic activity by inhibition of endothelial cell proliferation induced by fibroblast growth factor (FGF-2) and vascular endothelial growth factor (VEGF). PSA has also been shown to release angiostatin-like fragments from plasminogen. PSA has certain sequence homology with other serine proteases, such as g-nerve growth factor and epidermal growth factor binding protein. This indicates that PSA may have functions similar to those growth factors.

Free and complex

The predominant molecular form present in blood plasma is the 80-90 kDa complex of PSA with a1-antichymotrypsin (ACT), free PSA (f-PSA) represents a small but variable proportion of the total PSA (t-PSA) concentration. Minor fractions of PSA exist as complexes with other protease inhibitors, predominantly alpha-2-macroglobulin (A_2M), inter-a-trypsin inhibitor (IATI) and, at high concentrations, apha-1-antitrypsin (AAT). In seminal plasma, some 5% of the PSA is complexed with protein C inhibitor (PCI), a55kDa extracellular protease inhibitor.

While all commercial assays for PSA recognize f-PSA the PSA-ACT and PSA-AAT complexes, and possibly also the PSA-IATI

complex, none recognize the PSA- A_2M complex in which the PSA molecule is interiorized by the altered conformational structure of the inhibitor. This results in an incorrect calculation of the level of PSA when released from prostate into blood. f-PSA is mainly in the biologically inactive or "nicked" form, with cleavage of the polypeptide chain between residues 85-86, 145-146 or 182-183. Additional sites for internal cleavage of the molecule have been described between residues 54-55, 57-58 and 146-147 in PSA derived from BPH nodular fluids.

Prostate specific antigen epitopes

Prostate specific antigen has five major epitope domains. Four of these domains have been tentatively localized to amino-acid residues 1-13, 53-64, 80-91 and 151-164, with anti- bodies specific to f-PSA binding to the kallikrein loop between residues 84-91. The latter domain incorporates the active site and is blocked by complexing with ACT. In the PSA-ACT complex a further domain is affected by a variable degree of steric hindrance to antibody binding [111]. This phenomenon affects the molar response of free and complexed PSA, with the result that some antibodies (and assays) show a marked preference for binding f-PSA and give relatively less signal per unit mass of complexed PSA. The PSA-ACT complex covers all of the epitope domains and is therefore not detected by conventional immunoassays. Assays formulated with two (monoclonal) antibodies directed against epitope domains unaffected by ACT binding will give an equivalent signal with both free and complexed PSA and this is considered as equimolar.

The gradual adoption by industry of a 90:10 mixture of PSA-ACT complex and f-PSA as the calibration for t-PSA assays has minimized the differences that were previously apparent between equimolar and non-equimolar assays in clinical samples.

Clinical implications of PSA in men

PSA was first used as a semen marker to identify rape victims.. After it was discovered in serum, preliminary clinical studies indicated that increases in serum PSA were related to certain pathologic changes in the prostate, especially to prostate cancer. Although PSA is not a disease-specific marker, a large number of studies have demonstrated that it is a valuable tumor marker for prostate cancer diagnosis, staging, and prognosis, as well as for monitoring patient response to treatment and detecting recurrent or metastatic disease Clinical studies also suggest that measuring the percentage of free PSA in total PSA may help distinguish between prostate cancer and benign prostatic hyperplasia (BPH).

. Extraprostatic PSA

Prostate specific antigen is expressed in nonprostatic tissues [118,119]. Since the levels expressed in the prostate are so much higher than in other tissues this has not proved an obstacle to the use of PSA as a prostate cancer marker. However, it might suffer from false positives due to this extraprostatic expression, but this view has been vigorously opposed.

Prostate specific antigen has been reported to be present at low levels in several biological fluids apart from serum, for example in ascitic fluid, pleural effusions, and cerebrospinal fluid. Since levels were unaffected by the sex of the patient these observations cannot be explained by spillover derived from prostatic sources. Furthermore, PSA expression has also been reported in a wide variety of tumors, like metastatic melanoma, both lung tumors and lung tissue, pituitary tumors, colon adenocarcinoma, pancreas salivary gland, and non-Hodgkin lymphoma of the kidny.

PSA in women

The first question that arises in a discussion of PSA in women is whether women have a prostate. Based on gross anatomy, the answer is obviously no. However, microscopically, the answer may be different. Histologic studies have shown that the female paraurethral gland (Scene's gland) tests positive for immunohistochemical staining of PSA and prostatic acid phosphatase (PAP), another marker for prostatic tissue. Structurally, this female gland resembles the underdeveloped male prostate prior to stimulation from androgens. An animal experiment has suggested that, in a male hormonal environment, the paraurethral tissue from female rats can develop into a prostate-like structure. Thus, the paraurethral gland in women has been regarded as a female prostate that is not fully developed due to the lack of androgen stimulation.

The female paraurethral gland does not, however, appear to be the key organ producing PSA. Studies show that breast tissue is a major source of PSA in women. Initially, PSA was detected by accident in cancerous breast tissue. After extensive characterization using biochemical and molecular techniques, PSA in cancerous breast tissue has been proven to be the same molecule as that synthesized in the male prostate [136]. The molecular size of female PSA is identical to that of PSA in seminal plasma, as well as to free PSA in male serum. PSA is also found in normal breast tissue and tissue from women with benign breast diseases, as well as in breast fluids, including milk from lactating women, cyst fluid, and nipple aspirate fluid.

The breast and paraurethral gland are not the only tissues that are able to produce PSA in women. PSA has also been found in other female tissues, including endometrium , ovary , beside that existing in extraprostatic male tissue, and various cancerous tissues .

PSA in female serum and body fluid

During the early years of PSA research, several studies reported that PSA was detectable in some female sera .. Since the understanding of female PSA was limited at that time, little attention was paid to those reports, and the findings were interpreted as a cross-reaction of polyclonal anti-PSA antibodies to some nonspecific antigens. After extensive research on serum PSA, it is now known that PSA is indeed present in female serum. More than 50% of female sera are found to contain a tiny but detectable amount of PSA when extremely sensitive PSA assays are used. Since steroid hormones play a role in stimulating the production of PSA, serum levels of PSA in women vary with the menstrual cycle and are elevated during pregnancy .

Women with excess androgens, such as those suffering from hirsutism, also have elevated serum PSA . In addition to milk and blood, amniotic fluid and urine contain PSA . Levels of PSA in amniotic fluid are low in early pregnancy and are gradually elevated with gestational age. Concentration of PSA reaches the highest level at about gestational week 20, declining slowly thereafter. PSA in amniotic fluid also affects PSA levels in serum. The variation of serum PSA concentration with gestation matches the PSA variation in amniotic fluid . Other body fluids in women that are found to have a detectable amount of PSA include saliva , ascitic fluid , pleural effusion, and cerebrospinal fluid.

PSA in normal and breast tumor

Prostate specific antigen could be detected in breast tumor cytosols , and was detected in 30% of breast tumors .

PSA expression has now been shown in benign breast disease such as fibroadenoma and in normal breast tissue as well as in tumors .

Because of these observations PSA levels have been measured in fluids associated with breast tissue. However, with the advent of highly sensitive assays it has become clear that there are low but detectable levels of PSA in the circulation of women . The source of this PSA has not yet been completely determined but it seems highly likely that it is derived from the normal breast. In women with detectable circulating PSA, levels varied with the menstrual cycle, peaking in the mid to late follicular phase.

Circulating levels of PSA are not elevated in women with breast cancer ; however, PSA levels in the serum of some women with fibroadenomas or with breast cysts can attain the same levels as seen in men with prostate cancer, reaching 55 ng/ml in one case . This PSA is largely in the free form.

Function

While PSA in seminal fluid is known to be proteolytically active, the situation is less clear for extraprostatic PSA. However, the presence of PSA in breast tumor cytosol is a favourable prognostic factor, which is incompatible with these explanations. Attempts have been made to suggest a role for PSA in the breast consistent with a favourable prognosis. It has been suggested that PSA might generate bioactive peptides from the BRCA1 gene product which might be growth inhibitory . It has also been observed that PSA added to the hormone-dependent human breast cancer cell line MCF-7 is growth inhibitory and stimulates conversion by the cells of estradiol to the less potent estrogen, estrone, Since the cells are hormone-dependent this action of PSA could mediate its growth inhibitory effect on the cell line. In confirmation of this hypothesis PSA had no effect on the growth of the hormone-independent cell line MDA-MB-231.

Regulation

The presence of PSA in breast cancer cytosols is significantly associated with the expression of progesterone receptors in the same cytosols . There is also an association with estrogen receptor expression but this is secondary to a known association of progesterone and estrogen receptors in breast tumours . Not surprisingly, PSA positivity in breast tumours was not univocally associated with estrogen and progesterone receptor positivity in fact, most PSA-producing BCs are positive for steroid hormone receptor, but not all tumours that are positive for steroid hormone receptor produce PSA .

The expression of PSA in the prostate is regulated by steroids, for example constitutive secretion of PSA from prostate epithelial cells is stimulated by testosterone . This raises the question of whether the tw ocorticosteroids they are stimulated to produce PSA in the culture medium. In the

T47-D cells this stimulation is blocked by estrogens. It is known that androgen, progestin and glucocorticoid receptors share a common hormone response element (HRE) which differs from that of the estrogen receptor. Presumably it is this common HRE which is associated with the PSA gene. The mRNA coding for PSA appeared within 2 h of progestin stimulation of T47-D cells and the PSA protein appeared after 4-8 h. Only androgens and progestins are active at nanomolar concentrations.

Clinical implications of PSA in women

Clinical studies show that PSA status in breast cancer (being positive or negative based on the detection limit of a PSA assay) is associated with the presence of estrogen and progesterone receptors (ER, PR). The ER-positive or PR-positive cancers are more likely to express PSA. Clinical studies also demonstrate PSA positivity to be inversely correlated with clinical stage and tumor size, suggesting that the presence of PSA in cancerous breast tissue may indicate a favorable prognosis. In fact, survival analysis does show that PSA is a favorable prognostic marker for breast cancer. Women with PSA-positive cancer have better disease-free survival and overall survival than those with PSA-negative cancer. A reduction of 40% to 60% in the risks for relapse and death has been demonstrated in those clinical studies.

Why PSA is a favorable marker for prognosis in breast cancer remains unknown. Because PSA expression is stimulated by androgens, which counteract the effect of estrogen, it has been speculated that the presence of PSA in breast tissue may indicate the suppression of estrogen in the tissue [173]. It may also be possible that PSA degrades IGFBP-3, inhibiting its function. It has been suggested that IGFBP-3 may play a role in breast cancer progression.

Two small studies have evaluated PSA concentrations in nipple aspirate fluid in relation to breast cancer risk. The results of a study by Sauter and colleague's suggest that low levels of PSA in nipple aspirate are associated with a significantly increased risk of breast cancer, whereas those of a study by Foretova et al do not support such an association. The molecular forms of PSA may also differ between women with and without breast cancer; one study demonstrated that breast cancer patients tend to have more free PSA in their sera and healthy women have more ACT-bound PSA . Using serum PSA to assess androgen excess in women, two studies showed that PSA concentrations are elevated significantly in hirsute women . In comparison to another serum androgen marker, 3a-androstanediol glucuronide, however, PSA does not provide better sensitivity and specificity in the diagnosis of hirsutism .

Another preliminary clinical study by Melegos and colleagues suggests that low levels of PSA in amniotic fluid may be associated with certain abnormal pregnancies. Women pregnant with a down syndrome fetus may have higher serum PSA than those with a normal pregnancy ..

Prostate specific antigen (PSA) is a 33-34 KD, single- chain glycoprotein of 240 amino acid residues with four glucidic side chains .The molecular features   7-10% carbohydrate contents, 3.1 S sedimentation coefficient and isoelectric point at pH 6.8-7.5. It is secreted by the epithelial cells lining the acini and ducts of the prostate gland, and found exclusively in the tissue of normal, benign hyperplastic, and cancerous prostate gland as well as in seminal plasma and prostatic fluid.

This antigen which was first identified in 1978, is a member of the human **kallikrein** family, and nowadays is widely used as the eminal fluid by **Hara et al** in 1974, and the **protein p30**, also isolated from plasma seminal fluid by **Sensabaugh et al** in 1978 .

Prostate specific antigen is known to be a serine protease enzyme with chymotrypsin and trypsin like activity . Its physiological role is believed to be liquefaction of the seminal coagulum forming after ejaculation.

A percent of 60-80% of the PSA secreted in the seminal plasma have chymotrypsin and trypsin like activity, while the remaining 20-40% are enzymatically inactive, presumably due to an internal bond cleavage between two lysine residues. But when PSA is released in the blood, 70-90% of its total amount (t-PSA) is inactivated by the major extracellular serine protease inhibitors: *$α_1$ - antichymotrypsin* ($α_1$**ACT**), *$α_2$-macroglobulin* ($α_2$**M**), and other acute phase proteins. In addition, about 10-30% of it may exist in a free uncomplexed form   (f-PSA)[1].

Total PSA, (i.e.: all of the immunologically detectable forms), consisting primarily of (PSA-ACT) and (f-PSA), has served as an excellent indicator of prostate disease when the concentration exceeds 4.0 ng/ml in serum. While the portion of PSA in the blood inactivated by (α_2M) is non–immunoreactive because it is probably engulfed in the (α_2M) structure. The serum concentration is also increased with the prostate volume and the stage of the prostate cancer.

Lilja et al and *Stenman et al* demonstrated that the proportion of (PSA-ACT) complex increases as a function of (t-PSA) concentration, they also showed that this proportion in BPH is somewhat less than that of PCA. Accordingly many studies have suggested that the ratio of (f-PSA/t-PSA), (f-PSA%) allows a better discrimination between BPH and PCA.

Although previously thought to be produced exclusively by the epithelial cells of the prostate, PSA at present is considered a wide spread biochemical marker present in many non–prostatic tissues and fluids. Many studies indicate that the PSA expression in both physiological and pathological conditions, is not organ or sex – specific, but rather a steroid hormone mediated response.

Isolation and Characterization of PSA of Benign and Malignant Breast Tumors

Introduction

Chromatography encompasses a diver and important group of separation methods that permit the scientist to separate, isolate and identify related components of complex mixtures. The purification of biomolecules is the most popular use of gel chromatography due to the ability of a gel to fractionate molecules on the basis of size.

Mannello et al[)] separate f-PSA and c-PSA with out any additional peaks from breast cyst fluid (BCF) using sephacryl S-300.

The purpose of this part of thesis is to isolate f-PSA and c-PSA from each of benign and malignant breast tumor tissues using sephadex G-200 and to finding of the optimum reaction conditions for ^{125}I-anti total PSA antibody with isolated f-PSA and c-PSA in breast tumor homogenate.

Patients

The two groups of breast tumor patients were included for prostate specific antigen PSA analysis. Benign and postmenopausal malignant breast tumor homogenates that showed maximal binding in the preliminary test were collected, pooled and used for isolation and characterization of PSA.

Gel preparation

1. The pre-swollen gel was suspended in PBS buffer pH8 equivalent to approximately three times the volume of settles gel.

2. The gel was allowed to settle and decanting the excess buffer.

3. The step 1 and 2 was repeated two more times to allow preliminary equilibration of the gel with the eluent buffer.

4. The gel was resuspended in a volume of PBS buffer pH8 approximately equal to the settled volume of gel.

5. The gel slurry was degassed by suction for 60 minutes.

. Preparation of the column.

Dimensions of the column

The dimensions of the column were chosen according to the following equation [224].

$$\text{Diameter} = \sqrt{\frac{m}{10}}$$

Where:

m= amount of protein in mg.

L = 30 x diameter

Where:

L: length of the column in centimeters

In view of the results of such calculation, a 0.96x29cm column has been used.

Bed packing

1. The column was mounted and packing reservoir vertically.

2. The column out let was opened and filled with eluent buffer to a height about 5cm and turn off column flow.

3. The degassed slurry was carefully mixed before pouring into the column using a glass rod attached to the inner surface of the column.

4. After the gel has settled for 5 min., the column outlet was opened.

5. Packing was continued until the gel reached a stable bed height (29cm). Then the column was equilibrated with PBS buffer pH8 for 24hr. at 4°C with dimensions (0.96x29cm) and a total bed volume of 21ml.

Void Volume Determination

The elution volume of blue dextran 2000 is equal to the column void volume (V_o), and was determined as follows:

A fresh solution of Blue Dextran (2mg.ml-1) was prepared in the eluent buffer in a sample volume of 1-2% of the total bed volume, and then applied to the column with a flow rate of 5 ml.hr-1. Fractions of 1 ml were collected and their absorbance were measured at $\lambda = 600$ nm.

Separation procedure

1. Sixty microliters (300μg protein) of 125I- anti total PSA antibody was added to 50μl (400μg protein) of benign Fibroadenoma breast tumor homogenate and the volumes were completed to (500μl) with TEBP buffer pH (8).

2. Two additional tubes, containing 60μl of 125I- anti total PSA antibody only for total activity computation, were set-asides until counting.

3. The mixtures were incubated for 180 minute at 4°C while continuously shaking.

4. At the end of incubation, the mixture was applied to the surface of sepharose CL-6B gel filtration column (0.96x29cm) equilibrated with PBS pH8 with a flow rate one-half that used for column elution.

5. Immediately 1ml of PBSS pH8 was added after the mixture to aid in sharp mixture application.

6. Elution was carried out using the PBS buffer pH8, to separate 125I- anti total PSA antibody and (125I- anti total PSA antibody/PSA) complexes, with a flow rate of 5ml/hr., and fraction volumes of 1ml.

7. The radioactivity of each fraction was counted in a gamma counter for 1 min.

8. The experiment was repeated using 50μl (250μg protein) of 125I- anti total PSA antibody and 50μl (400μg protein) of postmenopausal malignant infiltrative ductal carcinoma (IDC) breast tissue homogenate.

Calculations

1. The radioactivity (c.p.m) of each eluted fraction was plotted against fraction number.

2. The resultant fractions under peak were determined by the sum of the radioactivity in (c.p.m) of the fractions at that.

3. The percent radioactivity of (^{125}I- anti total PSA antibody/ PSA) complexes to total ^{125}I- anti total PSA antibody was calculated from the values area under peak of each species.

. Determination of the molecular weight (M.Wt) of free 125I- anti total PSA antibody and 125I- anti total PSA antibody/ PSA) complexes of benign breast tumor

A sepharose CL-6B gel was prepared for this purpose as described . A sepharose CL-6B column was used .

The M.Wt was determined as follows:

1. The column was calibrated by gel filtration kit, purchased from pharmacia fine chemicals, which contained standard proteins. Standard protein solutions were prepared according to the manufacturer instructions, then applied through three portions (500µl), proteins 1 and 4 in the first portion, proteins 2 and 5 in the second portion, proteins 3 and 6 in the third portion.

2. Elution was carried out with PBS buffer at a flow rate of 5 ml.hr^{-1}.

3. The absorbance of the fractions collected was measured at 280 nm to evaluate the elution volume (Ve) of the standard protein.

Table (68): Standard proteins and their molecular weights (All other details are explained in text).

Protein	M.wt (kDa)	Conc. mg.mL-1
Thyroglobulin	669	7.0
Ferritin	440	1.0
Catalase	322	7.0
Aldolase	158	7.0
Albumin	67	7.0
Ovalbomin	43	7.0

Calculations

1. The K_{av} values of the proteins eluted were determined using the following equation:

$$K_{av} = \frac{V_e - V_o}{V_t - V_o}$$

Where:

V_o = Void volume

V_e = Elution volume of each protein

V_t = Total gel - bed volume.

The calibration curve of K_{av} values vs. log M.wt. of the proteins were plotted.

Isolation of free prostate specific antigen from complex prostate specific antigen in tissue of breast tumors by gel filtration technique using sephadex G-200.

Gel preparation

1. With gentle mixing, a weight of 3g sephadex G-200 was slowly added to excess volume of PBS buffer pH8 (40ml of buffer per gram of gel).

2. The gel as allowed swelling at 4°C for 72 h. without stirring to equilibrate with buffer.

3. After equilibrated, the buffer was decanted and the gel was resuspended in excess volume of eluent buffer equal three times to settled volume of the gel and allowed it to settled for approximately 20 minutes.

4. The excess buffer and fine particles were removed by suction.

5. The gel was resuspended in a volume of PBS buffer pH8 approximately equal to the settled volume of gel.

6. The gel slurry was degassed by suction for 1hr.

Preparation of the column

Dimensions of the column

The Dimensions of the column was prepared as mentioned .

Bed packing

The column was packing with dimensions (1.25x37.5cm) and a total bed volume of 46ml.

Void Volume Determination

The column void volume was determined.

Isolation procedure

1. Five hundred microliters (3750µg protein) of benign Fibroadenoma breast tumor homogenate was applied to the surface of sephadex G-200 gel filtration column (1.25x37.5cm) equilibrated with PBS pH8 with a flow rate one-half that used for column elution.

2. Immediately 1ml of PBSS pH8 was added after the homogenate to aided in sharp homogenate application.

3. Elution was carried out using the PBS buffer pH8, to isolate free prostate specific antigen and complex prostate specific antigen with a flow rate of 6ml/hr., and fraction volumes of 1ml.

4. The elution of each fraction was determined at 280nm.

5. Twenty microliters (100µg protein) of 125I- anti total PSA antibody was added to 50µl of each fraction number of benign fibroadenoma breast tumor homogenate and the volumes were completed to (500µl) with TEBP buffer pH (8).

6. Two additional tubes, containing 20µl of 125I- anti total PSA antibody only for total activity computation, were set-asides until counting.

7. The mixtures were incubated for 180 minute at 4°C while continuously shaking.

8. The (125I-anti-total PSA antibody/PSA) complex was estimated 9.The experiment was repeated using 500µl (9000µg protein) of postmenopausal malignant infiltrative ductal carcinoma (IDC) breast tissue homogenate.

Calculations

1. The fraction number was determined at 280nm.

2. The values of B/T ratio for the eluted fractions were calculated

3. The values of B/T ratio and the absorbancies at 280nm were plotted against the fraction number.

4. The plotted B/T ratio against the fraction number for all the experiments gave a 2peaks profile. The first peak represents the ^{125}I-Ab bound to c-PSA while the second peak represents the ^{125}I-Ab bound to f-PSA

5. The value of B/T ratio of f-PSA was determined by dividing the sum of the radioactivity of the fractions under peak 2 by the total radioactivity.

$$\text{f-PSA\%} = \frac{\text{The radioactivity under peak2}}{\text{Total radioactivity}} \times 100$$

6. The percentage of free-PSA /total-PSA was determined by dividing the sum of the radioactivity of the fractions under peak 2 by the sum of radioactivity of two peaks appeared in the profile:

$$\text{f-PSA/t-PSA} = \frac{\text{The radioactivity under peak2}}{\text{The sum of radioactivity under (peak1+peak2)}} \times 100$$

7. The percentage of f-PSA /c-PSA was determined by dividing the sum of the radioactivity of the fractions under peak 2 by the sum of radioactivity of the fractions under peak 1 appeared in the profile:

$$\text{f-PSA/c-PSA} = \frac{\text{The radioactivity under peak2}}{\text{The radioactivity under peak1}} \times 100$$

Determination of the molecular weight (M.Wt) of free prostate specific antigen and complex prostate specific antigen of benign breast tumor

A sephadex G-200 gel was prepared for this purpose as described in section (2.4.1.). A sephadex G-200 column was used as mentioned in section (2.4.2.). The M.Wt was obtained as mentioned in section (2.3.5.)

Table (69): Standard proteins and their molecular weights (All other details are explained in text).

Protein	M.wt (kDa)	Conc. mg.mL-1
Catalase	322	7.0
Aldolase	158	7.0
Albumin	**67**	**7.0**
Ovalbomin	**43**	**7.0**
Cymotrypsinogen	**25**	**7.0**
Ribonuclease	**13.7**	**7.0**

Calculations

The M.Wt was obtained as mentioned .. **Dialysis for Concentration**

After preparing dialysis tube, the fractions that contained high levels of the binding activity were collected, pooled and concentrated by dialyzing against sucrose at 4 °C for 2hrs to get the required concentration to be used in the next experiments

Preliminary test of the binding of isolated PSA in breast tumor tissues with 125I-anti total PSA antibody

1. **Thirty microliters (30) μl (150) μg of 125I- anti total PSA antibody was added to 50 μl of (200μg) of each isolated f-PSA and c-PSA of benign breast tumor respectively and the volumes were completed to (500μl) with TEBP buffer pH 8.**

2. **Two additional tubes, containing 30μl of 125I- anti total PSA antibody only for total activity computation, were set-asides until counting.**

The (125I- anti-total PSA antibody/ f-PSA) and (125I- anti-total PSA antibody/ c-PSA) complexes were estimated .

Calculations

The B/T% was computed for each tube as mentioned

Most appropriate conditions of the binding of isolated free prostate specific antigen (f-PSA) and complex prostate specific antigen (c-PSA) of benign breast tumor with 125I- anti total PSA antibody

Effect of protein amounts of the isolated free prostate specific antigen and complex prostate specific antigen of benign breast tumor on the binding

1. Fifty microliters of increasing amounts (50, 100, 150, 200, 250μg) of isolated f-PSA and c-PSA of benign breast tumor were each added to (25 and 30) μl (125 and 150) μg of 125I- anti total PSA antibody respectively and the volumes were completed to (500μl) with TEB buffer pH 8

2. The (125I- anti-total PSA antibody/ f-PSA) and (125I- anti-total PSA antibody/ c-PSA) complexes were estimated.

Calculations

1. The B/T% was computed for each tube

2. The percent of binding values B/T% were plotted against the increasing amounts of protein of isolated f-PSA and c-PSA of benign breast tumor

Effect of 125I- anti total PSA antibody on the binding

1. Increasing volumes (10, 15, 20, 25, 30, 40 and 50 μl) of 125I- anti total PSA antibody containing (50, 75, 100, 125, 150, 200 and 250 μg protein) respectively were each added to 50 μl (100 and 150) μg of isolated f-PSA and c-PSA of benign breast tumor respectively and the volumes were completed to (500μl) with TEBP buffer pH 8

2. The (125I- anti-total PSA antibody/f-PSA) and (125I- anti-total PSA antibody/c-PSA) complexes were estimated.

Calculations

1. The B/T% was computed for each tube.

2. The concentration of the bound 125I- anti total PSA antibody was plotted against the concentration of 125I- anti total PSA antibody.

Effect of pH on the 125I- anti total PSA antibody binding

1. Fifty microliters (100 and 150) μg of isolated f-PSA and c-PSA of benign breast tumor were each added to (20 and 40) μl (100 and 200) μg of 125I- anti total PSA antibody respectively and the volumes were completed to (500μl) with TEBP buffer of different pH (6.8, 7, 7.2, 7.4, 7.6, 7.8, 8, 9).

2. The (125I- anti-total PSA antibody/f-PSA) and (125I- anti-total PSA antibody/c-PSA) complexes were estimated

Solution

Calculations

1. The B/T % values were determined.

2. The percent of binding values B/T% were plotted against their corresponding pH value.

Temperature dependency of the binding

1. Fifty microliters (100 and 150) μg of isolated f-PSA and c-PSA of benign breast tumor were each added to (20and 40) μl (100 and 200) μg of 125I- anti total PSA antibody respectively and the volumes were completed to (500μl) with TEBP buffer pH (7.8 and 8) respectively.

2.The (125I- anti-total PSA antibody/f-PSA) and (125I- anti-total PSA antibody/c-PSA) complexes were estimated.

3.The experiment was repeated at different temperatures (20, 37 and 45°C)

Calculations

1. The B/T % values were determined at each temperature.

2. The percent of binding values were plotted against the different temperatures of incubation.

. Time dependency of the binding

1. Fifty microliters (100 and 150) μg of isolated f-PSA and c-PSA of benign breast tumor were each added to (20and 40) μl (100 and 200) μg of 125I- anti total PSA antibody respectively and the volumes were completed to (500μl) with TEBP buffer pH (7.8 and 8) respectively.

2. Four additional tubes, two containing 20μl and the others containing

40μl of 125I- anti total PSA antibody only for total activity computation were set-asides until counting.

3. The mixtures were incubated at 4°C at different time (30, 60, 90, 120, 180 and 240 minute) while continuously shaking.

4. The (125I- anti-total PSA antibody/f-PSA) and (125I- anti-total PSA antibody/c-PSA) complexes were estimated.

Calculations

1. The B/T % values were determined.

2. The percent of binding values were plotted against the different time of incubation.

Result and Discussion

Separation of free 125I- anti total PSA antibody from (125I- anti total PSA antibody/ PSA) complexes

Figure (118) shows the elution profile of the incubation solution, which had contained PSA of benign and postmenopausal malignant breast tumor tissue homogenates respectively, using sepharose CL-6B (0.96 × 29 cm) column. Also, the elution profile of ^{125}I- anti total PSA antibody was shown in Figure (119)

The void volume of this column was 7 ml as predicted from the elution profile of the blue dextran as shown in Figure (118).

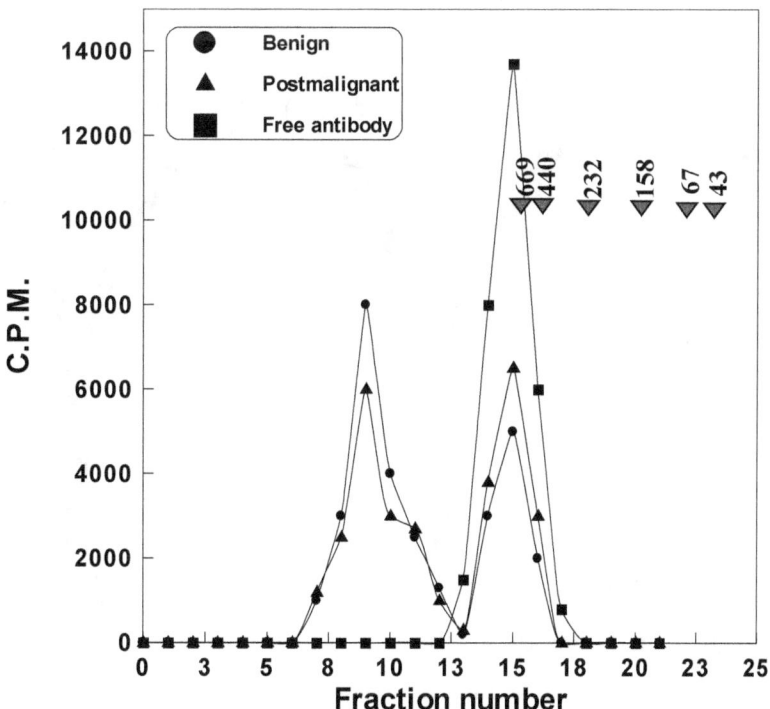

Figure (118): The elution profile of the complex of ^{125}I- anti total PSA antibody with isoforms of PSA in tissue each of benign and postmenopausal malignant breast tumor using sepharose CL-6B. (All other details are explained in the text).

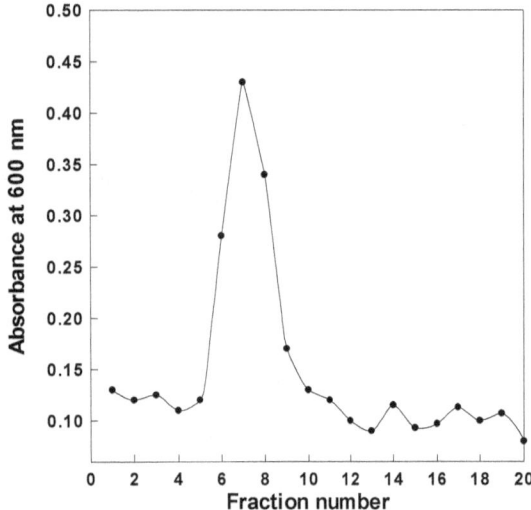

Figure (119): The elution profile of blue dextran 2000 by gel filtration chromatography using sepharose CL-6B. (All other details are explained in the text).

In each case two peaks of radioactivity appeared. The broad one or peak (1) represents the (^{125}I-anti total PSA antibody/PSA) complex with different forms of PSA. Figure (120) shows some of these suggested complexes [225].

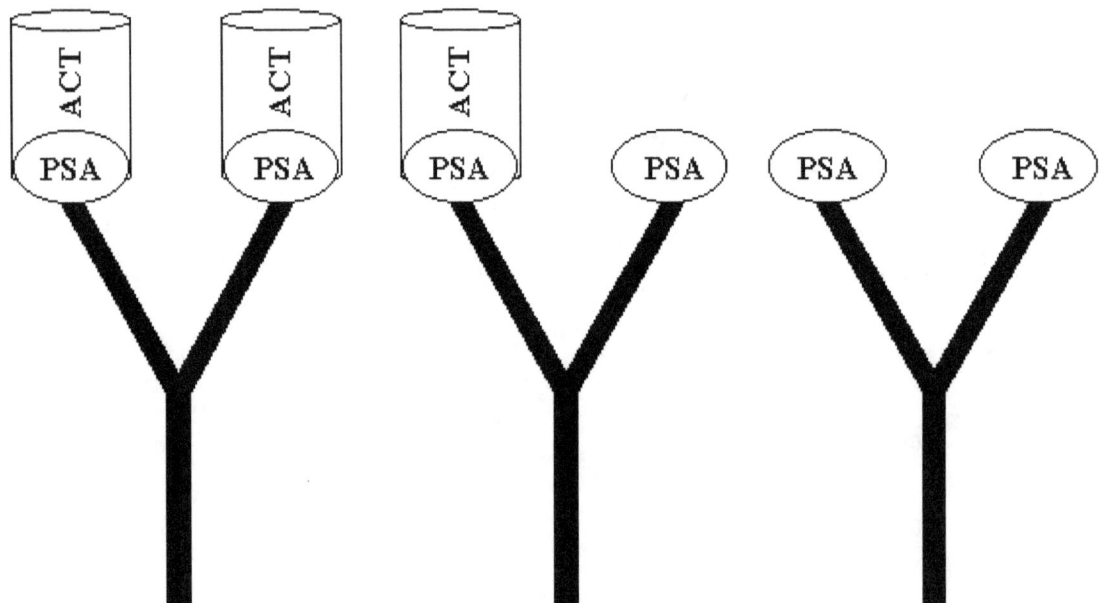

Figure (120): The suggested complexes between PSA and ^{125}I- anti total PSA antibody. (All other details are explained in the text).

Moreover, the high molecular weight of the first peak, which was expected from its low retention volume (9 ml) may attribute to the (^{125}I- anti total PSA antibody/PSA) polymerization and aggregation reaction, which may indicate the bivalency or multivalency of PSA).

Also the idea of bivalency of PSA could be supported by the finding of Lilja et al), they found that the PSA-ACT complex (for example) covers three of the epitopes of PSA leaving at least two epitopes available to react with antibody, so that PSA may be connected from both epitopes to produce a chain reaction and cause this enhancement in molecular weight.

The second peak (2) in figure (118) may represent the free ^{125}I- anti total PSA antibody. The high retention volume (15ml) indicates the presence of a relatively low molecular weight, approximately 127 kDa, which was obtained from the calibration curve in figure 121).

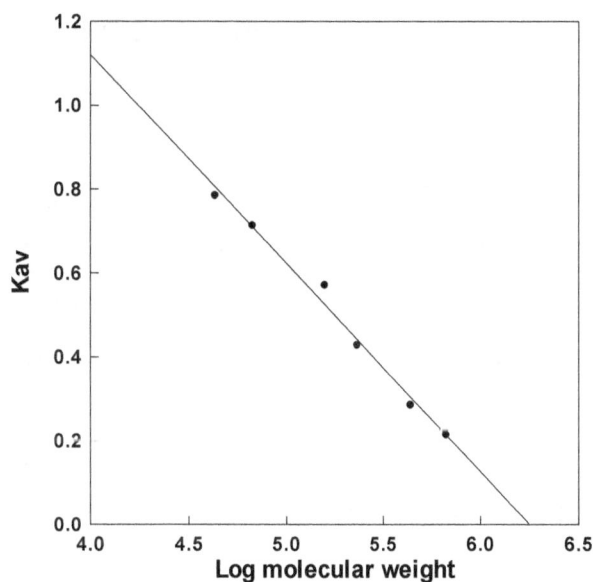

Figure (121): Calibration curve for determination of M.wt of complexes of PSA isoforms with ^{125}I- anti total PSA antibody by gel filtration chromatography using sepharose CL-6B. (All other details are explained in the text).

Isolation of f- PSA from c-PSA

Isolation of f-PSA from c-PSA of benign and postmenopausal malignant breast tumor tissue homogenates was performed by gel exclusion chromatography technique. Benign and malignant homogenates were applied to sephadex G-200 (1.25 × 37.5 cm) column. The void volume of this column was 8 ml as predicted from the elution profile of the blue dextran as shown in Figure (122).

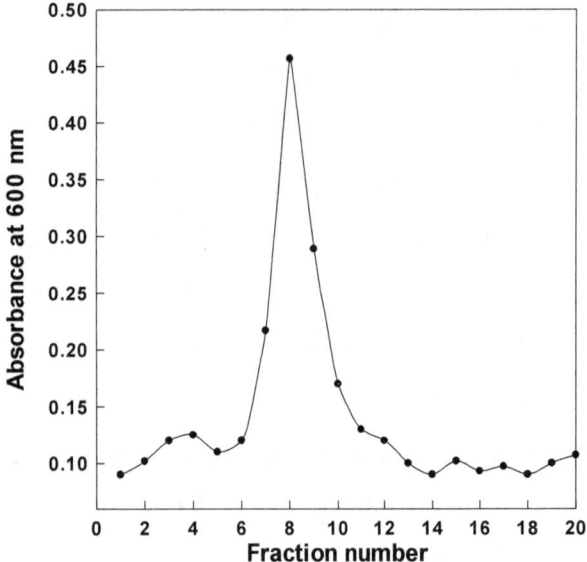

Figure (122): The elution profile of blue dextran 2000 by gel filtration chromatography using sephadex G-200. (All other details are explained in the text).

The resultant fractions of each homogenate type (benign and malignant) were measured at 280nm gave three peaks as shown in figure (123 A & B) respectively and the resultant fractions of each peak of benign and malignant were incubated with ^{125}I- anti total PSA antibody. According to the optimum conditions of those reactions gave two peaks may represents (^{125}I- anti total PSA antibody/c-PSA) complex at fraction nur (21) and (^{125}I- anti total PSA antibody/f-PSA) complex at fraction number (30).

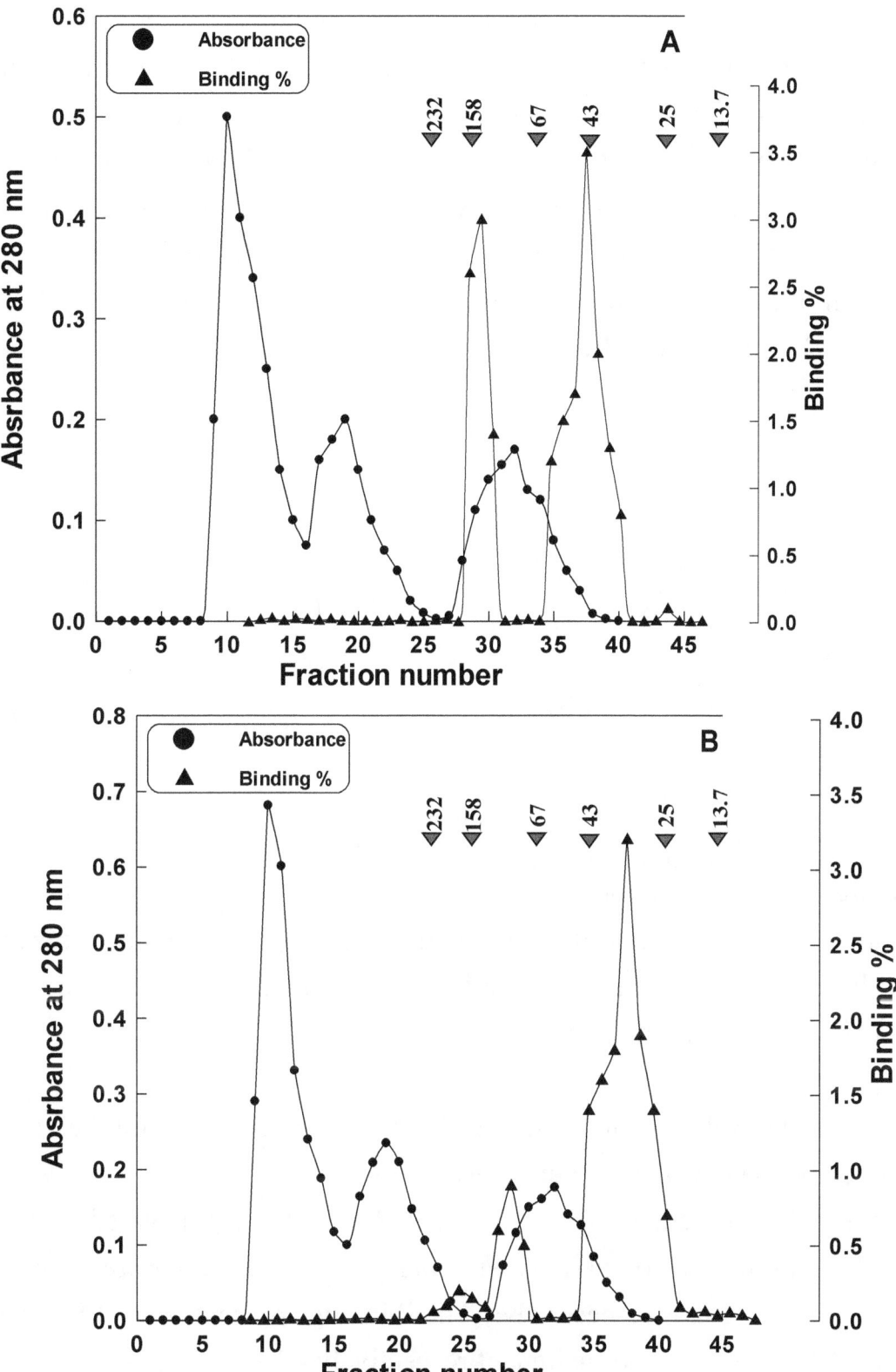

Figure (123): The elution profile of human PSA and complexes from:
A. Benign breast tumor tissue homogenates
B. Postmenopausal malignant breast tumor tissue homogenates
(All other details are explained in the text).

These results indicate that there are two immunologically active isoforms of PSA antigen, one of them was c-PSA (peak 1) and the other was f-PSA (peak 2). From these results it was concluded that these components are capable of binding to the ^{125}I- anti total PSA antibody with different affinity.

The chromatographic isolation was performed depending on the significant differences in molecular weight that was obtained by the calibration of the gel column with standard proteins as shown in figure (124) was 32.5 kDa for f-PSA and 101 kDa for c- PSA.

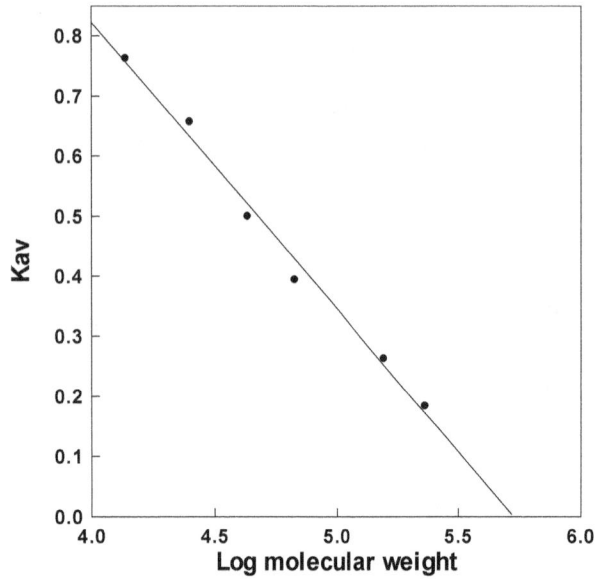

Figure (124): Calibration curve for determination of M.wt of isolated f-PSA and c-PSA from each of benign and postmenopausal malignant breast tumor by gel filtration chromatography using sephadex G-200. (All other details are explained in the text).

The obtained values of molecular weights were consistent with the values that were obtained by many other investigators[.

The binding percent of f-PSA and f-PSA/t-PSA from isolated PSA of benign breast tumor in comparison with malignant breast tumor are shown in figure (2-6 A & B) respectively. The binding percent of f-PSA was 12 and 12.1 for benign and malignant respectively whereas the f-PSA/t-PSA percent was 63% and 86% for benign and malignant respectively.

These results are nearly similar to these obtained previously by (Black et al), there is no significant differences between the percentage of benign breast tumors and malignant breast tumors with detectable f-PSA, but the proportion of f-PSA/t-PSA in malignant breast tumors is significantly higher in comparison to patients with benign breast tumors. Another studies suggested that the predominant form (>50%) of PSA in a significant proportion of females with breast cancer is f-PSA. In another study revealed f-PSA, as the predominant form and is highly specific for breast cancer in comparison with benign breast disease and normal tissue.

Also the binding portion of f-PSA/c-PSA from isolated PSA in benign breast tumors in comparison with malignant breast tumors is shown in figure (123 A & B) respectively. The binding portion of f-PSA/c-PSA were 1.7 and 6.0 for benign and malignant respectively

The present results suggest the measurement of f-PSA/t-PSA or f-PSA/c-PSA in the tissues of patients with benign and malignant breast tumors to have potential clinical applicability as a tumor marker either alone or in combination with other diagnostic markers.

Lopez-Otin and Diamandis) suggest that serum f-PSA, which is an enzymatically inactivation of PSA is over expressed in patients with benign and malignant breast disease. The mechanism of such changes is unknown, but the data in contrast to changes in prostate cancer where serum c-PSA increases and f-PSA decreases in cancer patients in comparison to patients with benign prostatic hyperplasia.

According to these studies the isolated f-PSA and c-PSA of benign breast tumor homogenate were used as the source in all subsequent experiments, since they give the highest binding.

The resultant fractions of each peak of benign breast tumor homogenate that gave binding were pooled, concentrated and then subjected to protein determination.

. Preliminary test of the binding of isolated PSA in breast tumor tissues with 125I- anti total PSA antibody

Each isolated f-PSA and c-ACT of benign breast tumor homogenate was detected through the incubation with ^{125}I- anti total PSA antibody for 180 min at 4°C with TEBP buffer pH (8) as a medium to complete the reaction. The (^{125}I- anti total PSA antibody/f-PSA) and (^{125}I- anti total PSA antibody/ c-PSA) complexes formed were separated from the unbound particulate by centrifugation at 1500xg for 30 min. This centrifugal speed was sufficient to precipitate the complex. After centrifugation the tubes were decanted, in order to get rid of the unbound antibody or antigen, while (^{125}I-anti total PSA antibody/f-PSA) complex or (^{125}I- anti total PSA antibody/c-PSA) complex remained as pellet in the bottom of the tube.

The preliminary conditions used in this experiment show the amount of binding B/T% for isolated f-PSA and c-PSA of benign breast tumor homogenate to be (12 and 7)%. The data revealed that the B/T% of f-PSA is more than c-PSA (>50%).

Most appropriate conditions of the binding of isolated PSA of benign breast tumor with 125I- anti total PSA antibody

Effect of protein amounts of the isolated PSA

Figure (125) shows the effect of increasing amounts of the isolated f-PSA and c-PSA of benign breast tumor on the binding with 125I- anti total PSA antibody. The results revealed that (25 and 40) □g protein were the most appropriate concentrations for the binding of f-PSA and c-PSA respectively. From these results, it could be concluded that the binding of 125I- anti total PSA antibody with its isolated f-PSA needed lower amounts of these antigens to get the equilibrium compared with the amounts required for isolated c-PSA. This may be due to cell hyper proliferation or increase in these antigen affinities for antibodies.

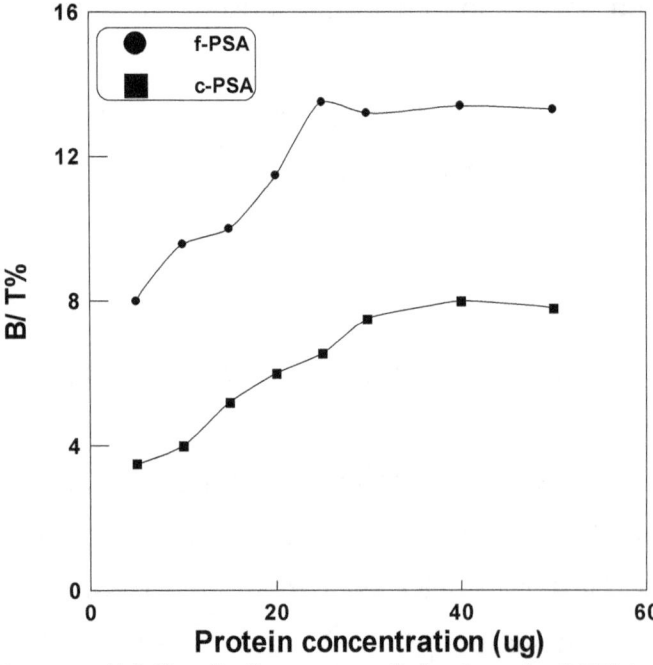

Figure (125): Influence of isolated f-PSA and c-PSA concentration (protein) in benign breast tumor homogenates on the binding with ^{125}I- anti total PSA antibody. (All other details are explained in the text).

According to these results, in all subsequent experiments, (25 and 40μg protein) of isolated f-PSA and c-ACT antigen respectively were used, since they give the highest binding.

Influence of 125I- anti total PSA antibody on the binding

Figure (126) shows that isolated f-PSA and c-PSA of benign breast tumor were saturated with ^{125}I- anti total PSA antibody concentrations equal to (100 and 200) □g respectively. From these results, it was found that isolated f-PSA fractions were saturated with smaller concentrations of ^{125}I- anti total PSA antibody than those required for isolated c-PSA. Thus it was concluded that isolated f-PSA had higher affinity at low concentrations toward ^{125}I- anti total PSA antibody than isolated c-PSA.

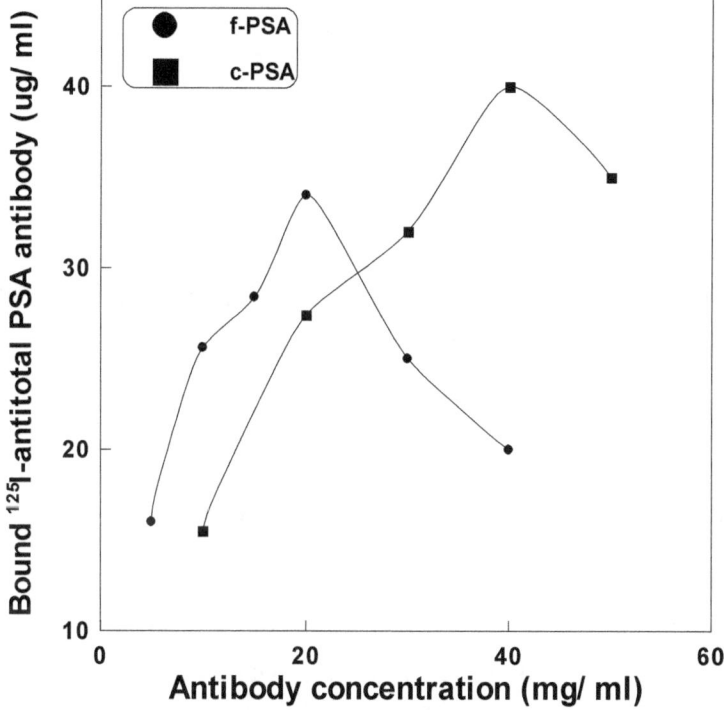

Figure (126): Effect of different concentrations of ^{125}I- anti total PSA antibody on the binding with each of isolated f-PSA and c-PSA in benign breast tumor homogenates. (All other details are explained in the text).

According to these results, in all subsequent experiments, (100 and 200μg protein) of ^{125}I- anti total PSA antibody respectively were used, since they give the highest binding.

Effect of pH on the 125I- anti total PSA antibody binding

Figure (127) shows the effect of increasing pH on the binding of ^{125}I- anti total PSA antibody to its isolated f-PSA and c-PSA of benign breast tumor. The results revealed that the optimum pH for isolated f-PSA and c-PSA fractions for the binding with ^{125}I- anti total PSA antibody was 7.8 and 8 respectively.

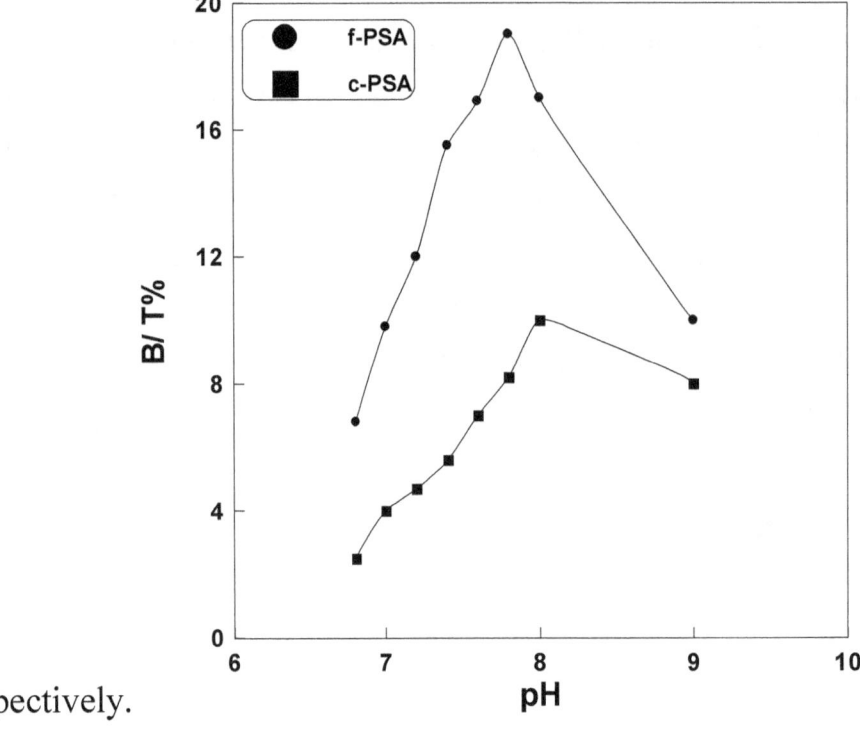

Figure (127): Effect of pH on the binding of ^{125}I- anti total PSA antibody with each of isolated f-PSA and c-PSA in benign breast tumor homogenates. (All other details are explained in the text).

These differences in the optimum pHs may suggest the differences in the binding sites of these isolated antigens.

According to the results obtained, the pH of the buffer used in all subsequent experiments were adjusted to 7.8 and 8.0 for isolated f-PSA and c-ACT antigen respectively.

Temperature dependency of the binding

The temperature dependency of the ^{125}I- anti total PSA antibody binding to its isolated antigens was investigated. Figure (128) shows that the optimum temperature of the binding of ^{125}I- anti total PSA antibody was 4°C with each isolated f-PSA and c-ACT antigen.

Figure (128): Effect of temperature on the binding of ^{125}I- anti total PSA antibody with each of isolated f-PSA and c-PSA in benign breast tumor homogenates. (All other details are explained in the text).

In general the loss of binding activity above the optimum temperature of each isolated f-PSA and c-ACT antigens may be due to degradation of these antigen molecules or to the irreversible dissociation of the antibody-antigen complexes.

In view of these results, the temperature used in all subsequent experiments was 4 °C

Time dependency of the binding

Figure 129) shows that at 4°C the apparent equilibrium of the ^{125}I- anti total PSA antibody binding reached within 180minutes for both isolated f-PSA and c-ACT antigens.

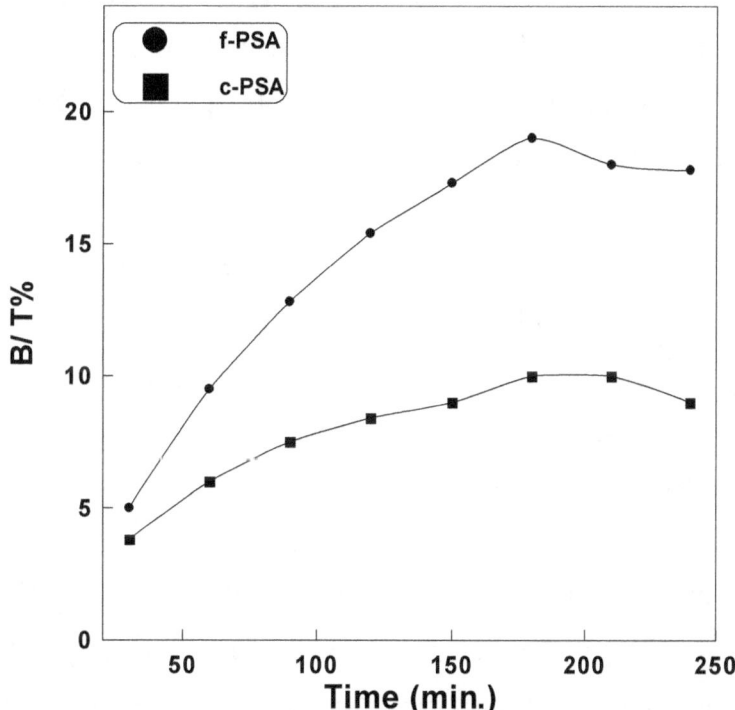

Figure (129): Effect of time on the binding of ^{125}I- anti total PSA antibody with each of isolated f-PSA and c-PSA in benign breast tumor homogenates. (All other details are explained in the text).

Spectroscopic Studies of PSA and its Complex

Introduction

Molecules absorb light; the wavelengths that are absorbed and efficiency of absorption depend on both the structure and environment of the molecule making absorption spectroscopy a useful tool for characterizing both small and large molecule.

The wavelength range (210-300) nm is the most important one for the protein characterization. This region is related to tryptophan, tyrosine, phenylalanine and histidine amino acid residues, which considered to be chromophores . Changes in the charge or the environment of these chromophores can lead to alteration in the absorption spectrum, and the conformational changes of a protein may also involve environmental changes of its chromophoric groups .

The electronic transitions for these chromophors come from $n\pi \rightarrow \pi^*$ and $\pi \rightarrow \pi^*$. These transitions are affected by many factors, which in turn affected the electronic environment that surround these transitions. Among these factors is the pH of the solvent, which determines the ionization state of ionizable chromophore. Also, the solvent polarity or perturbing agent affect the chromophore electronic transition where the λ_{max} for $\pi \rightarrow \pi^*$ transition occurs at shorter wavelength (blue shift) in polar erotic solvents (H_2O, alcohol) than in longer wavelength (red shift). The shift may or may not be accompanied by a change in intensity of the spectrum). Thus absorbance measurements can give an idea of the location of particular amino acids in a protein. Furthermore, information about the amino acids that are in a binding site is invariably valuable to the enzymologist to determine the reaction mechanism of an enzyme.

Although several new immunochemical techniques were developed to study such interactions , UV spectral remain as one of the most important methods in immunology because it provides a sensitive and quantitative measurements for the study of antibody structure and its specific ligand binding.

The U.V Spectrum of PSA, 125I-anti total PSA antibody and (125I-anti total PSA antibody / PSA) Complex

. The U.V spectrum of human PSA

1. One hundred microliters of human PSA (5.0 ng/ml) provided by (total PSA IRMA kit from Immunotech- Beckman Coulter company Czech Republic) was completed to 500µl with Tris/HCl buffer pH 7.2. Then placed in a 0.5 cm cuvette in sample beam

2. The absorption spectrum was immediately measured against the same buffer in reference beam in the area of (200-320 nm).

. U.V Spectrum of 125I - anti total PSA anti body

1. Ten microliters ^{125}I-anti total PSA antibody (50µg protein) was completed to 500µl with Tris/HCl buffer pH 7.2. Then placed in a 0.5 cm cuvette in sample beam

2. The absorption spectrum was immediately measured against the same buffer in reference beam in the area of (200-320 nm).

U.V Spectrum of (125I - anti total PSA anti body / PSA) complex

1. **One hundred microliters (500µg protein) of 125I- anti total PSA antibody was added to 100µl (100ng/ml) of standard PSA provided by (total PSA IRMA kit from Immunotech- Beckman Coulter company Czech Repoblic) and the volumes were completed to (500µl) with Tris/HCl buffer pH 7.2.**

The (125I-anti total PSA antibody / PSA) complex was estimated by following the steps 2, 3, 4, 5, 6 and 7 in section (3.3.4.) with the kit conditions.

The fractions that contained high levels of the binding activity of (^{125}I-anti total PSA antibody/ PSA) complex represented in peak 1 were collected and pooled.

One hundred microliters of (^{125}I-anti total PSA antibody/PSA) complex was completed to 500µl with Tris/HCl buffer pH 7.2. Then placed in a 0.5 cm cuvette in sample beam

Calculations

1. The radioactivity (c.p.m) of each eluted fraction was plotted against fraction number.

2. The absorption spectrum was measured against the same buffer in reference beam in the area of (200-320 nm).

. Factors Effecting the Absorption Properties of PSA and (125I-anti total PSA antibody / PSA) Complex

pH Effect

1. One hundred microliters of human PSA (5.0 ng/ml) provided by (total PSA IRMA kit from Immunotech- Beckman Coulter company Czech Repoblic) was completed to 500µl with Tris/HCl buffer at different pH (4, 7.2, 9 and 12.5). The pH was adjusted using 1N HCL Then placed in a 0.5 cm cuvette in sample beam

2. The absorption spectrum was immediately measured against the same buffer in reference beam in the area of (200-320 nm).

3. The experiment was repeated using 100µl of (^{125}I-anti total PSA antibody/ PSA) complex.

The effect of solvent polarity

1. One hundred microliters of human PSA (5.0 ng/ml) provided by (total PSA IRMA kit from Immunotech- Beckman Coulter company Czech Repoblic) was completed to 1000µl with Tris/HCl buffer pH 7.2 containing 20% ethanol Then placed in a 0.5 cm cuvette in sample beam.

2. The step 1 was repeated using 20% of each of ethylen glycol, glycerol and Dimethylsulphoxide (DMSO)

4. The absorption spectrum was immediately measured against the same buffer in reference beam in the area of (200-320 nm).

5. The absorption of human PSA in Tris/HCl buffer pH 7.2 was immediately measured against the λ$_{max}$ of human PSA in each solvent.

6. The experiment was repeated using 100µl of (^{125}I-anti total PSA antibody/ PSA) complex.

3.3.3. Spectrophotometic pH Titration of PSA and (125I-anti total PSA antibody / PSA) Complexes

1. One hundred microliters of human PSA (5.0 ng/ml) provided by (total PSA IRMA kit from Immunotech-Beckman Coulter company Czech Republic) was completed to 500µl with different buffers at different pH ranging 4 to 8. The maximum absorbency of each sample was measured at a wavelength of 211nm.

2. One hundred microliters of human PSA (5.0 ng/ml) provided by (total PSA IRMA kit from Immunotech-Beckman Coulter company Czech Republic) were completed to 500µl with different buffers at different pH ranging 9 to 12.5. The maximum absorbency of each sample was measured at a wavelength of 295nm.

3. The experiment was repeated using 100µl of (^{125}I-anti total PSA antibody/ PSA) complex.

Calculations

The absorbance of λ_{max} at each pH value was plotted versus corresponding pH.

. Results and Discussion

The U.V Spectrum of PSA, 125I-anti total PSA antibody and (125I-anti total PSA antibody / PSA) Complexes

The UV spectra of h-PSA and ^{125}I-anti total PSA antibody complex were scanned from 200-320 nm to determine the absorption spectra, and the alternation in the UV spectra as a results of their interaction.

. The U.V spectrum of human PSA

Figure (5-1) illustrates the U.V spectrum of human PSA (provided by total PSA IRMA kit from Immunotech- Beckman Coulter company Czech Republic) at pH 7.2. The spectrum shows that the λ_{max} for PSA is consisted of 2 peaks, at 229 nm and at 277 nm. Both peaks may be assigned to tyrosine residue and located in a way, that large part of it is on the surface of the protein molecule while the other part is buried.

As a result human PSA has a characteristic spectrum and can be identified by its peaks.

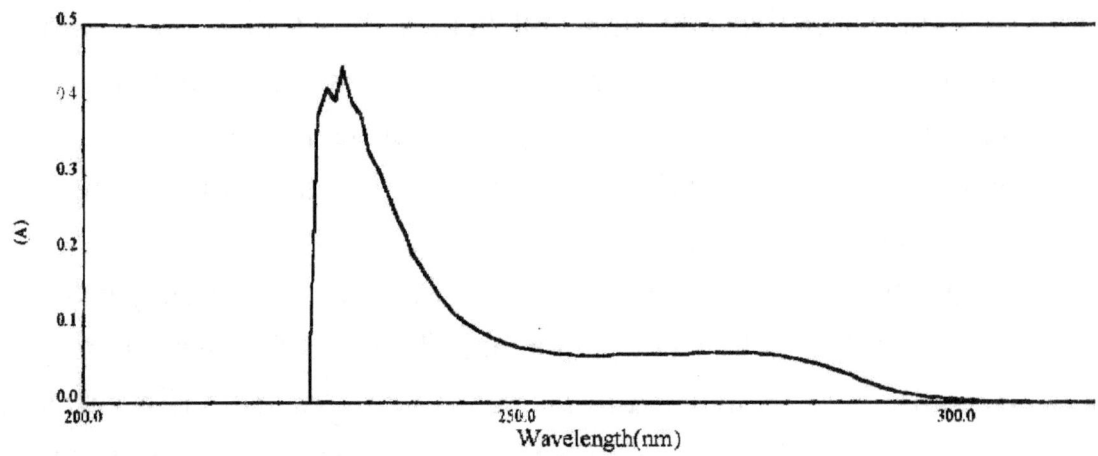

Figure (130): The U.V spectrum of human PSA at pH 7.2. (All other details are explained in the text).

The U.V Spectrum of 125I - anti total PSA anti body

Figure (131) illustrates the U.V spectrum of ^{125}I-anti total PSA antibody at pH 7.2. The spectrum shows that the λ_{max} for ^{125}I-anti total PSA antibody is

consisted of 2 peaks, at 230nm and at 270 nm. Which are assigned to the tyrosine and phenylalanine respectively. It seems that each of tyrosine and phenylalanine residue in the ^{125}I - anti total PSA anti body molecule is located in a way, that part of it is on the surface of the protein molecule while the other part is buried.

As a result ^{125}I - anti total PSA anti body has a characteristic spectrum and can be identified by its peaks.

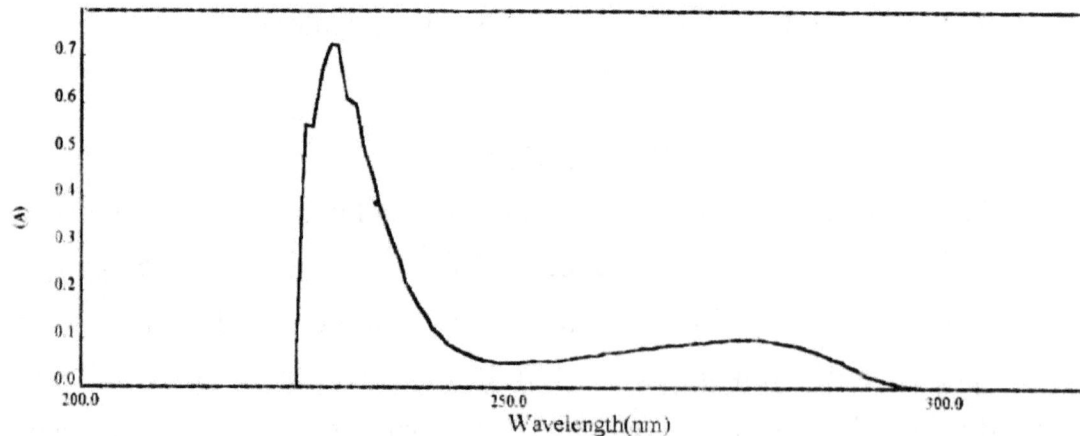

Figure (131): The U.V spectrum of ^{125}I - anti total PSA anti body at pH 7.2. (All other details are explained in the text).

The U.V Spectrum of (125I - anti total PSA anti body / PSA) complex

Figure (132) shows the elution profile of the incubation solution, which had contained PSA (provided by total PSA IRMA kit from Immunotech-Beckman Coulter company Czech Republic), using sepharose CL-6B (0.96 × 29 cm) column.

Figure (132) shows two peaks of radioactivity were appeared. The broad one or peak (1) represents the (^{125}I-anti total PSA antibody/PSA) complex.

Moreover, the high molecular weight of the first peak, which was expected from its low retention volume (10 ml) may be attributed to the same reason.

The second peak (2) in figure (132) may represent the free ^{125}I- anti total PSA antibody.

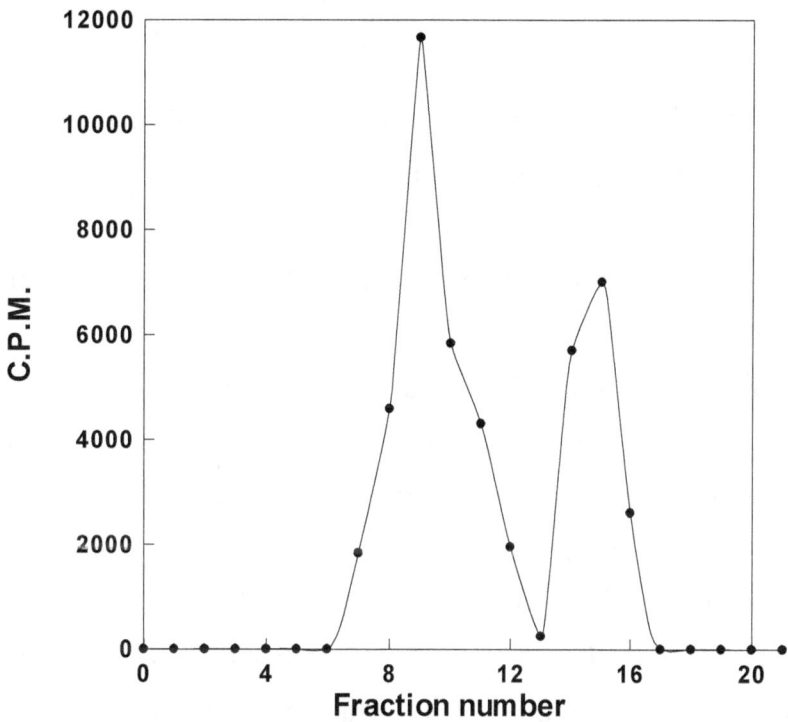

Figure (132): The elution profile of the complex of ^{125}I- anti total PSA antibody with human PSA. (All other details are explained in the text).

Figure (133) illustrates the U.V spectrum of ^{125}I-anti total PSA antibody/PSA complex at pH 7.2. The spectrum shows that the λ_{max} for ^{125}I-anti total PSA antibody is consisted of 2 peaks, at 234nm and at 283 nm. Which are assigned to the tyrosine and tryptophan respectively. It seems that each of tyrosine and tryptophan residues in the (^{125}I - anti total PSA anti body/PSA) complex molecule is located in a way, that a small part of tyrosine and large part of tryptophan residues are on the surface of the protein molecule while a large part of tyrosine and small part of tryptophan residues are buried.

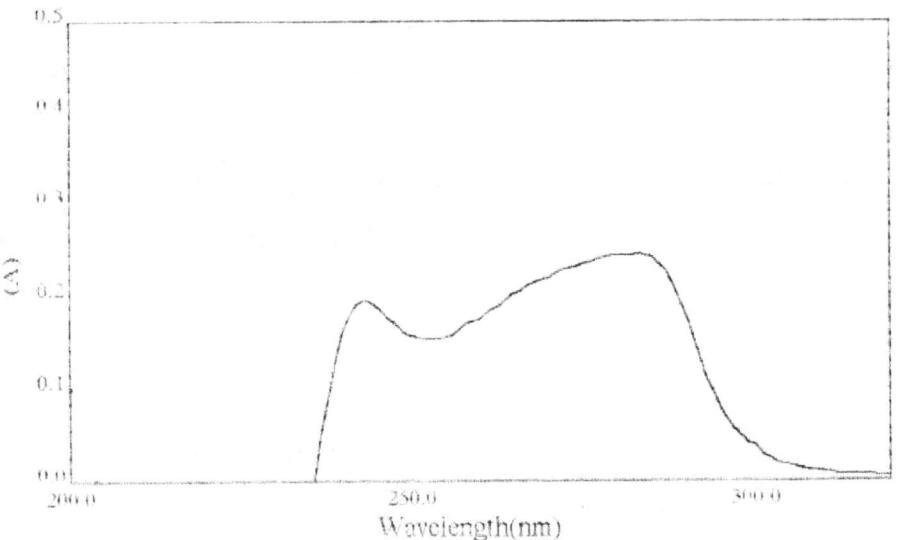

Figure (133): The U.V spectrum of (^{125}I - anti total PSA anti body/PSA) complex at pH 7.2. (All other details are explained in the text).

As a result (^{125}I-anti total PSA antibody/PSA) complex has a characteristic spectrum and can be identified by its peaks.

The spectrum of (^{125}I-anti total PSA antibody/PSA) complex as comparison with PSA indicated the disappearance of one tyrosine peak for PSA and appearance of the tryptophan peak for the (^{125}I - anti total PSA antibody/PSA) complex. Conformation changes may occur that effect the absorption of the PSA. These changes are due to the fitting of antibody to its antigen to form (^{125}I - anti total PSA antibody/PSA) complex. These results are agreed with finding of Seinerman et al who found that the surface of protein interactions are polar as well and the complex formation lead to the burial of charged and polar residues.

. Factors Effecting the Absorption Properties of PSA and (125I-anti total PSA antibody / PSA) Complex

The absorption spectrum of a chromophore is primarily determined by the chemical structure of the molecule. However, a large number of environmental factors produce detectable changes in λ_{max} and ε.

The general features of these environmental effects are the following:

pH Effect

The pH of the solvent determines the ionization state of the ionizable chromophore in the protein molecule. The UV spectrum of PSA and (^{125}I-anti total PSA antibody/PSA) complex were determined at different pH (4, 7.2, 9 and 12.5) and λ_{max} was obtained as shown in table (70).

In a neutral pH (7.2), Two λ_{max} were obtained for each PSA and (^{125}I-anti total PSA antibody/PSA) complex, λ_{max1} and λ_{max2} of PSA were (229 and 277) nm respectively whereas λ_{max1} and λ_{max2} of (^{125}I-anti total PSA antibody/PSA) complex were (234 and 283) nm respectively.

In an acidic pH (4), there were decreases in both λ_{max1} of each PSA and (^{125}I-anti total PSA antibody/PSA) complex and λ_{max2} of PSA with decreasing in pH. The blue shift has been observed in absorption of tyrosine residue may be attributed to conformational changes and chromophore in native PSA and (^{125}I-anti total PSA antibody/PSA) complex are on the surface of their. On the other hand the blue shift may be due to the increasing of hydrogen bond formed in the presence of highly positively charged state [260]. The λ_{max2} of the (^{125}I-anti total PSA antibody/PSA) complex remain almost constant. This result prove that the absorbance is related to tryptophan residue where the change in pH will no affect the ionization state of this residue and certainly the absorbance will no be changed.

When the pH value was increased from (7.2 to 9), there were an increase in both λ_{max1} of each PSA and (^{125}I-anti total PSA antibody/PSA) complex and λ_{max2} of PSA. The red shift has been observed in absorption of tyrosine residue, and this certainly related to the ionization of side chain of the tyrosine and this led to availability of the lone pair on the oxygen atom to be happened easier and at lower energy level (red shift). The λ_{max2} of the (^{125}I-anti total PSA antibody/PSA) complex remain almost constant. This result prove that the absorbance is related to tryptophan residue where the change in pH will no affect the ionization state of this residue and certainly the absorbance will no be changed.

At pH 12.5 no absorbance was recorded. The disappearance of both λ_{max1} of each PSA and (^{125}I-anti total PSA antibody/PSA) complex and λ_{max2} of PSA of tyrosine residue and λ_{max2} of the (^{125}I-anti total PSA antibody/PSA) complex of tryptophan residue due to conformational changes and chromophore in native complex were buried in the interior of their complexes.

Table (70): Effect of increasing pH on the λmax of PSA and (^{125}I - anti total PSA antibody/PSA) complex. (All other details are explained in the text).

pH	λ_{max} (nm)	
	Standard PSA	(^{125}I - anti total PSA antibody/PSA) complex
4	225, 274	227, 284
7.2	229, 277	234, 283
9	232, 279	235, 284
12.5	--	--

It must be noted that the spectral shifts of proteins produced by pH cannot be simply attributed to the inductive effects of vicinal charges, such spectral changes must therefore be attributed mainly to rearrangements of secondary and tertiary structure, although the possibility of field effects due to unusually close conjunction of charges to aromatic groups is not excluded.

The Effect of Solvent Polarity

Table (71) shows the effect of 20% ethanol and ethylene glycol at neutral pH on the human PSA and (^{125}I-anti total PSA antibody/PSA) complex spectra.

The data obtained previously show that the λ_{max} of PSA and (^{125}I-anti total PSA antibody/PSA) complex at neutral pH were (229 and 277) nm and (234 and 283) nm respectively. The λ_{max} value of tyrosine in PSA and (^{125}I-anti total PSA antibody/PSA) complex was shifted towards longer wavelengths (red shift) in 20% of each ethanol and ethylene glycol due to the hydrogen bonding of the OH groups of tyrosines with the solvent or with the π-electron system of the benzene ring where tyrosine was functioned as a hydrogen donor, while the λ_{max} value of tryptophan in (^{125}I-anti total PSA antibody/PSA) complex was shifted towards shorter wavelengths in 20% of each ethanol and ethylene glycol, this shift was attributed to $\pi \rightarrow \pi^*$ transitions. These two shifts in λ_{max} were accompanied with an increase in the absorbency of tryptophan and a decrease in the absorbency of tyrosine, these findings could be attributed to a change in the protein structure that bring the tryptophan residues to the surface of the protein while tyrosine residues were partly embedded in a hydrophobic region of the protein molecule.

The λ_{max} of tyrosine residues in PSA and (^{125}I-anti total PSA antibody/PSA) complex were shifted towards longer wavelengths in 20% glycerol without affecting the structure of these PSA and (^{125}I-anti total PSA antibody/PSA) complex, while the λ_{max} of tryptophan residues in (^{125}I-anti total PSA antibody/PSA) complex was not affected. The shift indicates that at 20% glycerol, the exposed tyrosines become solvated with glycerol (dipole–dipole interaction)).

Also table (71) shows the λ_{max} of PSA and (^{125}I-anti total PSA antibody/PSA) complex in 20% dimethylsulfoxide at neutral pH. It was found that two newer λ_{max} were appeared for PSA, at (263.2 and 293) nm which were assigned to phenylalanine and tryptophan residues respectively, while in (^{125}I-anti total PSA antibody/PSA) complex, a newer λ_{max} was appeared at 266.4nm which was assigned to histidine residue and the λ_{max} value of tryptophan was shifted towards longer wavelengths (red shift).

The appearance of these new λ_{max} values indicates that the protein was defolded due to change in the secondary and tertiary structure of the protein that bring the phenylalanine and tryptophan residues for PSA and histidin residues for (^{125}I-anti total PSA antibody/PSA) complex to expose to absorbance while tyrosine residue was buried inside each of PSA and (^{125}I-anti total PSA antibody/PSA) complex molecule, also it was found that PSA and (^{125}I-anti total PSA antibody/PSA) complex are highly sensitive to change in the polarity of the solvent.

Table (71): Effect of solvent polarity on the λmax of PSA and (^{125}I - anti total PSA antibody/PSA) complex. (All other details are explained in the text).

Solvent	λ_{max} (nm)

	Standard PSA	(^{125}I - anti total PSA antibody/PSA) complex
Ethanol	233.4, 280.1	238.24, 281.55
Ethylene glycol	235.41, 281.09	240.22, 278.68
Glycerol	236.02, 282.81	237.12, 283.3
DMSO	263.2, 293	266.4, 291.3

Spectrophotometric pH titration of PSA and (125I-anti total PSA antibody / PSA) complex

Spectrophotometric pH titration is the following of the change in absorbance of the chromophore with increasing pH [261]**. Many studies of protein structure require the determination of pk values for proton dissociation from ionizable amino acid side chains, because these values give an indication of the location of the amino acid in the protein. This can often be done spectrophotometrically because dissociation often changes the spectrum of one of the chromophores, the observation of tyrosine dissociation was performed by measuring the absorption at 295 nm (λ_{max} for the ionized form of tyrosine), and the observation of histidine dissociation was carried out by measuring the absorption at 211 nm.**

Figure (134 A&B) shows the pH titration curves of PSA and (^{125}I-anti total PSA antibody/PSA) complex for tyrosine and histidine respectively.

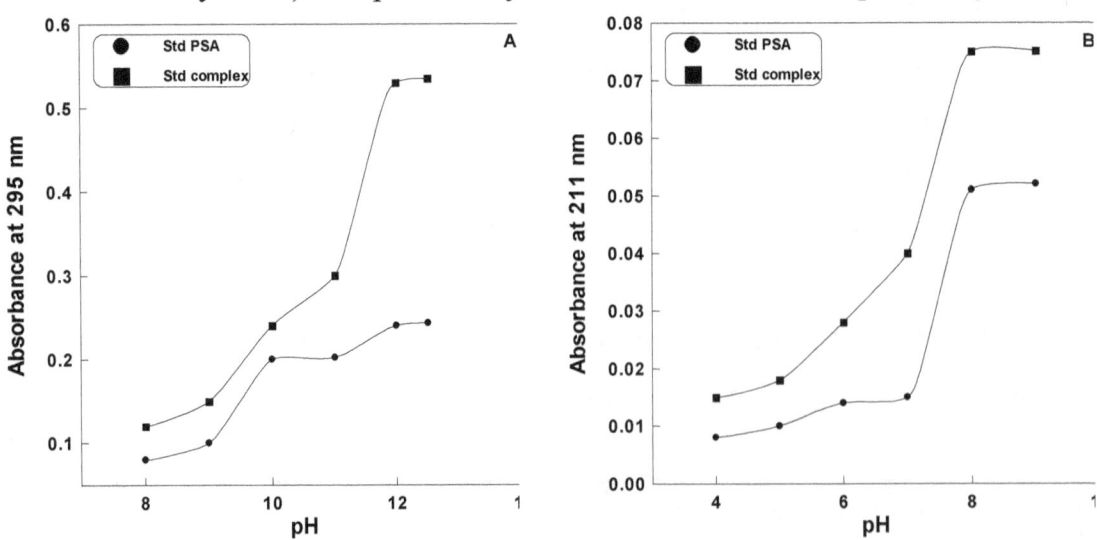

454

Figure (134): Spectrophotometric pH titration of PSA and (^{125}I-anti total PSA antibody / PSA) Complex for:

        A. Tyrosine residue
        B. Histidine residue
(All other details are explained in the text).

The (A) curves show that the pk$_a$ values for tyrosine are 9.2 and 11.2 for PSA and (^{125}I-anti total PSA antibody/PSA) complex respectively, while the pk$_a$ values for histidine in (B) curves were equal to 5.5 and 7.3 for PSA and (^{125}I-anti total PSA antibody/PSA) complex respectively. From the same figure, it was found that:

1. About 83.3 and 45.2% of tyrosine residues are located on the surface of the PSA and (^{125}I-anti total PSA antibody/PSA) complex molecule respectively.

2. About 16.7 and 54.8% of tyrosine residues are buried interior the folded structure of the PSA and (^{125}I-anti total PSA antibody/PSA) complex respectively.

3. About 27.4 and 37.3 % of histidine residues is located on the surface of the PSA and (^{125}I-anti total PSA antibody/PSA) complex molecule respectively.

4. About 72.6 and 62.7% of histidine residues is embedded in the interior region of the PSA and (^{125}I-anti total PSA antibody/PSA) complex molecule respectively.

5. In PSA the tyrosine residues were largely present on the surface of the molecule and the internal tyrosines are in a strongly nonpolar environment, while the internal tyrosine residues in (^{125}I-anti total PSA antibody/PSA) complex were in a strongly polar environment (e.g. a tyrosine surrounded by carboxyl groups). On the other hand, the histidine residues are slightly present on the molecular surface of PSA and (^{125}I-anti total PSA antibody/PSA) complex and the internal histidine residues of PSA are in a nonpolar environment whereas the internal histidine residues of (^{125}I-anti total PSA antibody/PSA) complex are likely to be in a strongly polar environment.

6. Finally, the percent of external tyrosine residues in PSA was greater than that of (^{125}I-anti total PSA antibody/PSA) complex and the percent of internal tyrosine in (^{125}I-anti total PSA antibody/PSA) complex was

greater than that of PSA. On the other hand the percent of internal histidine in PSA was greater than that of (^{125}I-anti total PSA antibody/PSA) complex.

Solvent perturbation studies

The determination of whether an amino acid is internal or external by measuring the spectra of a protein in polar and nonpolar solvents is called the solvent perturbation method. In fact, proteins are rarely studied in completely nonpolar solvents because most proteins are either insoluble or denatured in these solvents; therefore, mixtures of 80% water and 20% of reduced polarity solvent were used . Solvents alter the peak positions and intensities by altering the energy and probability of electronic transitions and this alteration arises from a difference in the salvation energies of the ground state and the first excited singlet state .

Table (72) shows the λ_{max} and ΔA values in the presence of different perturbants (20% ethanol, 20% ethylene glycol, 20% glycol and 20% DMSO).

Table (72): Solvent perturbation on PSA and (^{125}I-anti total PSA antibody / PSA) Complex. (All other details are explained in the text).

Perturbent substance	PSA			Complex		
	λ_{max}	A	ΔA	λ_{max}	A	ΔA
Ethanol	233.4	1.35	0.84	238.24	0.5	0.35
	280.1	0.9	0.83	281.55	0.57	0.37
Ethylene glycol	235.41	1.15	0.85	240.22	0.55	0.35
	281.09	0.94	0.84	278.68	0.43	0.38

Glycerol	236.02	1.15	0.81	237.12	0.6	0.34
	282.81	087	0.82	283.3	0.48	0.32
DMSO	263.2	0.25	0.15	266.4	0.37	0.27
	293	0.34	0.3	291.3	0.45	0.4

From the results listed in these Tables it was found that several spectral changes were obtained in the presence of these perturbants, like the alteration of the λ_{max} positions and intensities of PSA and (^{125}I-anti total PSA antibody/PSA) complex spectra, and the appearance of new chromophores on the surface of each the PSA and (^{125}I-anti total PSA antibody/PSA) complex molecule. These chromophores were embedded in an interior region of the protein in the absence of solvent.

From the solvent perturbation studies, the following remarks could be drawn:

About 0.83, 0.35 of tyrosine residues, 0.15 and 0.27 of phenylalanine residues and 0.3 and 0.4 of tryptophan residues are on the surface of PSA and (^{125}I-anti total PSA antibody/PSA) complex molecule respectively.

From spectrophotometic pH titration and solvent perturbation studies we conclude that

About 83,30, 15 and 25% of tyrosine, tryptophan, phenylalanine and histidine residues are on the surface of PSA molecule respectively, whereas about 39, 40, 27 and 36 % of tyrosine, tryptophan, phenylalanine, histidine and tryptophan residues are on the surface of (^{125}I-anti total PSA antibody/PSA) complex molecule respectively.

Finally if assumed that human PSA that used in spectrophotometric studies consisted only f-PSA that composed 4, 7, 6 and 11 of tyrosine, tryptophan, phenylalanine and histidine residues respectively [75], then about 3, 2, 1 and 3 of tyrosine, tryptophan, phenylalanine and histidine residues are on the surface of PSA molecule respectively, while about 1, 5, 5 and 10 of tyrosine, tryptophan, phenylalanine and histidine residues are buried interior the folded structure of the PSA molecule. Accordingly, from figure (5-2A&B) by dividing the maximum absorbency of PSA from the maximum absorbency of (^{125}I-anti total PSA antibody/PSA) complex and multiplied by the number of tyrosine and histidine residues in f-PSA concluding that (^{125}I-anti total PSA antibody/PSA) complex was composed 9 and 16 of tyrosine and histidine residues respectively, then about 4, and 6 of tyrosine and histidine residues are on the surface of PSA molecule respectively, while about 5, and 10 of tyrosine and histidine residues are buried interior the folded structure of the PSA molecule.

Isolation and Characterization of PSA of Benign and Malignant Breast Tumors

Introduction

Prostate specific antigen exists in different forms, free PSA with molecular weight approximately 32 kDa and complexes with protease inhibitors, predominantly alpha-1- antichemotrypsine (ACT) with molecular weight approximately 90 kDa.

Black *et al*, showed that there was no significant differences between the percentage of benign breast tumors and malignant breast tumors with detectable f-PSA, but the proportion of f-PSA/t-PSA in malignant breast tumors is significantly higher in comparison to patients with benign breast tumors. Another studies suggested that the predominant form (>50%) of PSA in a significant proportion of females with breast cancer is f-PSA. In another study revealed f-PSA, as the predominant form is highly specific for breast cancer in comparison with benign breast disease and normal tissue. According to these perturbation results studies, further work is required to investigate which of these studies should be supported.

Chromatography encompasses a diver and important group of separation methods that permit the scientist to separate, isolate and identify related components of complex mixtures [223]. The purification of biomolecules is the most popular use of gel chromatography due to the ability of a gel to fractionate molecules on the basis of size.

Mannello *et al*. separate f-PSA and c-PSA with out any additional peaks from breast cyst fluid (BCF) using sephacryl S-300.

The purpose of this part of thesis is to isolate f-PSA and c-PSA from each of benign and malignant breast tumor tissues using sephadex G-200 and to finding of the optimum reaction conditions for ^{125}I-anti total PSA antibody with isolated f-PSA and c-PSA in breast tumor homogenate.

Patients

The two groups of breast tumor patients described were included for prostate specific antigen PSA analysis. Benign and postmenopausal malignant breast tumor homogenates that showed maximal binding in the preliminary test were collected, pooled and used for isolation and characterization of PSA.

. Separation of free 125I- anti total PSA antibody from (125I- anti total PSA antibody/ PSA) complexes in tissue of breast tumors by gel-filtration technique using sepharose CL-6B

. Gel preparation

    1. The pre-swollen gel was suspended in PBS buffer pH8 equivalent to approximately three times the volume of settles gel.
    2. The gel was allowed to settle and decanting the excess buffer.
    3. The step 1 and 2 was repeated two more times to allow preliminary equilibration of the gel with the eluent buffer.
    4. The gel was resuspended in a volume of PBS buffer pH8 approximately equal to the settled volume of gel.
    5. The gel slurry was degassed by suction for 60 minutes.

Preparation of the column

2.3.2.1. Dimensions of the column

The dimensions of the column were chosen according to the following equation [224].

$$\text{Diameter} = \sqrt{\frac{m}{10}}$$

Where:

m= amount of protein in mg.

L = 30 x diameter

Where:

L: length of the column in centimeters

In view of the results of such calculation, a 0.96x29cm column has been used.

Bed packing

1. The column was mounted and packing reservoir vertically.
2. The column out let was opened and filled with eluent buffer to a height about 5cm and turn off column flow.
3. The degassed slurry was carefully mixed before pouring into the column using a glass rod attached to the inner surface of the column.
4. After the gel has settled for 5 min., the column outlet was opened.
5. Packing was continued until the gel reached a stable bed height (29cm). Then the column was equilibrated with PBS buffer pH8 for 24hr. at 4°C with dimensions (0.96x29cm) and a total bed volume of 21ml.

. Void Volume Determination

The elution volume of blue dextran 2000 is equal to the column void volume (V_o), and was determined as follows:

A fresh solution of Blue Dextran (2mg.ml-1) was prepared in the eluent buffer in a sample volume of 1-2% of the total bed volume, and then applied to the column with a flow rate of 5 ml.hr-1. Fractions of 1 ml were collected and their absorbance were measured at $\lambda = 600$ nm.

Separation procedure

1. Sixty microliters (300µg protein) of 125I- anti total PSA antibody was added to 50µl (400µg protein) of benign Fibroadenoma breast tumor homogenate and the volumes were completed to (500µl) with TEBP buffer pH (8).

2. Two additional tubes, containing 60μl of 125I- anti total PSA antibody only for total activity computation, were set-asides until counting.

3. The mixtures were incubated for 180 minute at 4°C while continuously shaking.

4. At the end of incubation, the mixture was applied to the surface of sepharose CL-6B gel filtration column (0.96x29cm) equilibrated with PBS pH8 with a flow rate one-half that used for column elution.

5. Immediately 1ml of PBSS pH8 was added after the mixture to aid in sharp mixture application.

6. Elution was carried out using the PBS buffer pH8, to separate 125I- anti total PSA antibody and (125I- anti total PSA antibody/PSA) complexes, with a flow rate of 5ml/hr., and fraction volumes of 1ml.

7. The radioactivity of each fraction was counted in a gamma counter for 1 min.

8. The experiment was repeated using 50μl (250μg protein) of 125I- anti total PSA antibody and 50μl (400μg protein) of postmenopausal malignant infiltrative ductal carcinoma (IDC) breast tissue homogenate.

Calculations

1. The radioactivity (c.p.m) of each eluted fraction was plotted against fraction number.

2. The resultant fractions under peak were determined by the sum of the radioactivity in (c.p.m) of the fractions at that.

3. The percent radioactivity of (^{125}I- anti total PSA antibody/ PSA) complexes to total ^{125}I- anti total PSA antibody was calculated from the values area under peak of each species.

Determination of the molecular weight (M.Wt) of free 125I- anti total PSA antibody and 125I- anti total PSA antibody/ PSA) complexes of benign breast tumor

A sepharose CL-6B gel was prepared for this purpose as described A sepharose CL-6B column was used.

The M.Wt was determined as follows:

4. The column was calibrated by gel filtration kit, purchased from pharmacia fine chemicals, which contained standard proteins. Standard protein solutions were prepared according to the manufacturer instructions, then applied through three portions (500µl), proteins 1 and 4 in the first portion, proteins 2 and 5 in the second portion, proteins 3 and 6 in the third portion.

5. Elution was carried out with PBS buffer at a flow rate of 5 ml.hr^{-1}.

6. The absorbance of the fractions collected was measured at 280 nm to evaluate the elution volume (Ve) of the standard protein.

Table (73): Standard proteins and their molecular weights (All other details are explained in text).

Protein	M.wt (kDa)	Conc. mg.mL-1
Thyroglobulin	669	7.0
Ferritin	440	1.0
Catalase	322	7.0
Aldolase	158	7.0
Albumin	67	7.0
Ovalbomin	43	7.0

Calculations

1. The K$_{av}$ values of the proteins eluted were determined using the following equation:

$$K_{av} = \frac{V_e - V_o}{V_t - V_o}$$

Where:
V$_o$= Void volume
V$_e$=Elution volume of each protein

V_t = Total gel - bed volume.

The calibration curve of K_{av} values vs. log M.wt. of the proteins were plotted.

Isolation of free prostate specific antigen from complex prostate specific antigen in tissue of breast tumors by gel filtration technique using sephadex G-200.

Gel preparation

1. With gentle mixing, a weight of 3g sephadex G-200 was slowly added to excess volume of PBS buffer pH8 (40ml of buffer per gram of gel).

2. The gel as allowed swelling at 4°C for 72 h. without stirring to equilibrate with buffer.

3. After equilibrated, the buffer was decanted and the gel was resuspended in excess volume of eluent buffer equal three times to settled volume of the gel and allowed it to settled for approximately 20 minutes.

4. The excess buffer and fine particles were removed by suction.

5. The gel was resuspended in a volume of PBS buffer pH8 approximately equal to the settled volume of gel.

6. The gel slurry was degassed by suction for 1hr.

Preparation of the column

Dimensions of the column

The Dimensions of the column was prepared as mentioned.

Bed packing

The column was packing as mentioned in section (2.3.2.2) with dimensions (1.25x37.5cm) and a total bed volume of 46ml.

Void Volume Determination

The column void volume was determined .

Isolation procedure

1. Five hundred microliters (3750µg protein) of benign Fibroadenoma breast tumor homogenate was applied to the surface

of sephadex G-200 gel filtration column (1.25x37.5cm) equilibrated with PBS pH8 with a flow rate one-half that used for column elution.

2. Immediately 1ml of PBSS pH8 was added after the homogenate to aided in sharp homogenate application.

3. Elution was carried out using the PBS buffer pH8, to isolate free prostate specific antigen and complex prostate specific antigen with a flow rate of 6ml/hr., and fraction volumes of 1ml.

4. The elution of each fraction was determined at 280nm.

5. Twenty microliters (100μg protein) of 125I- anti total PSA antibody was added to 50μl of each fraction number of benign fibroadenoma breast tumor homogenate and the volumes were completed to (500μl) with TEBP buffer pH (8).

6. Two additional tubes, containing 20μl of 125I- anti total PSA antibody only for total activity computation, were set-asides until counting.

7. The mixtures were incubated for 180 minute at 4°C while continuously shaking.

9. The (125I-anti-total PSA antibody/PSA) complex was estimated 9.The experiment was repeated using 500μl (9000μg protein) of postmenopausal malignant infiltrative ductal carcinoma (IDC) breast tissue homogenate.

Calculations

1. The fraction number was determined at 280nm.

2. The values of B/T ratio for the eluted fractions were calculated as .

3. The values of B/T ratio and the absorbancies at 280nm were plotted against the fraction number.

4. The plotted B/T ratio against the fraction number for all the experiments gave a 2peaks profile. The first peak represents the ^{125}I-Ab bound to c-PSA while the second peak represents the ^{125}I-Ab bound to f-PSA

5. The value of B/T ratio of f-PSA was determined by dividing the sum of the radioactivity of the fractions under peak 2 by the total radioactivity.

$$f\text{-PSA}\% = \frac{\text{The radioactivity under peak2}}{\text{Total radioactivity}} \times 100$$

6. The percentage of free-PSA /total-PSA was determined by dividing the sum of the radioactivity of the fractions under peak 2 by the sum of radioactivity of two peaks appeared in the profile:

$$f\text{-PSA}/t\text{-PSA} = \frac{\text{The radioactivity under peak2}}{\text{The sum of radioactivity under (peak1+peak2)}} \times 100$$

7. The percentage of f-PSA /c-PSA was determined by dividing the sum of the radioactivity of the fractions under peak 2 by the sum of radioactivity of the fractions under peak 1 appeared in the profile:

$$f\text{-PSA}/c\text{-PSA} = \frac{\text{The radioactivity under peak2}}{\text{The radioactivity under peak1}} \times 100$$

Determination of the molecular weight (M.Wt) of free prostate specific antigen and complex prostate specific antigen of benign breast tumor

A sephadex G-200 gel was prepared for this purpose as described . A sephadex G-200 column was used as and The M.Wt was obtained .

Table (74): Standard proteins and their molecular weights (All other details are explained in text).

Protein	M.wt (kDa)	Conc. mg.mL-1
Catalase	322	7.0
Aldolase	158	7.0
Albumin	**67**	**7.0**
Ovalbomin	**43**	**7.0**
Cymotrypsinogen	**25**	**7.0**
Ribonuclease	**13.7**	**7.0**

Calculations

The M.Wt was obtained as mentioned.

Dialysis for Concentration

After preparing dialysis tube, the fractions that contained high levels of the binding activity were collected, pooled and concentrated by dialyzing against sucrose at 4 °C for 2hrs to get the required concentration to be used in the next experiments.

Preliminary test of the binding of isolated PSA in breast tumor tissues with 125I-anti total PSA antibody

3. Thirty microliters (30) µl (150) µg of 125I- anti total PSA antibody was added to 50 µl of (200µg) of each isolated f-PSA and c-PSA of benign breast tumor respectively and the volumes were completed to (500µl) with TEBP buffer pH 8.

2. Two additional tubes, containing 30µl of 125I- anti total PSA antibody only for total activity computation, were set-asides until counting.

The (125I- anti-total PSA antibody/ f-PSA) and (125I- anti-total PSA antibody/ c-PSA) complexes were estimated by following the steps 2, 3 and 4 in section (2.4.3.6).

Calculations

The B/T% was computed for each tube as mentioned.

Most appropriate conditions of the binding of isolated free prostate specific antigen (f-PSA) and complex prostate specific antigen (c-PSA) of benign breast tumor with 125I- anti total PSA antibody

. Effect of protein amounts of the isolated free prostate specific antigen and complex prostate specific antigen of benign breast tumor on the binding

1. Fifty microliters of increasing amounts (50, 100, 150, 200, 250µg) of isolated f-PSA and c-PSA of benign breast tumor were each added to (25 and 30) µl (125 and 150) µg of 125I- anti total PSA antibody

respectively and the volumes were completed to (500µl) with TEB buffer pH 8

2. The (125I- anti-total PSA antibody/ f-PSA) and (125I- anti-total PSA antibody/ c-PSA) complexes were estimated.

Calculations

1. The B/T% was computed for each tube

2. The percent of binding values B/T% were plotted against the increasing amounts of protein of isolated f-PSA and c-PSA of benign breast tumor

. Effect of 125I- anti total PSA antibody on the binding

3. Increasing volumes (10, 15, 20, 25, 30, 40 and 50 µl) of 125I- anti total PSA antibody containing (50, 75, 100, 125, 150, 200 and 250 µg protein) respectively were each added to 50 µl (100 and 150) µg of isolated f-PSA and c-PSA of benign breast tumor respectively and the volumes were completed to (500µl) with TEBP buffer pH 8

4. The (125I- anti-total PSA antibody/f-PSA) and (125I- anti-total PSA antibody/c-PSA) complexes were estimated.

Calculations

1. The B/T% was computed for each tube.

2. The concentration of the bound 125I- anti total PSA antibody was plotted against the concentration of 125I- anti total PSA antibody.

Effect of pH on the 125I- anti total PSA antibody binding

1. Fifty microliters (100 and 150) µg of isolated f-PSA and c-PSA of benign breast tumor were each added to (20 and 40) µl (100 and 200) µg of 125I- anti total PSA antibody respectively and the volumes were completed to (500µl) with TEBP buffer of different pH (6.8, 7, 7.2, 7.4, 7.6, 7.8, 8, 9).

2. The (125I- anti-total PSA antibody/f-PSA) and (125I- anti-total PSA antibody/c-PSA) complexes were estimated

Solution

TEBP buffer of different pH (6.8, 7, 7.2, 7.4, 7.6, 7.8, 8, 9) were prepared.

Calculations

1. The B/T % values were determined.

2. The percent of binding values B/T% were plotted against their corresponding pH value.

Temperature dependency of the binding

1. Fifty microliters (100 and 150) μg of isolated f-PSA and c-PSA of benign breast tumor were each added to (20 and 40) μl (100 and 200) μg of 125I- anti total PSA antibody respectively and the volumes were completed to (500μl) with TEBP buffer pH (7.8 and 8) respectively.

2. The (125I- anti-total PSA antibody/f-PSA) and (125I- anti-total PSA antibody/c-PSA) complexes were estimated.

3. The experiment was repeated at different temperatures (20, 37 and 45°C)

Calculations

1. The B/T % values were determined at each temperature.

2. The percent of binding values were plotted against the different temperatures of incubation.

Time dependency of the binding

1. Fifty microliters (100 and 150) μg of isolated f-PSA and c-PSA of benign breast tumor were each added to (20 and 40) μl (100 and 200) μg of 125I- anti total PSA antibody respectively and the volumes were completed to (500μl) with TEBP buffer pH (7.8 and 8) respectively.

2. Four additional tubes, two containing 20μl and the others containing

40μl of 125I- anti total PSA antibody only for total activity computation were set-asides until counting.

3. The mixtures were incubated at 4°C at different time (30, 60, 90, 120, 180 and 240 minute) while continuously shaking.

4. The (125I- anti-total PSA antibody/f-PSA) and (125I- anti-total PSA antibody/c-PSA) complexes were estimated.

Calculations

1. The B/T % values were determined.

4. The percent of binding values were plotted against the different time of incubation.

Result and Discussion

. Separation of free 125I- anti total PSA antibody from (125I- anti total PSA antibody/ PSA) complexes

Figure (135) shows the elution profile of the incubation solution, which had contained PSA of benign and postmenopausal malignant breast tumor tissue homogenates respectively, using sepharose CL-6B (0.96 × 29 cm) column. Also, the elution profile of ^{125}I- anti total PSA antibody was shown in Figure (136)

The void volume of this column was 7 ml as predicted from the elution profile of the blue dextran as shown in Figure (137).

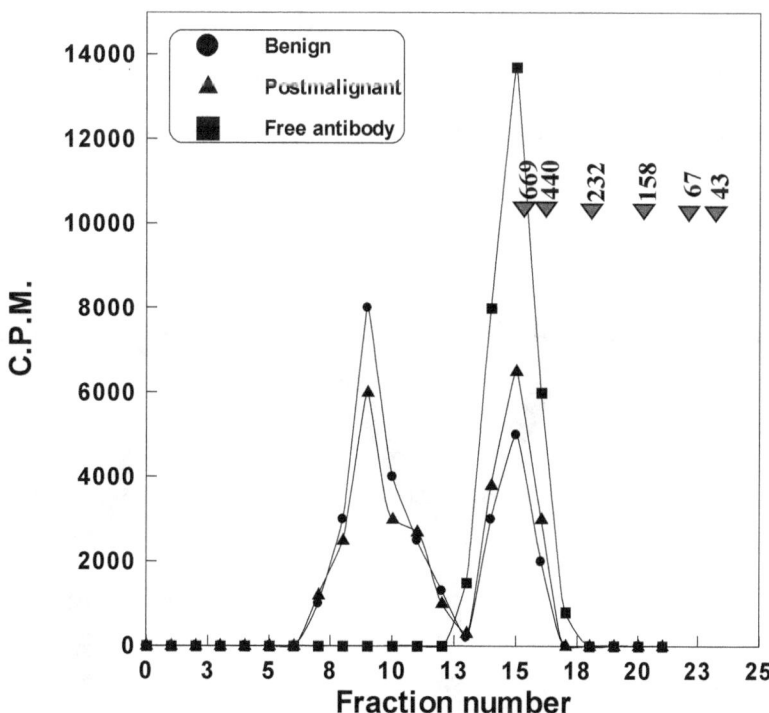

Figure (135): The elution profile of the complex of ^{125}I- anti total PSA antibody with isoforms of PSA in tissue each of benign and postmenopausal malignant breast tumor using sepharose CL-6B. (All other details are explained in the text).

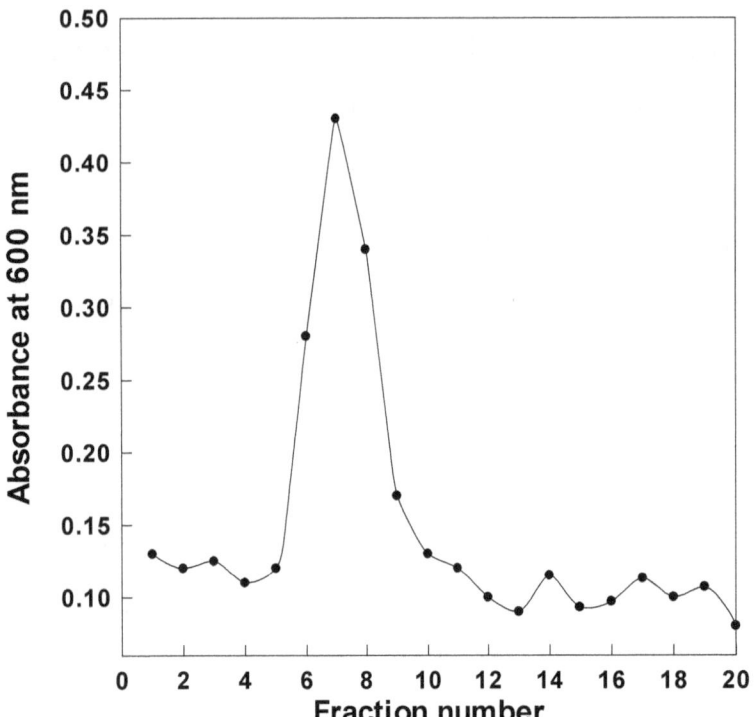

Figure (136): The elution profile of blue dextran 2000 by gel filtration chromatography using sepharose CL-6B. (All other details are explained in the text).

In each case two peaks of radioactivity appeared. The broad one or peak (1) represents the (^{125}I-anti total PSA antibody/PSA) complex with different forms of PSA. Figure (137) shows some of these suggested complexes.

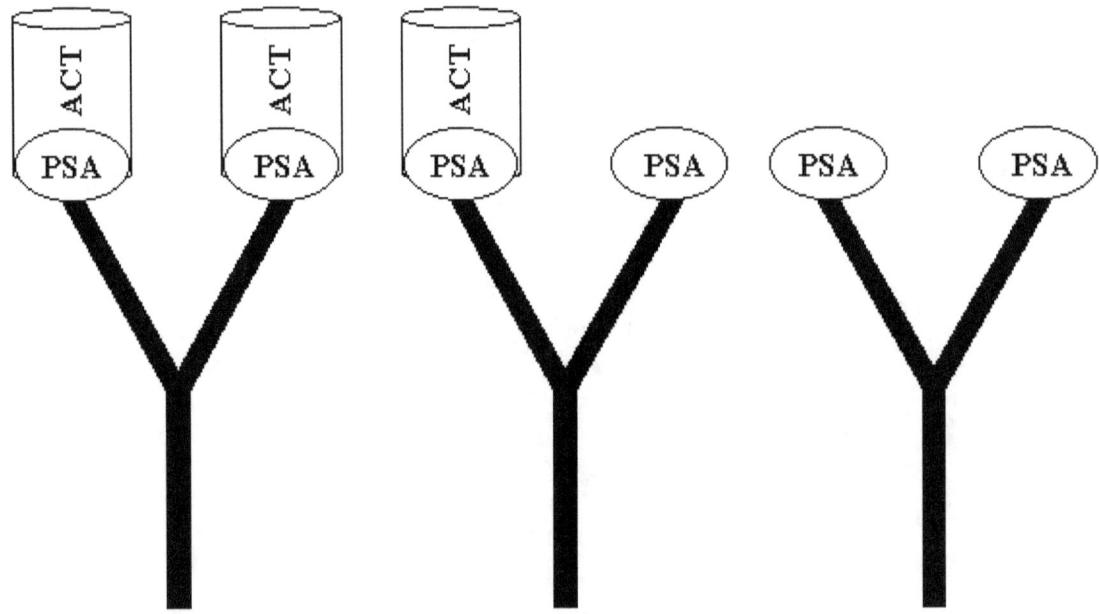

Figure (137): The suggested complexes between PSA and ^{125}I- anti total PSA antibody. (All other details are explained in the text).

Moreover, the high molecular weight of the first peak, which was expected from its low retention volume (9 ml) may attribute to the (^{125}I- anti total PSA antibody/PSA) polymerization and aggregation reaction, which may indicate the bivalency or multivalency of PSA.

Also the idea of bivalency of PSA could be supported by the finding of Lilja et al[)], they found that the PSA-ACT complex (for example) covers three of the epitopes of PSA leaving at least two epitopes available to react with antibody, so that PSA may be connected from both epitopes to produce a chain reaction and cause this enhancement in molecular weight.

The second peak (2) in figure (135) may represent the free ^{125}I- anti total PSA antibody. The high retention volume (15ml) indicates the presence of a relatively low molecular weight, approximately 127 kDa, which was obtained from the calibration curve in figure (138).

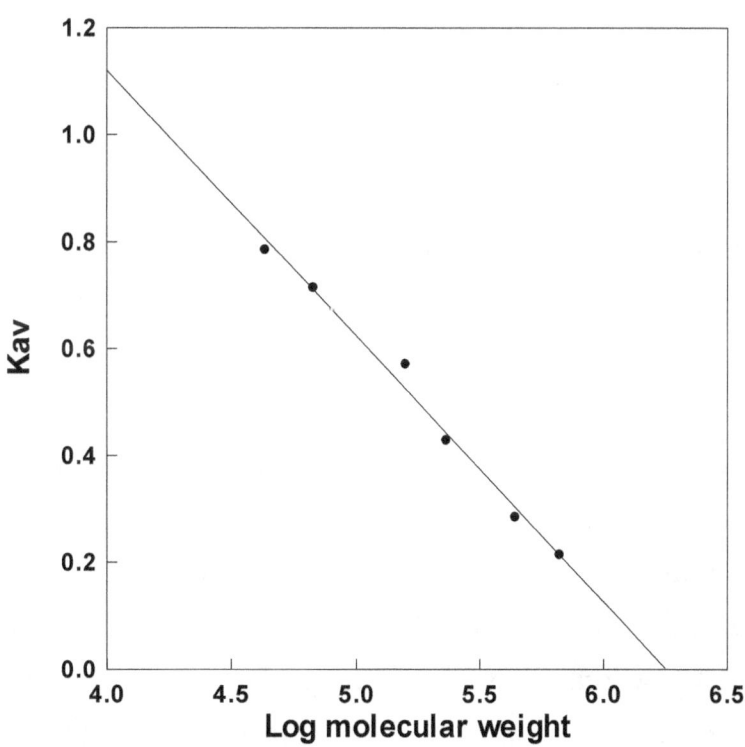

Figure (138): Calibration curve for determination of M.wt of complexes of PSA isoforms with ^{125}I- anti total PSA antibody by gel filtration chromatography using sepharose CL-6B. (All other details are explained in the text).

Isolation of f-PSA from c-PSA of benign and postmenopausal malignant breast tumor tissue homogenates was performed by gel exclusion chromatography technique. Benign and malignant homogenates were applied to sephadex G-200 (1.25 × 37.5 cm) column. The void volume of this column was 8 ml as predicted from the elution profile of the blue dextran as shown in Figure (139).

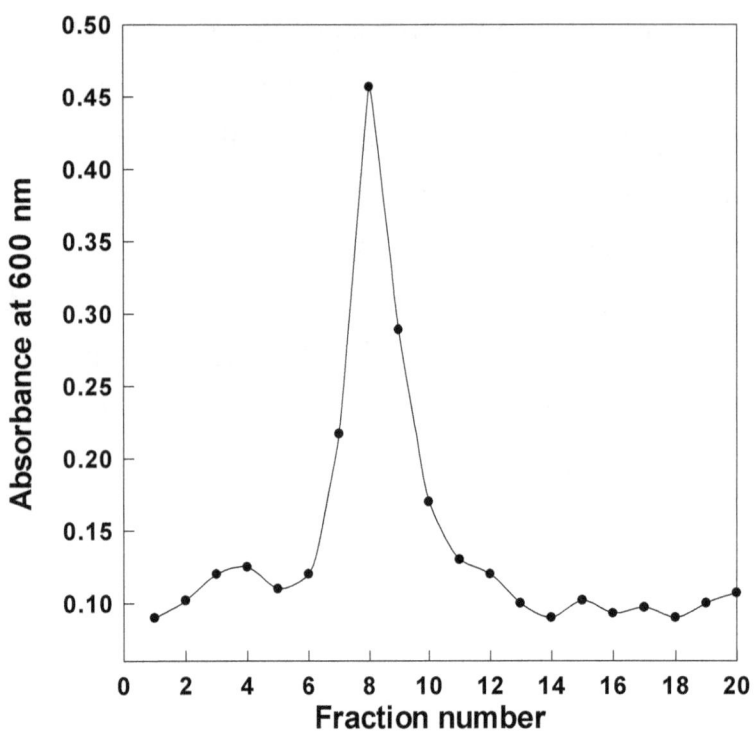

Figure (139): The elution profile of blue dextran 2000 by gel filtration chromatography using sephadex G-200. (All other details are explained in the text).

The resultant fractions of each homogenate type (benign and malignant) were measured at 280nm gave three peaks as shown in figure (140 A & B) respectively and the resultant fractions of each peak of benign and malignant were incubated with ^{125}I- anti total PSA antibody. According to the optimum conditions of those reactions gave two peaks may represents (^{125}I- anti total PSA antibody/c-PSA) complex at fraction number (21) and (^{125}I- anti total PSA antibody/f-PSA) complex at fraction number (30).

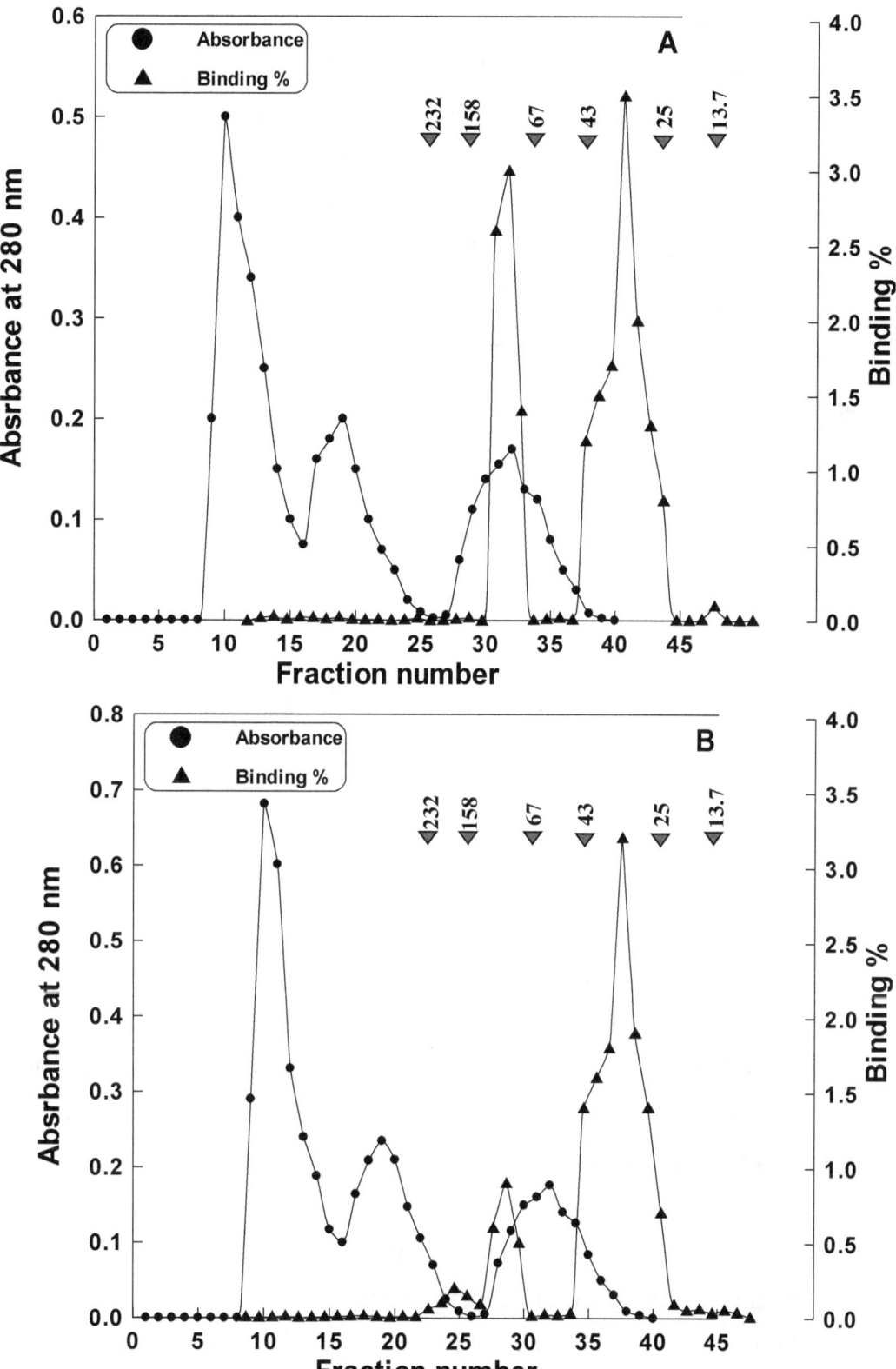

Figure (140): The elution profile of human PSA and complexes from:
A. Benign breast tumor tissue homogenates
B. Postmenopausal malignant breast tumor tissue homogenates
(All other details are explained in the text).

These results indicate that there are two immunologically active isoforms of PSA antigen, one of them was c-PSA (peak 1) and the other was f-PSA (peak 2). From these results it was concluded that these components are capable of binding to the ^{125}I- anti total PSA antibody with different affinity.

The chromatographic isolation was performed depending on the significant differences in molecular weight that was obtained by the calibration of the gel column with standard proteins as shown in figure (141) was 32.5 kDa for f-PSA and 101 kDa for c- PSA.

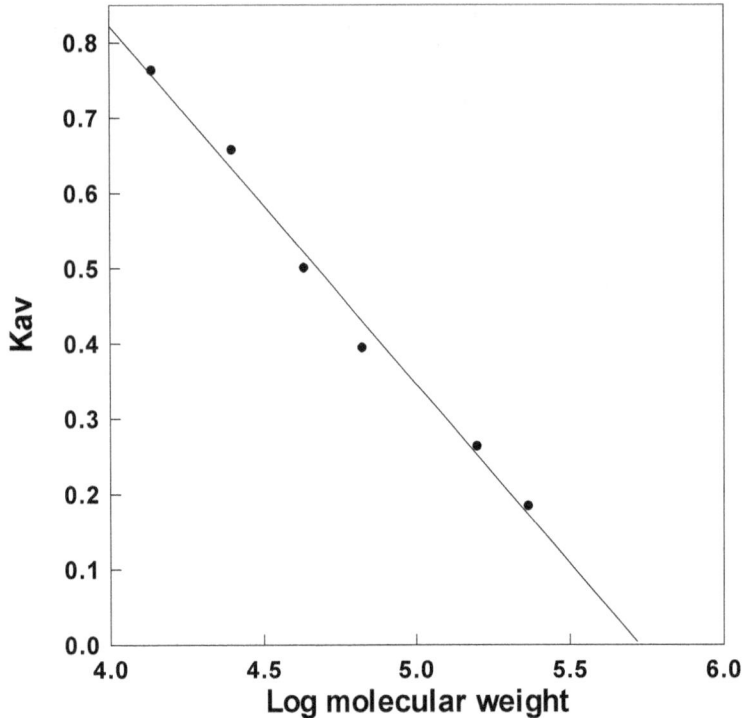

Figure (141): Calibration curve for determination of M.wt of isolated f-PSA and c-PSA from each of benign and postmenopausal malignant breast tumor by gel filtration chromatography using sephadex G-200. (All other details are explained in the text).

The obtained values of molecular weights were consistent with the values that were obtained by many other investigators.

The binding percent of f-PSA and f-PSA/t-PSA from isolated PSA of benign breast tumor in comparison with malignant breast tumor are shown in figure (2-6 A & B) respectively. The binding percent of f-PSA was 12 and 12.1 for benign and malignant respectively whereas the f-PSA/t-PSA percent was 63% and 86% for benign and malignant respectively.

These results are nearly similar to these obtained previously by (Black et al) [184], there is no significant differences between the percentage of benign breast tumors and malignant breast tumors with detectable f-PSA, but the proportion of f-PSA/t-PSA in malignant breast tumors is significantly higher in comparison to patients with benign breast tumors. Another studies suggested that the predominant form (>50%) of PSA in a significant proportion of females with breast cancer is f-PSA [157,174,184,222]. In another study revealed f-PSA, as the predominant form and is highly specific for breast cancer in comparison with benign breast disease and normal tissue [189].

Also the binding portion of f-PSA/c-PSA from isolated PSA in benign breast tumors in comparison with malignant breast tumors is shown. The binding portion of f-PSA/c-PSA were 1.7 and 6.0 for benign and malignant respectively

The present results suggest the measurement of f-PSA/t-PSA or f-PSA/c-PSA in the tissues of patients with benign and malignant breast tumors to have potential clinical applicability as a tumor marker either alone or in combination with other diagnostic markers.

Lopez-Otin and Diamandis suggest that serum f-PSA, which is an enzymatically inactivation of PSA is over expressed in patients with benign and malignant breast disease. The mechanism of such changes is unknown, but the data in contrast to changes in prostate cancer where serum c-PSA increases and f-PSA decreases in cancer patients in comparison to patients with benign prostatic hyperplasia .

According to these studies the isolated f-PSA and c-PSA of benign breast tumor homogenate were used as the source in all subsequent experiments, since they give the highest binding.

The resultant fractions of each peak of benign breast tumor homogenate that gave binding were pooled, concentrated and then subjected to protein determination.

Preliminary test of the binding of isolated PSA in breast tumor tissues with 125I- anti total PSA antibody

Each isolated f-PSA and c-ACT of benign breast tumor homogenate was detected through the incubation with ^{125}I- anti total PSA antibody for 180 min at 4°C with TEBP buffer pH (8) as a medium to complete the reaction. The (^{125}I- anti total PSA antibody/f-PSA) and (^{125}I- anti total PSA antibody/ c-PSA) complexes formed were separated from the unbound particulate by centrifugation at 1500xg for 30 min. This centrifugal speed was sufficient to precipitate the complex. After centrifugation the tubes were decanted, in order to get rid of the unbound antibody or antigen, while (^{125}I- anti total PSA antibody/f-PSA) complex or (^{125}I- anti total PSA antibody/c-PSA) complex remained as pellet in the bottom of the tube.

The preliminary conditions used in this experiment show the amount of binding B/T% for isolated f-PSA and c-PSA of benign breast tumor homogenate to be (12 and 7)%. The data revealed that the B/T% of f-PSA is more than c-PSA (>50%).

Most appropriate conditions of the binding of isolated PSA of benign breast tumor with 125I- anti total PSA antibody

Effect of protein amounts of the isolated PSA

Figure (142) shows the effect of increasing amounts of the isolated f-PSA and c-PSA of benign breast tumor on the binding with 125I- anti total PSA antibody. The results revealed that (25 and 40) µg protein were the most appropriate concentrations for the binding of f-PSA and c-PSA respectively. From these results, it could be concluded that the binding of 125I- anti total PSA antibody with its isolated f-PSA needed lower amounts of these antigens to get the equilibrium compared with the amounts required for isolated c-PSA. This may be due to cell hyper proliferation or increase in these antigen affinities for antibodies.

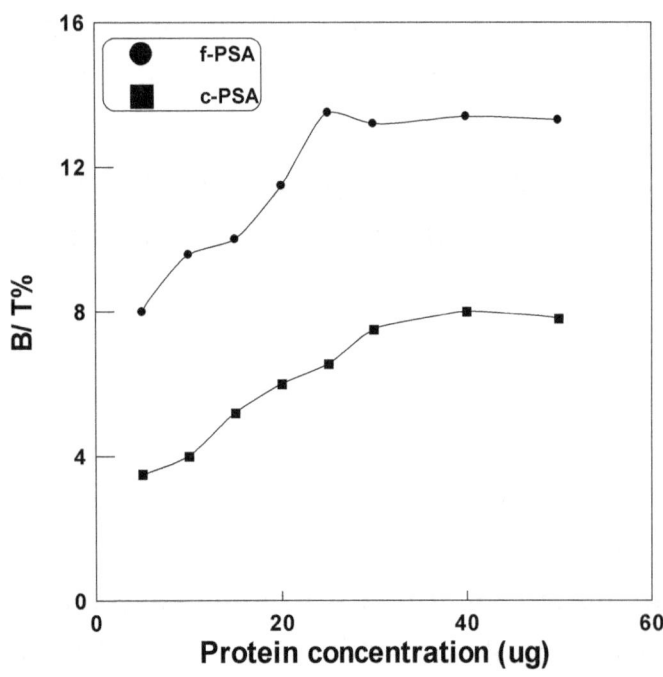

Figure (142): Influence of isolated f-PSA and c-PSA concentration (protein) in benign breast tumor homogenates on the binding with ^{125}I- anti total PSA antibody. (All other details are explained in the text).

According to these results, in all subsequent experiments, (25 and 40μg protein) of isolated f-PSA and c-ACT antigen respectively were used, since they give the highest binding.

Influence of 125I- anti total PSA antibody on the binding

Figure (143) shows that isolated f-PSA and c-PSA of benign breast tumor were saturated with ^{125}I- anti total PSA antibody concentrations equal to (100 and 200) μg respectively. From these results, it was found that isolated f-PSA fractions were saturated with smaller concentrations of ^{125}I- anti total PSA antibody than those required for isolated c-PSA. Thus it was concluded that isolated f-PSA had higher affinity at low concentrations toward ^{125}I- anti total PSA antibody than isolated c-PSA.

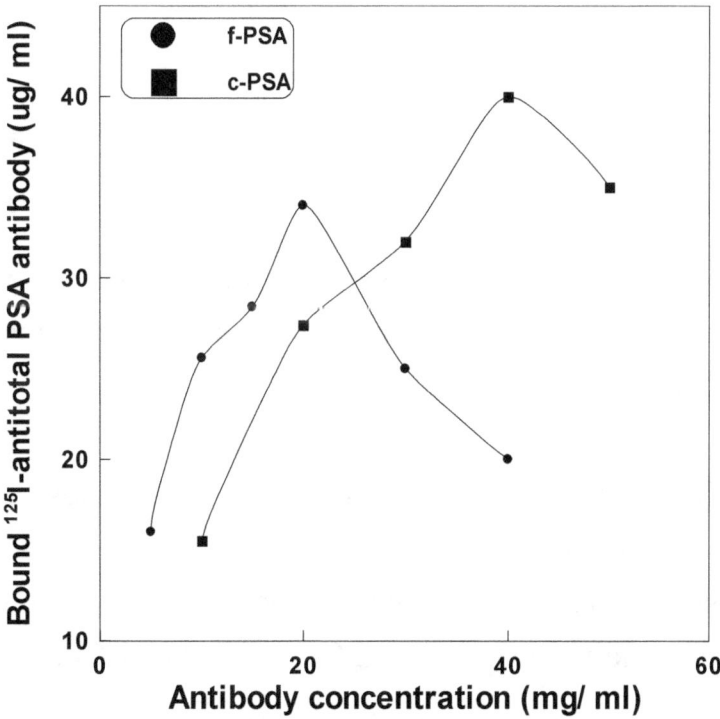

Figure (143): Effect of different concentrations of ^{125}I- anti total PSA antibody on the binding with each of isolated f-PSA and c-PSA in benign breast tumor homogenates. (All other details are explained in the text).

According to these results, in all subsequent experiments, (100 and 200μg protein) of ^{125}I- anti total PSA antibody respectively were used, since they give the highest binding.

Effect of pH on the 125I- anti total PSA antibody binding

Figure (144) shows the effect of increasing pH on the binding of ^{125}I- anti total PSA antibody to its isolated f-PSA and c-PSA of benign breast tumor. The results revealed that the optimum pH for isolated f-PSA and c-PSA fractions for the binding with ^{125}I- anti total PSA antibody was 7.8 and 8 respectively.

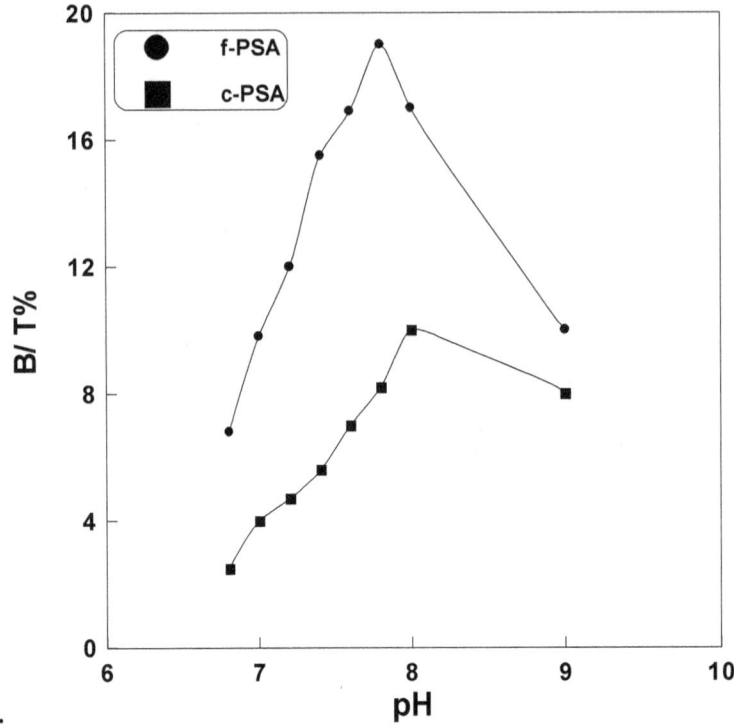

Figure (144): Effect of pH on the binding of ^{125}I- anti total PSA antibody with each of isolated f-PSA and c-PSA in benign breast tumor homogenates. (All other details are explained in the text).

These differences in the optimum pHs may suggest the differences in the binding sites of these isolated antigens.

According to the results obtained, the pH of the buffer used in all subsequent experiments were adjusted to 7.8 and 8.0 for isolated f-PSA and c-ACT antigen respectively.

Temperature dependency of the binding

The temperature dependency of the ^{125}I- anti total PSA antibody binding to its isolated antigens was investigated. Figure (145) shows that the optimum temperature of the binding of ^{125}I- anti total PSA antibody was 4°C with each isolated f-PSA and c-ACT antigen.

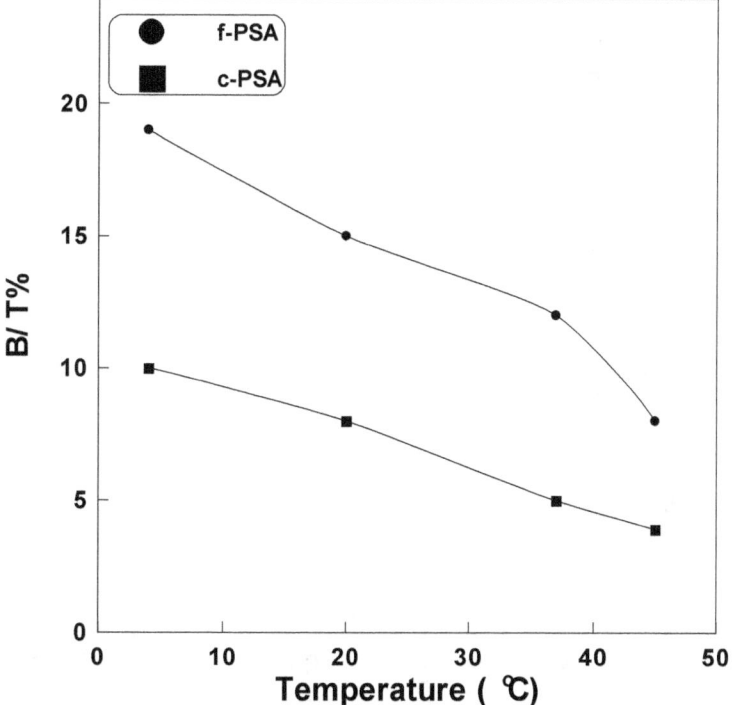

Figure (145): Effect of temperature on the binding of ^{125}I- anti total PSA antibody with each of isolated f-PSA and c-PSA in benign breast tumor homogenates. (All other details are explained in the text).

In general the loss of binding activity above the optimum temperature of each isolated f-PSA and c-ACT antigens may be due to degradation of these antigen molecules or to the irreversible dissociation of the antibody-antigen complexes.

In view of these results, the temperature used in all subsequent experiments was 4 °C

Time dependency of the binding

Figure (146) shows that at 4°C the apparent equilibrium of the ^{125}I- anti total PSA antibody binding reached within 180minutes for both isolated f-

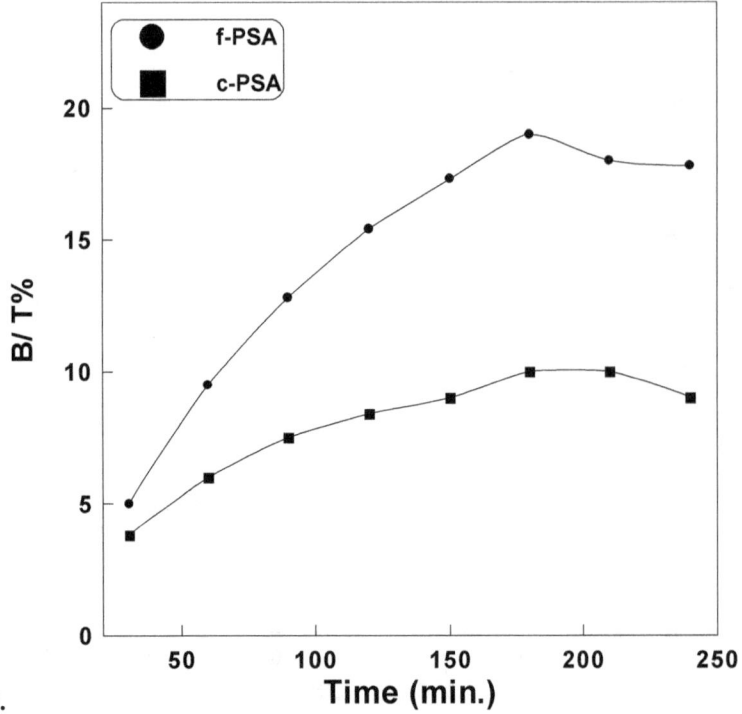

PSA and c-ACT antigens.

Figure (146): Effect of time on the binding of ^{125}I- anti total PSA antibody with each of isolated f-PSA and c-PSA in benign breast tumor homogenates. (All other details are explained in the text).

Spectroscopic Studies of PSA and its Complex

Introduction

Molecules absorb light; the wavelengths that are absorbed and efficiency of absorption depend on both the structure and environment of the molecule making absorption spectroscopy a useful tool for characterizing both small and large molecule.

The wavelength range (210-300) nm is the most important one for the protein characterization. This region is related to tryptophan, tyrosine, phenylalanine and histidine amino acid residues, which considered to be chromophores . Changes in the charge or the environment of these chromophores can lead to alteration in the absorption spectrum, and the conformational changes of a protein may also involve environmental changes of its chromophoric groups .

The electronic transitions for these chromophors come from $n\pi \to \pi^*$ and $\pi \to \pi^*$. These transitions are affected by many factors, which in turn affected the electronic environment that surround these transitions. Among these factors is the pH of the solvent, which determines the ionization state of ionizable chromophore. Also, the solvent polarity or perturbing agent affect the chromophore electronic transition where the λ_{max} for $\pi \to \pi^*$ transition occurs at shorter wavelength (blue shift) in polar erotic solvents (H_2O, alcohol) than in longer wavelength (red shift) [252]. The shift may or may not be accompanied by a change in intensity of the spectrum. Thus absorbance measurements can give an idea of the location of particular amino acids in a protein. Furthermore, information about the amino acids that are in a binding site is invariably valuable to the enzymologist to determine the reaction mechanism of an enzyme.

Although several new immunochemical techniques were developed to study such interactions, UV spectral remain as one of the most important methods in immunology because it provides a sensitive and quantitative measurements for the study of antibody structure and its specific ligand binding[256,250].

This work is planned to study the spectroscopic behavior of standard PSA equipped with the supplied kit and a complex prepared for this standard PSA antigen.

. Spectroscopic studies

. The U.V Spectrum of PSA, 125I-anti total PSA antibody and (125I-anti total PSA antibody / PSA) Complex

The U.V spectrum of human PSA

1. One hundred microliters of human PSA (5.0 ng/ml) provided by (total PSA IRMA kit from Immunotech- Beckman Coulter company Czech Republic) was completed to 500µl with Tris/HCl buffer pH 7.2. Then placed in a 0.5 cm cuvette in sample beam

2. The absorption spectrum was immediately measured against the same buffer in reference beam in the area of (200-320 nm).

U.V Spectrum of 125I - anti total PSA anti body

3. Ten microliters ^{125}I-anti total PSA antibody (50μg protein) was completed to 500μl with Tris/HCl buffer pH 7.2. Then placed in a 0.5 cm cuvette in sample beam

4. The absorption spectrum was immediately measured against the same buffer in reference beam in the area of (200-320 nm).

U.V Spectrum of (125I - anti total PSA anti body / PSA) complex

2. **One hundred microliters (500μg protein) of 125I- anti total PSA antibody was added to 100μl (100ng/ml) of standard PSA provided by (total PSA IRMA kit from Immunotech- Beckman Coulter company Czech Repoblic) and the volumes were completed to (500μl) with Tris/HCl buffer pH 7.2.**

The (125I-anti total PSA antibody / PSA) complex was estimated by following the steps 2, 3, 4, 5, 6 and 7 in section (3.3.4.) with the kit conditions.

The fractions that contained high levels of the binding activity of (^{125}I-anti total PSA antibody/ PSA) complex represented in peak 1 were collected and pooled.

One hundred microliters of (^{125}I-anti total PSA antibody/PSA) complex was completed to 500μl with Tris/HCl buffer pH 7.2. Then placed in a 0.5 cm cuvette in sample beam

Calculations

1. The radioactivity (c.p.m) of each eluted fraction was plotted against fraction number.

2. The absorption spectrum was measured against the same buffer in reference beam in the area of (200-320 nm).

Factors Effecting the Absorption Properties of PSA and (125I-anti total PSA antibody / PSA) Complex

pH Effect

7. One hundred microliters of human PSA (5.0 ng/ml) provided by (total PSA IRMA kit from Immunotech- Beckman Coulter company Czech Repoblic) was completed to 500μl with Tris/HCl buffer at different

pH (4, 7.2, 9 and 12.5). The pH was adjusted using 1N HCL Then placed in a 0.5 cm cuvette in sample beam

8. The absorption spectrum was immediately measured against the same buffer in reference beam in the area of (200-320 nm).

9. The experiment was repeated using 100μl of (^{125}I-anti total PSA antibody/ PSA) complex.

The effect of solvent polarity

2. One hundred microliters of human PSA (5.0 ng/ml) provided by (total PSA IRMA kit from Immunotech- Beckman Coulter company Czech Repoblic) was completed to 1000μl with Tris/HCl buffer pH 7.2 containing 20% ethanol Then placed in a 0.5 cm cuvette in sample beam.

2. The step 1 was repeated using 20% of each of ethylen glycol, glycerol and Dimethylsulphoxide (DMSO)

10. The absorption spectrum was immediately measured against the same buffer in reference beam in the area of (200-320 nm).

11. The absorption of human PSA in Tris/HCl buffer pH 7.2 was immediately measured against the λ_{max} of human PSA in each solvent.

12. The experiment was repeated using 100μl of (^{125}I-anti total PSA antibody/ PSA) complex.

Spectrophotometic pH Titration of PSA and (125I-anti total PSA antibody / PSA) Complexes

1. One hundred microliters of human PSA (5.0 ng/ml) provided by (total PSA IRMA kit from Immunotech-Beckman Coulter company Czech Republic) was completed to 500μl with different buffers at different pH ranging 4 to 8. The maximum absorbency of each sample was measured at a wavelength of 211nm.

2. One hundred microliters of human PSA (5.0 ng/ml) provided by (total PSA IRMA kit from Immunotech-Beckman Coulter company Czech Republic) were completed to 500μl with different buffers at different pH ranging 9 to 12.5. The maximum absorbency of each sample was measured at a wavelength of 295nm.

3. The experiment was repeated using 100μl of (^{125}I-anti total PSA antibody/ PSA) complex.

Calculations

The absorbance of λ_{max} at each pH value was plotted versus corresponding pH.

Results and Discussion

The U.V Spectrum of PSA, 125I-anti total PSA antibody and (125I-anti total PSA antibody / PSA) Complexes

The UV spectra of h-PSA and ^{125}I-anti total PSA antibody complex were scanned from 200-320 nm to determine the absorption spectra, and the alternation in the UV spectra as a results of their interaction.

The U.V spectrum of human PSA

Figure (147) illustrates the U.V spectrum of human PSA (provided by total PSA IRMA kit from Immunotech- Beckman Coulter company Czech Republic) at pH 7.2. The spectrum shows that the λ_{max} for PSA is consisted of 2 peaks, at 229 nm and at 277 nm. Both peaks may be assigned to tyrosine residue and located in a way, that large part of it is on the surface of the protein molecule while the other part is buried.

As a result human PSA has a characteristic spectrum and can be identified by its peaks.

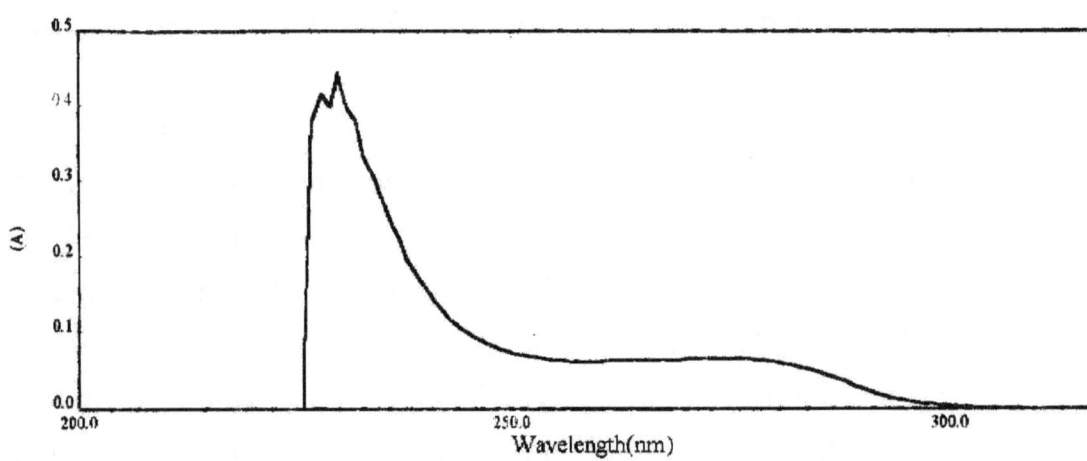

Figure (147): The U.V spectrum of human PSA at pH 7.2. (All other details are explained in the text).

The U.V Spectrum of 125I - anti total PSA anti body

Figure (148) illustrates the U.V spectrum of ^{125}I-anti total PSA antibody at pH 7.2. The spectrum shows that the λ_{max} for ^{125}I-anti total PSA antibody is

consisted of 2 peaks, at 230nm and at 270 nm. Which are assigned to the tyrosine and phenylalanine respectively. It seems that each of tyrosine and phenylalanine residue in the ^{125}I - anti total PSA anti body molecule is located in a way, that part of it is on the surface of the protein molecule while the other part is buried.

As a result ^{125}I - anti total PSA anti body has a characteristic spectrum and can be identified by its peaks.

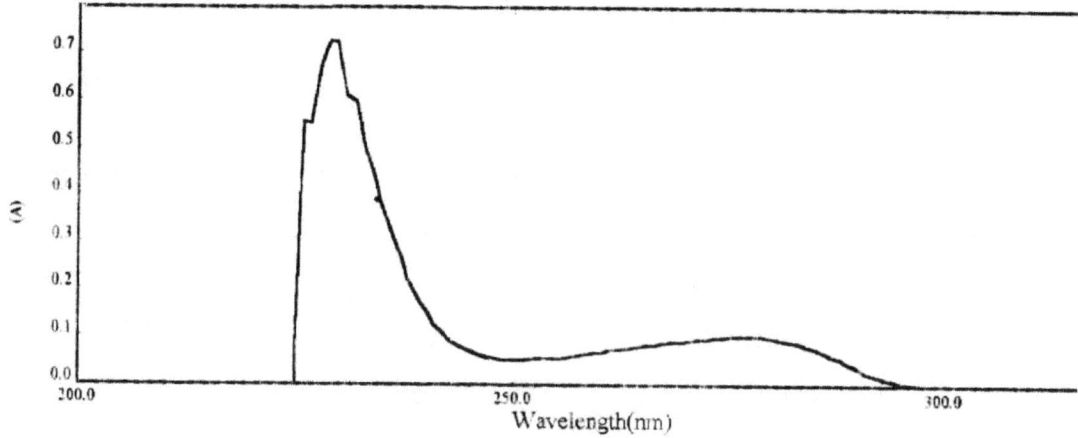

Figure (148): The U.V spectrum of ^{125}I - anti total PSA anti body at pH 7.2. (All other details are explained in the text).

. The U.V Spectrum of (125I - anti total PSA anti body / PSA) complex

Figure (149) shows the elution profile of the incubation solution, which had contained PSA (provided by total PSA IRMA kit from Immunotech-Beckman Coulter company Czech Republic), using sepharose CL-6B (0.96 × 29 cm) column.

Figure (149) shows two peaks of radioactivity were appeared. The broad one or peak (1) represents the (^{125}I-anti total PSA antibody/PSA) complex.

Moreover, the high molecular weight of the first peak, which was expected from its low retention volume (10 ml) may be attributed to the same reason.

The second peak (2) in figure (5-3) may represent the free ^{125}I- anti total PSA antibody.

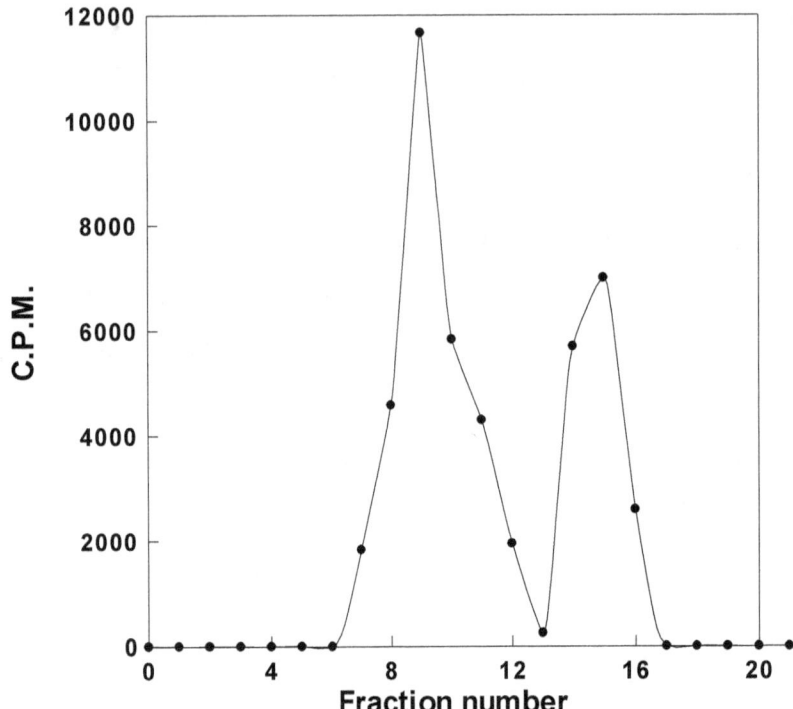

Figure (149): The elution profile of the complex of ^{125}I- anti total PSA antibody with human PSA. (All other details are explained in the text).

Figure (150) illustrates the U.V spectrum of ^{125}I-anti total PSA antibody/PSA complex at pH 7.2. The spectrum shows that the λ_{max} for ^{125}I- anti total PSA antibody is consisted of 2 peaks, at 234nm and at 283 nm. Which are assigned to the tyrosine and tryptophan respectively. It seems that each of tyrosine and tryptophan residues in the (^{125}I - anti total PSA anti body/PSA) complex molecule is located in a way, that a small part of tyrosine and large part of tryptophan residues are on the surface of the protein molecule while a large part of tyrosine and small part of tryptophan residues are buried.

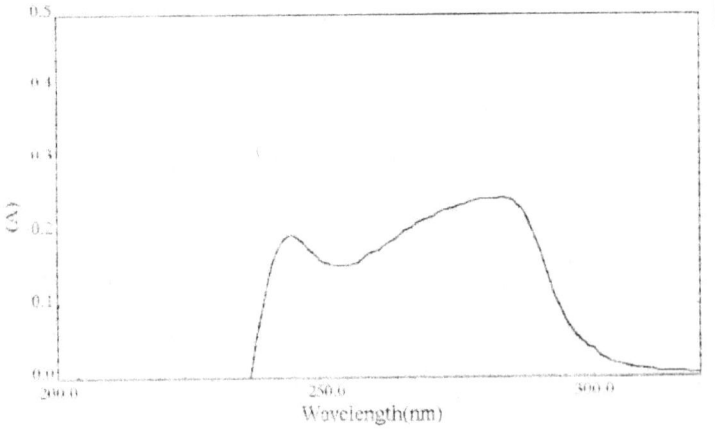

Figure (150): The U.V spectrum of (^{125}I - anti total PSA anti body/PSA) complex at pH 7.2. (All other details are explained in the text).

As a result (^{125}I-anti total PSA antibody/PSA) complex has a characteristic spectrum and can be identified by its peaks.

The spectrum of (^{125}I-anti total PSA antibody/PSA) complex as comparison with PSA indicated the disappearance of one tyrosine peak for PSA and appearance of the tryptophan peak for the (^{125}I - anti total PSA antibody/PSA) complex. Conformation changes may occur that effect the absorption of the PSA. These changes are due to the fitting of antibody to its antigen to form (^{125}I - anti total PSA antibody/PSA) complex. These results are agreed with finding of Seinerman et al who found that the surface of protein interactions are polar as well and the complex formation lead to the burial of charged and polar residues.

Factors Effecting the Absorption Properties of PSA and (125I-anti total PSA antibody / PSA) Complex

The absorption spectrum of a chromophore is primarily determined by the chemical structure of the molecule. However, a large number of environmental factors produce detectable changes in λ_{max} and ε.

The general features of these environmental effects are the following:

pH Effect

The pH of the solvent determines the ionization state of the ionizable chromophore in the protein molecule. The UV spectrum of PSA and (^{125}I-anti total PSA antibody/PSA) complex were determined at different pH (4, 7.2, 9 and 12.5) and λ_{max} was obtained as shown in table (75).

In a neutral pH (7.2), Two λ_{max} were obtained for each PSA and (^{125}I-anti total PSA antibody/PSA) complex, λ_{max1} and λ_{max2} of PSA were (229 and 277) nm respectively whereas λ_{max1} and λ_{max2} of (^{125}I-anti total PSA antibody/PSA) complex were (234 and 283) nm respectively.

In an acidic pH (4), there were decreases in both λ_{max1} of each PSA and (^{125}I-anti total PSA antibody/PSA) complex and λ_{max2} of PSA with decreasing in pH. The blue shift has been observed in absorption of tyrosine residue may be attributed to conformational changes and chromophore in native PSA and (^{125}I-anti total PSA antibody/PSA) complex are on the surface of their . On the other hand the blue shift may be due to the increasing of hydrogen bond formed in the presence of highly positively charged state [260]. The λ_{max2} of the (^{125}I-anti total PSA antibody/PSA) complex remain almost constant. This result prove that the absorbance is related to tryptophan residue where the change in pH will no affect the ionization state of this residue and certainly the absorbance will no be changed.

When the pH value was increased from (7.2 to 9), there were an increase in both λ_{max1} of each PSA and (^{125}I-anti total PSA antibody/PSA) complex and λ_{max2} of PSA. The red shift has been observed in absorption of tyrosine residue, and this certainly related to the ionization of side chain of the tyrosine and this led to availability of the lone pair on the oxygen atom to be happened easier and at lower energy level (red shift). The λ_{max2} of the (^{125}I-anti total PSA antibody/PSA) complex remain almost constant. This result prove that the absorbance is related to tryptophan residue where the change in pH will no affect the ionization state of this residue and certainly the absorbance will no be changed.

At pH 12.5 no absorbance was recorded. The disappearance of both λ_{max1} of each PSA and (^{125}I-anti total PSA antibody/PSA) complex and λ_{max2} of PSA of tyrosine residue and λ_{max2} of the (^{125}I-anti total PSA antibody/PSA) complex of tryptophan residue due to conformational changes and chromophore in native complex were buried in the interior of their complexes.

Table (75): Effect of increasing pH on the λmax of PSA and (^{125}I - anti total PSA antibody/PSA) complex. (All other details are explained in the text).

pH	λ_{max} (nm)	
	Standard PSA	(^{125}I - anti total PSA antibody/PSA) complex
4	225, 274	227, 284
7.2	229, 277	234, 283
9	232, 279	235, 284
12.5	--	--

It must be noted that the spectral shifts of proteins produced by pH cannot be simply attributed to the inductive effects of vicinal charges, such spectral changes must therefore be attributed mainly to rearrangements of secondary and tertiary structure, although the possibility of field effects due to unusually close conjunction of charges to aromatic groups is not excluded .

The Effect of Solvent Polarity

Table (76) shows the effect of 20% ethanol and ethylene glycol at neutral pH on the human PSA and (^{125}I-anti total PSA antibody/PSA) complex spectra.

The data obtained previously show that the λ_{max} of PSA and (^{125}I-anti total PSA antibody/PSA) complex at neutral pH were (229 and 277) nm and (234 and 283) nm respectively. The λ_{max} value of tyrosine in PSA and (^{125}I-anti total PSA antibody/PSA) complex was shifted towards longer wavelengths (red shift) in 20% of each ethanol and ethylene glycol due to the hydrogen bonding of the OH groups of tyrosines with the solvent or with the π-electron system of the benzene ring where tyrosine was functioned as a hydrogen donor, while the λ_{max} value of tryptophan in (^{125}I-anti total PSA antibody/PSA) complex was shifted towards shorter wavelengths in 20% of each ethanol and ethylene glycol, this shift was attributed to $\pi \rightarrow \pi^*$ transitions. These two shifts in λ_{max} were accompanied with an increase in the absorbency of tryptophan and a decrease in the absorbency of tyrosine, these findings could be attributed to a change in the protein structure that bring the tryptophan residues to the surface of the protein while tyrosine residues were partly embedded in a hydrophobic region of the protein molecule.

The λ_{max} of tyrosine residues in PSA and (^{125}I-anti total PSA antibody/PSA) complex were shifted towards longer wavelengths in 20% glycerol without affecting the structure of these PSA and (^{125}I-anti total PSA antibody/PSA) complex, while the λ_{max} of tryptophan residues in (^{125}I-anti total PSA antibody/PSA) complex was not affected. The shift indicates that at 20% glycerol, the exposed tyrosines become solvated with glycerol (dipole–dipole interaction).

Also table (76) shows the λ_{max} of PSA and (^{125}I-anti total PSA antibody/PSA) complex in 20% dimethylsulfoxide at neutral pH. It was found that two newer λ_{max} were appeared for PSA, at (263.2 and 293) nm which were assigned to phenylalanine and tryptophan residues respectively, while in (^{125}I-anti total PSA antibody/PSA) complex, a newer λ_{max} was appeared at 266.4nm which was assigned to histidine residue and the λ_{max} value of tryptophan was shifted towards longer wavelengths (red shift).

The appearance of these new λ_{max} values indicates that the protein was defolded due to change in the secondary and tertiary structure of the protein that bring the phenylalanine and tryptophan residues for PSA and histidin residues for (^{125}I-anti total PSA antibody/PSA) complex to expose to absorbance while tyrosine residue was buried inside each of PSA and (^{125}I-anti total PSA antibody/PSA) complex molecule, also it was found that PSA and (^{125}I-anti total PSA antibody/PSA) complex are highly sensitive to change in the polarity of the solvent.

Table (76): Effect of solvent polarity on the λmax of PSA and (^{125}I - anti total PSA antibody/PSA) complex. (All other details are explained in the text).

Solvent	λ_{max} (nm)	
	Standard PSA	(^{125}I - anti total PSA antibody/PSA) complex
Ethanol	233.4, 280.1	238.24, 281.55
Ethylene glycol	235.41, 281.09	240.22, 278.68
Glycerol	236.02, 282.81	237.12, 283.3
DMSO	263.2, 293	266.4, 291.3

Spectrophotometic pH titration of PSA and (125I-anti total PSA antibody / PSA) complex

Spectrophotometric pH titration is the following of the change in absorbance of the chromophore with increasing pH . Many studies of protein structure require the determination of pk values for proton dissociation from ionizable amino acid side chains, because these values give an indication of the location of the amino acid in the protein. This can often be done spectrophotometrically because dissociation often

changes the spectrum of one of the chromophores, the observation of tyrosine dissociation was performed by measuring the absorption at 295 nm (λ_{max} for the ionized form of tyrosine), and the observation of histidine dissociation was carried out by measuring the absorption at 211 nm.

Figure (151 A&B) shows the pH titration curves of PSA and (^{125}I-anti total PSA antibody/PSA) complex for tyrosine and histidine respectively.

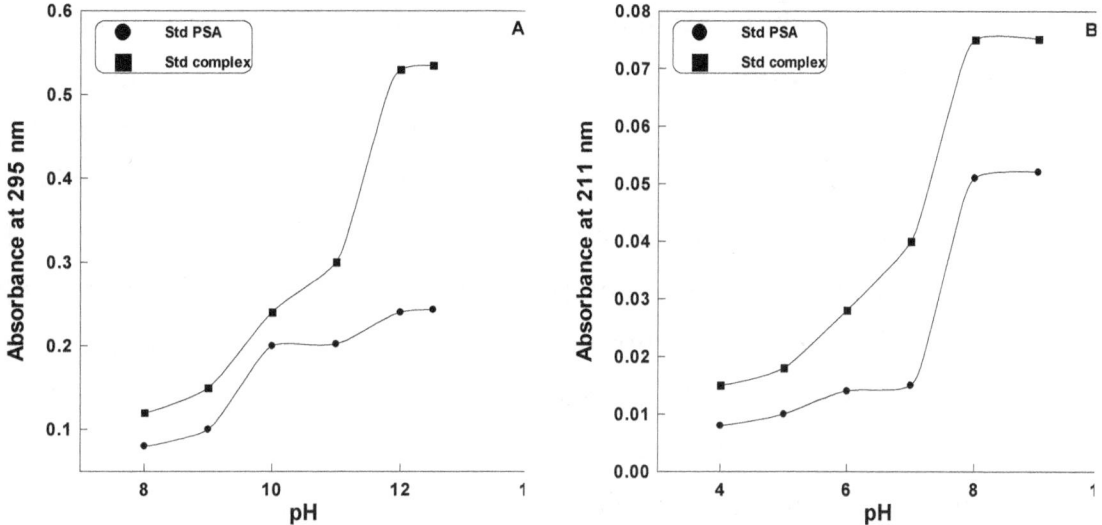

Figure (151): Spectrophotometric pH titration of PSA and (^{125}I-anti total PSA antibody / PSA) Complex for:

        A. Tyrosine residue

        B. Histidine residue

(All other details are explained in the text).

The (A) curves show that the pk$_a$ values for tyrosine are 9.2 and 11.2 for PSA and (^{125}I-anti total PSA antibody/PSA) complex respectively, while the pk$_a$ values for histidine in (B) curves were equal to 5.5 and 7.3 for PSA and (^{125}I-anti total PSA antibody/PSA) complex respectively. From the same figure, it was found that:

    1. About 83.3 and 45.2% of tyrosine residues are located on the surface of the PSA and (^{125}I-anti total PSA antibody/PSA) complex molecule respectively.

    2. About 16.7 and 54.8% of tyrosine residues are buried interior the folded structure of the PSA and (^{125}I-anti total PSA antibody/PSA) complex respectively.

3. About 27.4 and 37.3 % of histidine residues is located on the surface of the PSA and (^{125}I-anti total PSA antibody/PSA) complex molecule respectively.

4. About 72.6 and 62.7% of histidine residues is embedded in the interior region of the PSA and (^{125}I-anti total PSA antibody/PSA) complex molecule respectively.

5. In PSA the tyrosine residues were largely present on the surface of the molecule and the internal tyrosines are in a strongly nonpolar environment, while the internal tyrosine residues in (^{125}I-anti total PSA antibody/PSA) complex were in a strongly polar environment (e.g. a tyrosine surrounded by carboxyl groups). On the other hand, the histidine residues are slightly present on the molecular surface of PSA and (^{125}I-anti total PSA antibody/PSA) complex and the internal histidine residues of PSA are in a nonpolar environment whereas the internal histidine residues of (^{125}I-anti total PSA antibody/PSA) complex are likely to be in a strongly polar environment.

6. Finally, the percent of external tyrosine residues in PSA was greater than that of (^{125}I-anti total PSA antibody/PSA) complex and the percent of internal tyrosine in (^{125}I-anti total PSA antibody/PSA) complex was greater than that of PSA. On the other hand the percent of internal histidine in PSA was greater than that of (^{125}I-anti total PSA antibody/PSA) complex.

Solvent perturbation studies

The determination of whether an amino acid is internal or external by measuring the spectra of a protein in polar and nonpolar solvents is called the solvent perturbation method. In fact, proteins are rarely studied in completely nonpolar solvents because most proteins are either insoluble or denaturated in these solvents; therefore, mixtures of 80% water and 20% of reduced polarity solvent were used . Solvents alter the peak positions and intensities by altering the energy and probability of electronic transitions and this alteration arises from a difference in the salvation energies of the ground state and the first excited singlet state .

Table (77) shows the λ_{max} and ΔA values in the presence of different perturbants (20% ethanol, 20% ethylene glycol, 20% glycol and 20% DMSO).

Table (77): Solvent perturbation on PSA and (^{125}I-anti total PSA antibody / PSA) Complex. (All other details are explained in the text).

Perturbent substance	PSA			Complex		
	λ_{max}	A	ΔA	λ_{max}	A	ΔA
Ethanol	233.4	1.35	0.84	238.24	0.5	0.35
	280.1	0.9	0.83	281.55	0.57	0.37
Ethylene glycol	235.41	1.15	0.85	240.22	0.55	0.35
	281.09	0.94	0.84	278.68	0.43	0.38
Glycerol	236.02	1.15	0.81	237.12	0.6	0.34
	282.81	087	0.82	283.3	0.48	0.32
DMSO	263.2	0.25	0.15	266.4	0.37	0.27
	293	0.34	0.3	291.3	0.45	0.4

From the results listed in the Tables, it was found that several spectral changes were obtained in the presence of these perturbants, like the alteration of the λ_{max} positions and intensities of PSA and (^{125}I-anti total PSA antibody/PSA) complex spectra, and the appearance of new chromophores on the surface of each the PSA and (^{125}I-anti total PSA antibody/PSA) complex molecule. These chromophores were embedded in an interior region of the protein in the absence of solvent.

From the solvent perturbation studies, the following remarks could be drawn:

About 0.83, 0.35 of tyrosine residues, 0.15 and 0.27 of phenylalanine residues and 0.3 and 0.4 of tryptophan residues are on the surface of PSA and (^{125}I-anti total PSA antibody/PSA) complex molecule respectively.

From spectrophotometic pH titration and solvent perturbation studies we conclude that

About 83, 30, 15 and 25% of tyrosine, tryptophan, phenylalanine and histidine residues are on the surface of PSA molecule respectively, whereas about 39, 40, 27 and 36 % of tyrosine, tryptophan, phenylalanine, histidine and tryptophan residues are on the surface of (^{125}I-anti total PSA antibody/PSA) complex molecule respectively.

Finally if assumed that human PSA that used in spectrophotometric studies consisted only f-PSA that composed 4, 7, 6 and 11 of tyrosine, tryptophan, phenylalanine and histidine residues respectively [75], then about 3, 2, 1 and 3 of tyrosine, tryptophan, phenylalanine and histidine residues are on the surface of PSA molecule respectively, while about 1, 5, 5 and 10 of tyrosine, tryptophan, phenylalanine and histidine residues are buried interior the folded structure of the PSA molecule. Accordingly, from figure (5-2A&B) by dividing the maximum absorbency of PSA from the maximum absorbency of (^{125}I-anti total PSA antibody/PSA) complex and multiplied by the number of tyrosine and histidine residues in f-PSA concluding that (^{125}I-anti total PSA antibody/PSA) complex was composed 9 and 16 of tyrosine and histidine residues respectively, then about 4, and 6 of tyrosine and histidine residues are on the surface of PSA molecule respectively, while about 5, and 10 of tyrosine and histidine residues are buried interior the folded structure of the PSA molecule.

References

1. Disaia, P.J., and Creasman, W.T. (1989) In Breast Diseases and Colorectal Cancer in Clinical Gynecologic Oncology. Third ed., pp. 1-5, The C.V. Mosby Company, Toronto.

2. Parkin, D.M, Idara F., and Muir, C.S. (1988) Int. Cancer **41**, 184-97.

3. Coleman, M.P., Esteve, J., Damiecki, P., Arston, A., and Renard, H. (1993) Trends in cancer incidence and mortality, Lynos: International Agency for Research on Cancer (IARC. Scientific Publication No. 121).

4. Steven, A., Schroeder, Marcus, A., Krupp, Lawrence, M., Tierney, and Stephen, J. Mcphee (1989) In Current Medical Diagnosis & Treatment, pp 429-435.

5. Wooster, K., Neunausen, S.L., Mangion, J. et al (1994) Science **265**, 2088-2090.

6. Nowak, R. (1994) Science **265**, 1796-1799.

7. McPherson, K., Steel, C.M., and Dixon, J.M. (1994) BMJ **309**, 1003-1007.

8. Wealheral, Ledinglaic, and Warrel (1984) Oxford Text Book of Medicine, Oxford University press.

9. Anderson, D.E., and Badzioch, M.D. (1985) Cancer **56**, 383.

10. Gilbertsen, V.A. (1975) Semin. Oncol. **1**, 87-90.

11. Canellos, G.P., Hellman, S., and Veronesi, U. (1982) N. Engl. J. Med. **306**, 1430.

12. Calson, H.E. (1980) N. Engl. J. Med. **303**, 795-799.

13. Trialists Collaborative Group (1992) In: Early Breast Cancer Lancet 339, 1-15, 71-85.

14. Singleton, W.V., and Mecarty, K.S. (1987) Gyncol. Oncol. **26**, 271-275.

15. Gelber, R.D., and Goldhirsch, A. (1986) J. Clin. Oncol. **4,** 1696.

16. Donegan, W.L., and Spratt, J.S. (1979) In Cancer of the breast, W.B. Saunders, Philadelphia.

17. Horris, J.R, Leven, M.B., and Helman, S. (1987) Semin Oncol. **5**, 403.

18. Butta, A., Maclennan, K., Flanders, K.C., and et al. (1992) Cancer Res. **52**, 4261.

19. Slevin, M.L., Stubbs, L., Plant, H.J., and et al. (1990) Br. Med. J. **300,** 1458.

20. Wilson, J.D., Aiman, J., and MacDonald, P.C. (1980) Adv. Inter. Med. **25**,1.

21. Bailey & Love's, In Short Practice of Surgery, 21st edition (1991) pp. 806-810.

22. Helman, S. (1983) In Controversies in Breast Cancer, Conference Sponsored by the MD Aderson Hospital and Tumor Institute, Chicago, Year Book, Medical Publishers, Inc.

23. Chetty, U. (1979) Br. J. Surg. **67**, 789.

24. Humphrey, U. et al (1983) Contemp Surg. **23,** 97.

25. Cole, P., Elwood, J.M., and Kaplan, S.D. (1978) AM. J. Epidemiol. **108**, 112.

26. Oluwole, S.F., and Freeman, H.P. (1979) AM. J. Surg **37**, 786.

27. Rao, B.R.(1981) Cancer **47**, 2016.

28. Devitt, J.E. (1972) Surg. Gynecol Obstet. **134**, 803.

29. Yamamota, K. (1985) Ann. Rev. Genet. **19**, 209-252.

30. Reddik, K, and Holland, J.F. (1976) Proc. Natl. Acad. Sci. USA, **73**, 2308.

31. Wood, C.H., Varela, V., and Palmquist, M. et al (1977) J. Surg. Oncol. **5**, 251.

32. Secreto, G., Recchione, C., and Caralleri, A., et al (1983) Br. J. Cancer, **47**, 269.

33. Sharon, P. (1984) Scientific American **6**, 86-95.

34. Handas, S., and Nakamura, K. (1984) J. Biol. Chem. **95**, 1323-9.

35. Magdelenal, I.I. (1992) J. Immunol Meth. **150**, 133-143.

36. Suresh, M.R.(1944): Cancer markers In: The immunoassay handbook (Wild, D. ed.) pp. 441-460 NY. Stockton Press.

37. Suresh, M.R., Noujam, A.A., and Longeneeker, B.M. (1991): Recent developments in monoclonal antibodies In: biotechnology Current Progress, Vol. **1**, (Cheremisinoff, P.N., and Ferrante, L.M. eds), pp. 83-101, USA Technomic Publ.

38. Sano, T, Smith, L.S., and Cantor, C.R. (1992) Conjugates Science **258**, 120-122.

39. Suresh, M.R. (1996) Anticancer Research **16**, 2273-2278.

40. Suresh, M.R. (1991): Immunoassay for cancer-associated carbohydrate antigens In, Seminars in cancer Biol., Vol. **2**, pp. 367-377, London W.D. Saunders Publ.

41. American Society of Clinical Oncology (1996) J. Clin. Oncol. **14**, 2843-2877.

42. Greene, G.L., Sobel, N.B., and King, W.J. et al (1984) J. Steroid Biochem. **20**, 51-56.

43. Green, S., Gronemeyer, H., and Chambon, P. (1987) In Structure and function of steroid hormone receptors in Sluyser M. ed.: Growth Factors and Oncogenes in Breast Cancer, England, Mlis pp. 728, Horwood.

44. Walter, P., Green, S., and Green, G. et al. (1985) Proc. Nail. Acad. Sci. U.S.A. **82**, 7889-7893.

45. Orti, E., Bodwell, J.I., and Munck, A. (1992) Endocr. Rev. **13**, 105-128.

46. Ruh, T.S., Ruh, M.F., and Singh R.K. (1988) In Nuclear acceptor Sites: Interaction with estrogen-versus antiestrogen-receptor complexes, in Moudgil V.K. ed.: Steroid Receptors in Health and Disease, pp. 233-250, New York, N.Y. Plenum.

47. Bauer, K.D., Bagwell, C.B., and Giaretti, W. et al. (1993) Cytometry **4**, 486-491.

48. Tandon, A.K., Clark, G.M., and Chamness, G.C. et al. (1990) N. Engl. J. Med. **322**, 297-302.

49. Reid, P.E., Culling, C.F.A., and Dunn, W.L. (1978) J. Histochem. Cytochem. **26**, 187-192.

50. Culling, C.F.A., and Reid, P.E. (1979) J. Histochem. Cytochem. **27**, 1177-1179.

51. Bouchier, I.A.D., and Clamp, J.R. (1971) Clin. Chim. Acta., **35**, 219-224.

52. Horowitz M.I.(1977), "Gastrointestinal glycoproteins' In: "The glycoconjugates" eds., Horowitz M.I. and Pigman W., 1st ed. Vol. **1**, pp. 189-213 Academic Press.

53. Tuppy, H., and Gottschalk, A. (1972) "The structure of sialic acids and their quantitation" In: "Glycoproteins" ed. Gottschelk A., 2nd ed., pp. 403-449 Amsterdam.

54. Schauer, R. (1982) Adv. Carbohydr. Chem. Biochem., Vol. **40**, in press.

55. Schauer, R. (1978) Methods Enzymol. **50C**, 64-89.

56. Buscher, H.P., Casals-Stenzel, J., and Schauner, R. (1974) Eur. J. Biochem. **50**, 71-82.

57. Reuter, G., Vielgenthart, J.F.G., Wember, M., Schauer, R. and Howard, R.J. (1980) Biochem. Biophys. Res. Commun. **94**, 567-572.

58. Sheshadri, N. (1994) Annals. of Clinical and Laboratory Science, Vol. **24** No. 4, 376-384.

59. Warren, L. (1959) J. Biol. Chem. **234**, 1971-5.

60. Shukla, A.K., and Schauer, R. (1981) Physiol. Chem. **362**, 236-7.

61. Shamberger, R.J. (1986) Anticancer Res. **6**, 717-20.

62. Hara, S., Takemori, Y., Yamaguchi, M., Nakamura, M., and Ohkura, Y. (1987) Anal. Biochem. **164**, 138-45.

63. Pigman W., (1977): Blood group glycoproteins" In: "The glycoconjugates" eds. Horowitz M.I. and Pigman W., 1st ed., Vol. **1**, pp. 181-188 Academic press.

64. Schwick, H.G., Heide, K., and Haupt, H. (1977) "Plasma" In: "The glycoconjugates" eds., Horowitz M.I. and Pigman W., 1st ed., Vol. **1**, pp. 261-321 Academic press.

65. Patel, V. (1978) "Degradation of glycoprotein" In: "the glycoconjugates" eds., Horowitz M.I. and Pigman W., 1st ed., Vol. **2**, pp. 185-229 Academic press.

66. Baumann, H. and Doyle, D. (1984) "Determination of carbohydrate structures in glycoproteins and glycolipids", Molecular and chemical characterization of membrane receptors, pp. 125-160.

67. Horowitz M.I., (1978) "Immunological aspect and lectins" In: "The glycoconjugates" eds., Horowitz M.I. and Pigman W., 1st ed., Vol. **2**, pp. 387-425 Academic press.

68. Yogeeswaran, G., and Salk, P. (1981) Science **212**, 1514-6.

69. Bahl, O.P., and Shah, R.H. (1977) "Glycoenzymes and glycohormons" In: "The glycoconjugate" eds., Horowitz M.I. and Pigman W., 1st ed., Vol. **2**, pp. 385-422 Academic press.

70. Schauer, R. (1985) Trends Biochem. Sci. **10**: 357-60.

71. Reid, P.E., Culling, C.F.A., Dunn, W.L., and Clay M.G. (1978) J. Histochem. Cytochem. **26**, 1033-1041.

72. AL-Suhail, A., Reid, P.E., Yeung, M., Corret, S., Frolich, J., and Brooks, D.E. (1981) "Serum level of sialic acid: Effect of age, smoking and heart disease". C.S.C.C. 25th annual convention No. 5.

73. McNeil, C., Berrett, C.R., Lucia, Y.S.U., Trentelman, E.F., and Helmick, W.M. (1965) Am. J. Cli. Path. **43**, 130-133.

74. Carter, A., and Martin, N.H. (1962) J. Clin. Path. **15**: 69-72.

75. Erbil, K.M., Jones, J.D., and Klee, G.G. (1986) Cancer **57**, 1889.

76. Khanderia, V., and Keller, J. (1983) J. Surg. Oncol **23**, 163-166.

77. Stefnelli, N., Klotz, H., Engel, A., and Bauer, P. (1985) J. Cancer Res. Clin. Oncol. **109**, 55-59.

78. Shamberger, R.J. (1984) J. Clin. Chem. Clin. Biochem. **22**, 647-651.

79. Silver, H.K.B., Rangel, D.M, and Morton, D.L. (1978) Cancer **41**, 1497-1499.

80. Brozmanova, E., and Skrovina, B. (1972) Neoplasma **19**, 115-123.

81. Silver, H.K.B., karim, K.A., Archibald, E.A., and Salinas, F.A. (1979) Cancer Res. **39**, 5036-5042.

82. Moss, A.J., Bissada, N.K., Bayed, C.M., and Hunter, W.C. (1979) Urology **13**, 182-184.

83. Silver, H.K.B., Karim, K.A., Salinas, F.A., and Swenterton, K.D. (1981) Surg. Gynecol. Obstect. **153**, 203-213.

84. Harvey, H.A., Lipton, A., White, D., and Davidson, E. (1981) Cancer, **47**, 324-327.

85. Bradely, W.P., Blasco, A.P., Weiss, J.F., Alxander, J.C., Silverman, N.A., and Chretien, P.B. (1977) Cancer, **40**, 2264-2272.

86. Culling, C.F.A., Reid, P.E., Burton, J.D., and Dunn, W.L. (1975) J. Clin. Pathol. **28**, 650-656.

87. Culling, C.F.A., Reid, P.E., Worth, A.J., and Dunn, W.L. (1977) J. Clin. Pathol. , **30**, 1056.

88. Corfield, A.P., Michalski, J.C., and Schauer, R. (1981) "In perspective in inherited metabolic diseases", (Tettamanti G., Durand P. and Di Donato S., eds.), Vol. **4**, pp. 3-70, Edi Ermes Publ., Milano.

89. Varki, A., and Kornfeld, S. (1980) J. Exp. Med. **152**, 532-544.

90. Culling, C.F.A., Reid, P.E., Clay, M.G., and Dunn, W.L. (1973) J. Histochem. Cytochem. **22**, 826-831.

91. Casals-Stenzel, J., Buscher, H.P., and Schauer, R. (1975) Anal. Biochem. **65**, 501-507.

92. Erbil, K.M., Jones, J.D., and klee, G.G. (1985) Cancer **55**, 404-409.

93. Silver, H.K.B., Rangel, D.M., and Morton, D.L. (1978) Cancer **41**, 1497-1499.

94. Fukushima, K., Hirota, M., and Terasaki, P.I., et al. (1984) Cancer Res. **44**, 5279-5285.

95. Kessel, D., and Allen, J. (1975) Cancer Res. **35**: 670-672.

96. Berg, E.L., Robinson, M.K., and Mansson, O., et al. (1991) J. Biol. Chem. **266**, 14869-14872.

97. Lundblad, A. (1980) Scand. J. Clin. Lab. Invest. **40**, 3-11.

98. Reintgein, D.S., Cruse, C.W., Wells, K.E., Saba, H.I., and Fabri, P.J. (1992) Ann. Plast. Surg. **28**, 55-9.

99. Patel, P.S., Adhvaryn, S.G, and Baxi, R.B. (1991) Int. J. Biol. Markers **6**, 177-82.

100. Xing, R.D., Chen, R.M., Wang, Z.S., and Zhang, Y.Z. (1991) J. Oral Maxillofac. Surg. **49**, 843-7.

101. Horowitz, M.I., and Pigman, W. (1977) "The glycoconjugate" eds., Horowitz, M.I., and Pigman, W, 1st ed., Vol. **1**, pp. 1-10, Academic Press.

102. Martin, D.W. Jr. (1985) "Glycoproteins, proteoglycons, and glycosaminoglycans" In.: "Harper's review of biochemistry" eds., Martin, D.W.Jr., Mayer, P.A., Rodwell V.W., and Granner, D.K. 20th ed., pp. 464-479, Lange Medical Publications.

103. Filipe, M.I., and Fengar, C. (1979) Histochem. J. **11**, 277-287.

104. Shehan, D.G., and Jervis, H.R. 1(1976) Am. J. Anta. **146**, 103-132.

105. La Mont, J.T., Smith, B.F., and Moore, J.R.L. (1984) Hepatology **4**, 515-565.

106. Zinn, A.B., Plantner, J.J., and Carlson, D.M. (1977) "Nature of linkages between protein core and oligosaccharides": In "The glycoconjugates" eds., Horowitz, M.I., and Pigman, W. 1st ed., Vol. **1**, pp. 69-85, Academic Press.

107. Lenten, L.V., and Ashwell, G. (1970) J. Biol. Chem. **246**, 1889-1894.

108. Glick, M.C., and Flowers, H. (1978) "Surface membrane" In: "The glycoconjugates" eds., Horowitz, M.I., and Pigman, W. 1st ed., Vol. **2**, pp. 337-384, Academic Press.

109. Hoskins, L.C. (1978) "Degradation of mucus glycoproteins in the gastrointestinal tract" In: "The glycoconjugates" eds., Horowitz, M.I., and Pigman, W. 1st ed., Vol. **2,** pp. 235-253, Academic Press.

110. Holden, K.G., and Griggs, L. (1987) "Respiratory tract" In: "The glycoconjugates" eds., Horowitz, M.I., and Pigman, W. 1st ed., Vol. **1**, pp. 215-237, Academic Press.

111. Ferguson, R.N., Edelhoch, H., and Saroff, H.A. (1975) Biochem. **14**, 282-289.

112. Wasserman, R.L., and Capra, J.D. (1977) "Immunoglobulins" In: "The glycoconjugates" eds., Horowitz, M.I., and Pigman, W. 1st ed., Vol. **1**, pp. 323-348, Academic Press.

113. Glandemans, C.P.J. (1975) Adv. Carbohydr. Chem. Biochem. **31**, 313-346.

114. Pazur, J.H., and Forsberg, L.S. (1978) Carbohydr. Res. **60**, 167-178.

115. Stern, P.L., Willison, K.R., Lennox, E., Calfre, G., Milstein, C., Secher, D., Ziegler, A., and Springer, T. (1978) Cell **14**, 775-785.

116. Granner, D.K. (1985) "Hormone action" In: "Harper's review of biochemistry" eds., Martin, D.W., Mayes, P.A., Rodwell, V.W., and Granner, D.K. 20th ed., pp. 505-515, Lange Medical Publications.

117. Macbeth, R.A.L., and Bekesi, J.G. (1962) Cancer Res. **22**, 1170-1175.

118. Seibert, F.B., Seibert, M.V., Atno, A.J., and Campbell, H.W. (1974) J. Clin. Invest. **26**, 90-102.

119. Nigelson, G.L., and Postle, G. (1976) N. Engl. J. Med. **295**, 253-258.

120. Cunietti, E., Vaiani, G., Gandini, M, Monti, M, Locatelli, E., Gandini, R., and Reggiani, A. (1985) Cancer Detec. Prev. **8**, 222-232.

121. Schachter, H. (1978) "Glycoprotein biosynthesis" In: "The glycoconjugates" eds., Horowitz, M.I., and Pigman, W. 1st ed., Vol. **2**, pp. 87-181, Academic Press.

122. Both, S.N., King, J.P.G., Leonard, J.C., and Dykes, P.W. (1973) Gut **14**, 794-799.

123. Barondes, S.H., (1981) Annual Review of Biochemistry **50**, 207-231.

124. Toyoshima, S., Osawa, T., and Tonomura, A. (1970) Biochem. Biophys. Acta, **221**, 514-521.

125. Kornfeld, S., and Kornfeld, R. (1978) "Use of lectins in the study of mammalian glycoproteins" In: "The glycoconjugates" eds., Horowitz, M.I., and Pigman, W. 1st ed., Vol. **2**, pp. 437-449, Academic Press.

126. Powell, J. (1980) Biochem. J. **187**, 123.

127. Sammel, H. (1984) Science **223**, 4639.

128. Springer, G., and Desai, P. (1971) Biochemistry **10**, 3749.

129. Lis, H., and Sharon, N. (1981) Ann. Rev. Biochem. **50**, 207-31.

130. Miller, J.B., Hsu, R., Heinrikson, R., and Yachnin, S. (1975) Proc. Natl. Acad. Sci. USA, **72**, 1388-91.

131. Ceri, H., Kobiler, D., and Barondes, S.H. (1981) J. Biol. Chem. **256**, 390-94.

132. Pereira, M.E.A., and Kabat, E.A. (1979) Crit. Rev. Immunol. **1**, 1-73.

133. Barondes, S. (1981) Annu. Rev. Bioche. **50**, 207.

134. Boldt, D.H., et al. (1975) J. Immunol. **114**, 1532-1536.

135. Scott, R.E., and Rosenthal, A.S. (1977) J. Immunol. **119**, 143-148.

136. Alhadeff, J.A. (1989) Crit. Rev. Oncol. Hematol. **9**, 37-47.

137. Taner, O., Nursen, E., and Limrin, A. (1990) Clin. Chem. **36**, 393-397.

138. Shamberger, R.J. (1984) J. Clin. Chem. Clin. Biochem. **22**, 647-651.

139. Vegh, Zs., Kremmer, T., Boldizsar, M., Gesztes, K.A., and Szajani, B. (1991) Clin. Chim. Acta **203**, 259-268.

140. Raynes, J.G. (1983) Biomed. Pharmacother. **37**, 136-138.

141. Lipton, A., Harvey, H.A., and De Long, S. et al. (1979) Cancer **43**, 1766-1771.

142. Dnistria, A.M., Schwartz, M.K., and Katopodis, N. (1982) Cancer **50**, 9.

143. Katopodis, N., Hirshaut, Y., and Geller, N.L. (1982) Cancer Res. **42**, 5270-5.

144. Rothenberg, R.E., La Ruja, R.D., Mueller, O.T., and Pryce, E.H. (1994) Breast Dis. **7**, 3, 197-202.

145. Lowry, O.H., Rosebrough, N.J., Farr, A.L., and Randall, R.J. (1951) J. Biol. Chem. **193**, 265-75.

146. Wilkinson, L. SYSTAT. (1990) The system for Statistics Evanston, II: SYSTAT Inc.

147. Crook, M. (1993) Clin. Biochem. **26**, 31-38.

148. Cohen, S.L., Lincoln, S.T., and Rosen, S.T. (1986) Cancer Invest. **4**, 305-327.

149. Patel, P.S., Baxi, B.R., Desal, S.S., and Balar, D.B. (1990) Ind. J. Pathol. Microbiol. **33**, 124-128.

150. Fukushima, K. (1991) J. Exp. Med. **163**, 17-30.

151. Itai, S., Arii, S., Tobe, R., Kitahara, A., and Kim, Y.C. et al. (1988) Cancer **61**, 775-87.

152. Hakomori, S. (1985) Cancer Res. **45**, 2405-14.

153. Warren, L., Fuhrer, J.P., and Buck, C.A. (1972) Proc. Natl. Acad. Sci. **69**, 1838-42.

154. Schutter, E.M.J., Vissr, J.J., and Van kamp, G.J. et al. (1992) Tumor Biol. **13**, 121-32.

155. Kakari, S., Stirngou, E., Toumbis, M., Ferderigos, As., and Poulaki, I. et al. (1991) Anticancer Res. **11**, 2107-10.

156. Dnistrian, A.M., and Schwartz, M.K. (1981) Clin. Chem. **27(10)**, 1737-1739.

157. Mannello, F., Bocchiotti, G., Troccoli, R., and Gazzanelli, G. (1993) Breast Cancer Res. Treat. **24**, 167-170.

158. Patel, P.S., Baxi, B.R., Adhvaryu, S.G., and Balar, D.B. (1990) Cancer Lett. **51**, 203-208.

159. Haq, M., Haq, S., Tutt, P., and Crook, M. (1993) Ann. Clin. Biochem. **30**, 383-386.

160. Hansen, H.J., Snyder, J.J., Miller, E., Vandevoorde, J.P., Miller, O.N., Hines, L.R., and Burns, J.J. (1974) Hum. Pathol. **5**, 139-147.

161. Harshman, S., Reynolds, V.H., Neumaster, T., Patikas, T., and Worrall, T. (1974) Cancer **34**, 291.

162. Barlow, J.J., and Dillard, P.H. (1972) Obstet. Gynecol. **39**, 727.

163. Aronson. N.N., and De Duve. C. (1978) J. Biol. Chem. **243**, 4564.

164. Robert, A. (1962) Cancer Res. **22**, 1170.

165. Lawrence, M., Gerald, B., and Zoltan, A. (1977) Clin. Chem. **23**, 2055.

166. Yogeswarar, G. (1983) Advance Cancer Res. **38**, 289.

167. Weimer, and Mashin, J. (1965) Clin. Chem. Principle and Techniques.

168. Waalkes, P.T., Mrochek, J.E., Dinsmore, S.R., and Tormey, D.C. (1978) J. Natl. Cancer Inst. **6**, 703.

169. Bradley, W.P., Blasco, A.P., and Weiss, J.F. (1977) Cancer **40**, 2264-2272.

170. Bhuvarahamurthy, V., Balasubramanian, N., Subramanian, S., and Govindasamy, S. (1992) Bichem. Int. **28**, 105.

171. Scambia, G., Panici, B., Perrone, L., Sonsini, C., Giannelli, S., and Gallo, A. et al. (1990) Br. J. Cancer **62**, 147.

172. Sherblow, A.P., Buck, R.L., and Carraway, K.L. (1980) J. Biol. Chem. **255**, 783.

173. Yogeeswaran, G., and Salk, P.L. (1981) Science **212**, 1514.

174. Yamamooto, K. (1984) Eur. J. Biochem. **143**, 133.

175. Bolmer, S., and Davidson, E. (1981) Biochemistry **20**, 1047.

176. Yaskhiko T., Wataru I. and Mitsunon Y., Am. J. Nephrol 1988; **2**: 21.

177. McCord, J.M., and Fridovich, I. (1969) J. Biol. Chem. **244**: 6049-6055.

178. Fridovich, I. (1972) Acc. Chem. Res. **5**, 321-326.

179. Wever, R., Oudega, B., and Van, Gelder, B.F. (1972) Biochem. Biophys. Acta, **302**, 475-478.

180. Batalie, R., Klein, B., Durie, B., and Sany, J. (1989) Clin. Exp. Rheumatol. **7**, 319-28.

181. Sigureirsson, B., Lindelof, B., Edhag, O., and Allander, E. (1992) N. Engl. J. Med. **326**, 363-7.

182. Polivakova, J., Vosmikova, K., and Horak, L. (1992) Neoplasma **39(4)**, 233-6.

183. Borrello, S., De Leo, M.E., Wohirab, H., and Galeoti, T. (1992) FEBS Lett. **310**, 249-54.

184. Van Balgooy, J.N.A, and Roberts, E. (1979) Comp. biochem. Physiol. **62B**, 263-8.

185. Winterbourn, C.C., Hawkins, R.E., Brian, M., and Carrell, R.W. (1975) J. Lab. Clin. Med. **85(2)**, 337-341.

186. Knee, J.K., Mitidieri, E., and Affonso, O.R.b (1991) Cancer Lett **57**, 199-202.

187. Bolzan, A.D., Bianchi, M.S., and Bianchi, N.O. (1993) Cancer Res. **6**, 142-6.

188. Yoshimitsu, K., Kobayashi, Y., and Usui, T. (1984) Acta Paediatr. Scand., **73**, 92-6.

189. Abella, A., Clerc, D., Chalas, J., Baret, A., Leluc, R., and Lindenbaum, A. (1987) Ann. Biol. Clin. **45**, 152-5.

190. Tsurn, S., Nomoto, K., Aiso, S., Ogata, T., and Zinnaka, Y. (1983) Int. Arch. Allergy appl. Immunol. **71(1)**, 88-92.

191. Galeotti, T., Masotti, L., Borrello, S., and Casali, E. (1991) Xenobiotica **21(8)**, 1041-51.

192. Oberley, L., and Oberley, T.D. (1988) Mol. Cell Biochem. **84**, 147-53.

193. Wong, Y.F., Wong, W.S.H., and Fung, Y.H., et al. (1993) Med. Sci. Res. **21**, 397-8.

194. Mizuno, K., and Kozutsumi, T. (1981) J. Biol. Chem. **256**, 4247.

195. Briles, E., and Gregory, W. (1979) J. Cell Biol. **81**, 528.

196. Bishayee, S., and Dorai, D. (1980) Biochem. Biophys. Acta **623**, 89.

197. Bohlool, B.B., Schmidt, E.L. (1976) J. Bacteriol. **125**, 1188-94.

198. Burger, M., and Goldberg, A. (1967) Proc. Nat. Acad. Sci. USA **57**, 359.

199. Kaplan, R., Li, S., and Kehoe, J. (1977) Biochemistry **16**, 4297.

200. Hans J., Attila, B., Sigrun, G., and Michael, K. (1989) Biochem. Biophys. Res. Commun. **163**, 506.

201. Ronald, L., James, F., and Wayne, W. (1982) J. Biol. Chem. **257**, 7574.

202. Liener, I. (1955) Arch. Biochem. Biophys. **54**, 223.

203. Lis, H., and Sharon, N. (1972) Methods Enzymol. **28**, 360.

204. Gabius, H., Bandlow, G., Schirramacher, V., and Vehmeyer, K. (1987) Int. J. Cancer **39**, 634.

205. Kaplan, A. (1998) Clinical Chemistry Theory, Analysis and Correlation, 2nd edition pp. 180.

206. Ahmed, H., Chatterjee, B.P., Klem, S., and Schauer, R. (1986) Biol. Chem. Hoppeseyler **367**, 501-506.

207. Basu, S., Sarkar, M., and Mondal, C. (1986) Mol. Cell biochem. **71**, 149-157.

208. Varki, A., and Kornfeld, S. (1980) J. Exp. Med. **152**, 532-544.

209. Schauer, R., Shukla, A.K., Schroder, G., and Muller, E. (1984) Pure and Appl. Chem. **57**, 907-921.

210. Shukla, A.K., and Schauer, R. (1982) Hoppe-Seyler's and Physiol. Chem. **363**, 255-262.

211. Ravindranath, M.H., Higa, H.H., Copper, E.L., and Paulson, J.C. (1985) J. Biol. Chem. **260**, 8850-8856.

212. Elvin, A., and Manfred, M. (1967) Experimental Immuno Chemistry, Second edition, Illinois, USA.

213. Gottschalk, A. (1960) In: The chemistry and biology of sialic acids and related substances, Cambridge University Press, London.

214. Bakhtear, M. (1992) Ph.D. thesis, College of Science, Baghdad University.

215. Wild, J., Robinson, D., and Winchester, B. (1983) Biochem. J. **21**, 167.

216. Dolichos, R. (1983) Biochemistry **22**, 2741.

217. Finstand, C., Good, R., and Litman, C. (1974) Ann. N. Y. Acad. Sci. **234**, 170.

218. Dipti, G., Fred, C., and Brewer, J. (1994) Biochemistry **33**, 5526-5530.

219. Pemberton, R. (1970) Vax Song. **18**, 74.

220. Goldstein, I., and Hays, C. (1978) Adv. Carbohydr. Chem. Biochem. **35**, 127.

221. Goebel, W., et al. (1934) J. Exper. Med. **60**, 599.

222. Nassir M. (1995) Ph.D. thesis, College of Science, Baghdad University.

223. Nowak, T., and Barondes, S.(1975) Biochimica et Biophys. Acta **393**, 15.

224. Finstad, C., Litman, G., Finstand, J., and Good, R. (1972) J. Immunol. **108**, 1704.

225. Oppenheim, J., Nachbar, M., Salton, M., and Aull, F. (1974) Res. Commun. **58**, 1127.

226. Peters, B.P. et al. (1979) Biochemistry **18**, 5505-5511.

227. Roche, A.C. et al. (1975) FEBS Lett. **57**, 245-249.

228. Mohan et al. (1982) Biochem. J. **203**, 253-261.

229. Miller, R.L. (1982) J. Invertebr. Pathol. **39**, 210-214.

230. Babal, P. (1994) Biochem. J. **229(2)**, 341.

231. Kawagishi, H. (1994) FEBS Lett. **340**, 56.

232. Scopes, R. (1982) Protein purification principles and practice, Springer Verlag, pp. 162, New York Heidelber Berlin.

233. Pharmacies fine chemical: Gel filtration calibration kit instruction mannual for protein molecular weight determination.

234. Csizman, L. (1960) Proc. Soc. Exp. Biol. N. Y. **103**, 157.

235. Laemmli, U.K. (1970) Nature **227**, 680.

236. Hans, F. (1977) Application note 306 LKB-Produkter AB., pp. 1-15, Bromma, Sweden.

237. Huda, H. (1998) M.Sc. thesis, College of Science, Baghdad University.

238. Isamu, M., Heruko, K., Naoko, I., and Yukiko, S. (1986) Carb. Res. **151**, 261.

239. Goldstein, I.J., Hughes, R.C., Monsigny, M., Osawa, T., and Sharon, N. (1980) Nature **285**, 66.

240. Czech, M.P., Lynn, W.S. (1973) Biochem. Biophys. Acta **297**, 368-377.

241. Scatchard, G. (1949) Ann. N. Y. Acad. Sci. **51**, 660.

242. Emil, L. (1985) In: Principles of biochemistry, seven edition, pp. 289, International Student Edition.

243. Rae-Venter, B., and Dao, T. (1982) Biochem. Biophys. Res. Commun. **107**, 624.

244. Hussain, M. (1990) M.Sc. thesis, College of Science, Baghdad University.

245. Waelbroeck, M., Van Obeerghen, E., and Demeyts, P. (1979) J. Biol. Chem. **254**, 7736.

246. Blumenthar, D.K., and Stul, J.T. (1982) Biochemistry **21**, 2386.

247. Laport, D.C., Wierman, E.M., and Storm, D.I. (1980) Biochemistry **19**, 3814.

www.ingramcontent.com/pod-product-compliance
Lightning Source LLC
Chambersburg PA
CBHW080648190526
45169CB00006B/2032